Building Secure Automotive IoT Applications

Developing robust IoT solutions for next-gen automotive software

Dr. Dennis Kengo Oka

Sharanukumar Nadahalli

Jeff Yost

Ram Prasad Bojanki

Building Secure Automotive IoT Applications

Copyright © 2024 Packt Publishing

All rights reserved. No part of this book may be reproduced, stored in a retrieval system, or transmitted in any form or by any means, without the prior written permission of the publisher, except in the case of brief quotations embedded in critical articles or reviews.

Every effort has been made in the preparation of this book to ensure the accuracy of the information presented. However, the information contained in this book is sold without warranty, either express or implied. Neither the authors, nor Packt Publishing or its dealers and distributors, will be held liable for any damages caused or alleged to have been caused directly or indirectly by this book.

Packt Publishing has endeavored to provide trademark information about all of the companies and products mentioned in this book by the appropriate use of capitals. However, Packt Publishing cannot guarantee the accuracy of this information.

Group Product Manager: Preet Ahuja

Publishing Product Manager: Suwarna Rajput

Book Project Manager: Ashwin Kharwa

Senior Editor: Roshan Ravi Kumar

Technical Editor: Irfa Ansari

Copy Editor: Safis Editing

Proofreader: Roshan Ravi Kumar

Indexer: Tejal Soni

Production Designer: Nilesh Mohite

Senior DevRel Marketing Executive: Rohan Dobhal

First published: August 2024

Production reference: 1070824

Published by Packt Publishing Ltd.

Grosvenor House

11 St Paul's Square

Birmingham

B3 1RB, UK

ISBN 978-1-83546-550-9

www.packtpub.com

To my loving wife, Mai, for your support and patience, and to my wonderful children, Mia, Elina, and Alyssa, for your joy and inspiration.

– Dr. Dennis Kengo Oka

To my late father, Eshwarappa Nadahalli, who provided me with boundless love and care throughout my childhood.

– Sharanukumar Nadahalli

To my family. First and foremost, my wife, Stacie. Then, to my four "presidential" daughters: Madison, Reagan, McKinley, and Kennedy.

– Jeff Yost

To my son, Aakash, and daughter, Aaradhya, who are my constant teachers, bringing new lessons and joy into my life each day.

– Ram Prasad Bojanki

Foreword

In their book, Dr. Dennis Kengo Oka, Sharanu Nadahalli, Jeff Yost, and Ram Prasad Bojanki provide a comprehensive overview of developing secure automotive **internet of things** (**IoT**) applications, addressing the intricacies and increasing importance of this field as vehicles integrate into the IoT. Understanding the convergence of automotive and IoT domains coupled with cybersecurity is crucial as modern vehicles evolve into sophisticated, interconnected systems.

The development of **software-defined vehicles** (**SDVs**), **electric vehicles** (**EVs**), **connected vehicles** (**CVs**), and increased vehicle automation has led to enhanced features and use cases as well as expanded attack surfaces. For instance, recent API attacks have highlighted vulnerabilities that could potentially affect entire vehicle fleets, underscoring the critical need for robust cybersecurity measures both in the vehicle and the cloud.

This book provides an in-depth exploration of the end-to-end development of automotive IoT applications. It examines current and future automotive technologies, vehicle architectures, and real-world automotive IoT use cases. The authors also address cybersecurity topics and provide insights into standards and methodologies such as ISO/SAE 21434 and secure software development life cycle. Further, the book delves into the complexities of the automotive software supply chain and related cybersecurity risks.

The authors tackle many topics that are often overlooked but highly relevant in everyday work. For example, they provide guidance throughout the different development phases of an automotive IoT application, including system design, vehicle architecture and cloud integration, software component development, deployment, and maintenance within a DevSecOps framework. The book also discusses related aspects such as **Automotive SPICE** (**ASPICE**), functional safety, and embedded hardware and software technologies.

For system designers, software developers, security process owners, and practitioners, this book serves as an excellent guide to building secure automotive IoT applications. It offers practical guidance and actionable insights using workflows, tables, diagrams, and illustrations to help you get started. Also, it eases the transition for software engineers from other industries into the automotive sector. Whether your company is in the early stages of developing automotive systems or cloud applications for IoT use cases, this book is an invaluable resource for navigating the complex landscape of automotive IoT.

Dr. André Weimerskirch, VP Product Integrity

Contributors

About the authors

Dr. Dennis Kengo Oka is an automotive cybersecurity expert with over 15 years of experience. He has a Ph.D. in automotive security from Chalmers University of Technology in Sweden. He has worked for Volvo, Bosch Group, and held the role of head of engineering and consulting Asia-Pacific. Some highlights of his career include security research for remote diagnostics and over-the-air updates, co-launching the ESCRYPT automotive security practice in Japan, and standardizing cybersecurity testing. Dennis is on the advisory board for Block Harbor, and with over 70 publications, including his book, *Building Secure Cars: Assuring the Automotive Software Development Lifecycle*, he is also a frequent speaker at international automotive and cybersecurity conferences.

I would like to express my gratitude to the staff at Packt for their guidance throughout writing this book. My deepest thanks to my co-authors Ram, Sharanu, and Jeff for the amazing collaboration. I would also like to thank my parents, Sven and Etsuko, and my siblings, Alex and Linda, for their unwavering support and encouragement, and for providing an escape to appreciate what truly matters in life.

Sharanukumar Nadahalli is a software engineering manager at Panasonic Automotive Systems of America with over 16 years of experience in the automotive industry. He has held various roles, including SW developer, technical leader, product architect, project manager, and technical manager. Before that, he worked in embedded system software development for over a year. Sharanu is an enthusiastic learner currently focusing on cybersecurity. Sharanu earned a Stanford lead certification from the Stanford University Graduate School of Business. He also holds a master's degree in general business from Clayton State University, Georgia, and a bachelor's degree in computer science from Visvesvaraya Technological University, India. Sharanu has also served as an area director in Toastmasters International and remains an active member.

I want to thank Jeff Yost for providing me with the opportunity to work on this assignment. I also thank Ram, Dennis, and Packt for the great collaboration and guidance. I would like to thank Steve Barron for giving me a leadership opportunity at Panasonic Automotive and my mentor, Don Turner, for persistently guiding me on various topics. I also thank my mother, Kalawathi, and my wife, Nagaveena, who always support me in going the extra mile in all things.

Jeff Yost is an engineering manager at Panasonic Automotive Systems of America. He is a 20-year veteran in embedded development and has been in automotive for over 5 years. He has spent most of his career in software but also had a stint as a product line general manager. Jeff enjoys working with teams and developing new products. He graduated with a BSEE from Missouri Science and Technology University before moving to the desert of Tucson and earning a master's degree in electrical and computer engineering from the University of Arizona. He has 3 US patents and developed many industry-leading products. After enjoying 20 years in the beautiful Pacific Northwest, he relocated to the golf cart community of Peachtree City, Georgia, where he bikes to work daily.

I want to thank a few people who allowed me to get started in the automotive industry. Specifically, I thank Paul Beets who hired me within about two days of first seeing my resume and I thank Steve Barron who put a lot of trust in me in allowing me to lead the software activities on a large and critical program. I also thank my wife, Stacie, for her encouragement and support in all things.

Ram Prasad Bojanki is a seasoned software development professional with over two decades of experience, with a particular focus on the automotive industry for 15 years. Most recently, Ram led the charge at Panasonic North America, Smart Mobility, where he spearheaded the entire lifecycle of the OneConnect Cloud Platform (IoT), from conceptualization and development to delivery and operational management. This experience highlights his well-honed abilities in designing and developing complex products and platforms across diverse environments, including both embedded systems and cloud technologies. He holds a Bachelor's in Electronics and Instrumentation Engineering from Andhra University, India.

I am deeply grateful to Jeff and Sharanu for extending the invitation to co-author this book, and to Dennis for his invaluable collaboration. My heartfelt thanks also go to my wonderful wife Jeevani and my loving parents Satyanarayana and Lakshmi for their unwavering support.

About the reviewers

Jean Paul Talledo Vilela is a senior technology implementer at VTTI, where he designs, develops, and integrates various transportation technologies for applied automotive research. He also interfaces directly with sponsor technical and management teams to support research development and delivery goals. Some of his current design work includes ADS system integration for SAE Level 4/5 vehicle automation, connected vehicle-to-everything (C-V2X) technology deployment for the work zone, intersection scenarios, and cybersecurity and secured message transactions for V2X communications. In addition, Jean Paul provides technical oversight, quality assurance, and mentoring to developing engineers and strategic research projects.

Dr. Krishnendu Kar is a digital IoT leader who holds a Ph.D. in mechanical engineering from West Virginia University and an MBA from Simon School, University of Rochester. With a decade at General Motors, focusing on vehicle control and diagnostics, he transitioned to IoT 5 years ago, specializing in product development as well as coding AI, computer vision, the AWS cloud, device, and mobile app development for smart home IoT products. At Kidde, he leads the execution of smart home products. He authored *Mastering Computer Vision with TensorFlow* and created iOS apps *Nity: DashCam & AI Map* and *Stretch Tracker*. He enjoys an active lifestyle, engaging in running, home improvement, and social networking.

Balaji Balasubramanian is a passionate engineering professional with over 15 years of experience across different fields, including xEV design and development, automation, and machine learning. Currently, he works for Tata Consultancy Services as an assistant consultant. He handles different roles, ranging from vehicle-level simulation, energy management strategies, eVCU, and BMS software development. He describes himself as purpose-driven and an active member of ASAM, CoE : xEV design and development. In his free time, he codes for data analytics and machine learning algorithms on sports and IoT device data. He received a master's degree in power electronics from IIT Madras and a bachelor's degree in electrical and electronics from TCE, Madurai.

I'd like to thank my family and friends for their constant support.

Table of Contents

Preface — xvii

Part 1: Introduction to Automotive IoT

1

Automotive Technology Trends — 3

Overview of current automotive trends	3	Introduction to automotive IoT	11
CASE	4	Automotive IoT	11
SDV and SOA	6	Automotive IoT use case examples	12
Mobile apps and the cloud	7	Data management for automotive IoT use cases	13
Modern software development	8	**Summary**	**13**
Standards and regulations	9	**References**	**13**

2

Introducing Automotive IoT Use Cases — 15

Enhanced driver experience and safety	16	Driver performance monitoring	22
Connected car services	16	Predictive maintenance	23
Advanced driver-assistance systems	17	**Connected mobility revolution**	**24**
Personalized in-car experience	18	Smart parking solutions	24
Phone as a key	19	Vehicle-to-Everything (V2X) communication	25
Optimized fleet management	**22**	Connected supply chain and manufacturing	27
Real-time vehicle tracking and telematics	22	**Summary**	**28**
		References	**28**

Part 2: Vehicle Architectures

3

Vehicle Architecture and Frameworks 31

The scale of vehicle architecture	31	Standard frameworks to support vehicle architecture and IoT	36
Distributed architecture	32		
Centralized zonal domain architecture	32	A high-level overview of the domain controller	36
A central computer with multiple domain-specific SoCs	34	Summary	48
A central computer with a single SoC	35	References	48

4

Vehicle Diagnostics 49

UDS	49	Diagnostic service management in Adaptive AUTOSAR	64
UDS message structure	50		
DoIP	53	Reflecting on the application of remote diagnostics	66
DoIP message format	56		
DoIP example message flow	60	Summary	66
Diagnostic communication workflow in Classic AUTOSAR	62	References	66

5

Next Wave of Vehicle Diagnostics 67

Technical requirements	67	Example of an SOVD interface as part of applications on the server side	75
Needs beyond UDS	68		
SOVD	69	SOVD documentation and demo	77
REST	73	SOVD and UDS comparison	78
SOVD example, demo, and details	73	Summary	79
Example of a diagnostic message using UDS and SOVD	74	References	79

Part 3: Secure Development for Automotive IoT

6

Exploring Secure Development Processes for Automotive IoT — 83

An overview of security threats and the need for security and secure development processes — 84	**NIST Cybersecurity Framework, ISO 27001, SOC 2, and OWASP** — 98
New cybersecurity threats — 84	NIST Cybersecurity Framework — 98
Examples of recent attacks — 85	ISO 27001 — 100
Simplified threat model of automotive IoT ecosystem — 85	SOC 2 — 101
	OWASP — 103
ISO/SAE 21434 and ASPICE for Cybersecurity — 92	**DevSecOps and agile development** — 105
ISO/SAE 21434 Overview — 93	V-model — 105
ISO/SAE 21434 organizational-level requirements — 93	Agile — 106
ISO/SAE 21434 project-level requirements — 95	Scrum — 107
ASPICE for Cybersecurity overview — 96	DevSecOps — 107
ASPICE for Cybersecurity – security activities — 97	**Summary** — 109
	References — 109

7

Establishing a Secure Software Development Platform — 111

Activities in the SSDLC — 112	**Project inventory** — 114
TARA/threat model — 113	Project information and risk level — 115
Requirements review — 113	Cybersecurity assurance level and activities — 116
Design review — 113	Example project inventory — 118
Code review — 113	**Practical steps for establishing a secure software development platform** — 120
Static application security testing — 113	Purpose and need — 120
Vulnerability scanning — 114	Overview of the secure software development platform — 121
Fuzz testing — 114	Requirements, policies, and compliance — 122
Dynamic application security testing — 114	Vulnerability management — 123
Interactive application security testing — 114	AppSec tooling — 124
Penetration testing — 114	

Common AppSec tooling and test approaches	124
SAST	124
SCA	126
DAST	128

Fuzz testing	129
Penetration testing	131
Summary	**133**
References	**134**

8

Securing the Software Supply Chain — 137

Software supply chain and distributed development	138
Overview of the software supply chain	138
RASIC, vendor security assessments, and CIADs	**140**
RASIC	141
Vendor security assessments	143
CIADs	145
Managing risks with OSS	**146**
Security vulnerabilities	147
License compliance	149
Operational risk	151

SBOM	153
SBOM formats	153
Executive Order 14028	154
NTIA	155
OpenChain	155
Secure software supply chain risk management	**155**
Identifying the risks	156
Assessing the risks	156
Mitigating the risks	156
Summary	**158**
References	**158**

Part 4: Automotive IoT Application Life Cycle

9

System Design of an Automotive IoT Application — 163

System design process overview	164
UXDD	165
Use case – remote diagnostics	167
System components	169
Vehicle telematics gateway	169
Vehicle cloud platform	170
End-user mobile device	171

Gateway design considerations	172
GNSS receivers	172
Wireless communication	173
Wired communication	175
CAN	176
Sensors	177
SIM/eSIM	177
Gateway hardware	178

Cloud design considerations	179	Service-oriented vehicle diagnostics	184
Device management	180	Regulatory compliance	185
Connectivity management	180	Build versus buy	187
Remote diagnostics applications	182	Summary	188
Classic vehicle ECU diagnostics	183	References	189

10

Developing an Automotive IoT Application — 191

Cloud backend deployment and service models	191	Rule engine	206
Deployment models	192	Application Programming Interface (API) gateway	206
Service models	193	Connectivity management	209
Server-based and serverless computing	195	IAM	209
IoT application architecture	198	Vehicle telematics gateway	210
Cloud device gateway	199	Remote diagnostics application	213
Edge computing	200	Predictive maintenance	216
Stream processing	200	Development process	218
Device management	202	Summary	219
OTA solutions	203	References	219
Telemetry datastore	205		

11

Deploying and Maintaining an Automotive IoT Application — 221

The DevSecOps life cycle	222	CD	233
The plan stage	224	The release stage	233
CI	226	The deploy stage	233
The code stage	226	The operate stage	239
The build stage	228	The monitor stage	243
The test stage	230	Summary	247
		References	248

Part 5: Automotive Software Insights

12

Processes and Practices — 251

Introduction to processes and practices	251
ASPICE	253
SWE.1 – Software Requirements Analysis	257
SWE.2 – Software Architectural Design	257
SWE.3 – Software Detailed Design and Unit Construction	258
SWE.4 – Software Unit Verification	258
SWE.5 – Software Integration and Integration Test	259
SWE.6 – Software Qualification Test	259
Functional safety	260
Vocabulary	261
Risk classification system	261
Development process	262
Additional automotive processes and practices	266
DFMEA	267
5 Whys root cause analysis	268
Fishbone	270
A-B-A testing	271
Summary	271
Reference	272

13

Embedded Automotive IoT Development — 273

Embedded software development	273
Electrical engineering	274
Schematics/block diagrams	277
Datasheets, errata, and application notes	278
Device drivers	279
Hardware Abstraction Layer (HAL)	280
Additional aspects of embedded development	280
Automotive-focused aspects	281
Power state management	283
Operating systems	285
Hypervisors	287
Development tools	289
Life cycle management tools	290
Software development ecosystem	294
You and your customers	295
You and your co-suppliers	296
You and your suppliers	296
Summary	299
References	299

14

Final Thoughts — 301

Agile	301	**Security**	310
Agile+ASPICE	304	**Summary**	311
Automotive embedded testing	305	**References**	312
Types of testing	309		

Index — 313

Other Books You May Enjoy — 332

Preface

The integration of the **internet of things** (**IoT**) into the automotive industry is driving an era of unprecedented innovation and connectivity. This book, *Building Secure Automotive IoT Applications* is crafted to provide a thorough understanding of the technologies, architectures, security approaches, and development practices that define this evolving field. It is structured into five comprehensive parts, offering both theoretical knowledge and practical insights.

The journey begins with an exploration of current automotive trends and the shift towards IoT applications. Readers will gain insights into the technological advancements that are revolutionizing the industry and the essential infrastructure required for IoT.

The focus then shifts to the evolution of vehicle architectures. Here, the transition from traditional mechanical systems to sophisticated electronic and software-integrated systems is examined, alongside the modern tools and methods used for an example use case based on vehicle diagnostics.

Recognizing the critical importance of cybersecurity, the book delves into secure development practices for automotive IoT. It covers new cybersecurity threats, secure development methodologies, and practical steps for establishing secure development environments. Additionally, strategies for managing risks in the software supply chain are discussed in detail.

The book also provides a detailed look at the life cycle of automotive IoT applications. It covers the end-to-end process of designing, developing, deploying, and maintaining these applications, offering practical guidance and strategies for effective implementation and management.

Finally, the book synthesizes these insights, focusing on the unique aspects of automotive software development. It addresses essential engineering practices, offers guidance for engineers transitioning into the automotive domain, and discusses the collaborative nature of development, including regulatory considerations.

This book aims to be a definitive guide for professionals and enthusiasts alike, blending in-depth theoretical knowledge with practical advice. Whether you are an industry veteran or a newcomer, you will find valuable information to help you navigate and succeed in the dynamic world of automotive IoT.

Who this book is for

If you have been working as an automotive software engineer focused on embedded development, but want to learn about growing IoT development, this book is for you. If you are an IoT software developer but want to learn automotive development, this book is for you. This book is an excellent resource to help you grow your automotive software expertise and prepare for a new career in automotive IoT development.

What this book covers

Chapter 1, *Automotive Technology Trends*, introduces the reader to automotive trends and describes how the automotive industry is changing to support new use cases for automotive IoT. This chapter gives the reader an overview of the technology trends enabling IoT and introduces relevant terminology and concepts.

Chapter 2, *Introducing Automotive IoT Use Cases*, introduces several automotive IoT use cases that significantly enhance vehicle functionality and driver safety through connected car services, ADAS, and personalized in-car experiences. Some of these use cases will be referenced throughout the book in the different chapters to allow the reader to follow the various topics on end-to-end automotive IoT application development.

Chapter 3, *Vehicle Architecture and Framework*, covers the evolution of vehicle architecture, spanning more than two decades, tracing its journey from distributed systems to integrated approaches. We'll explore essential technologies and frameworks such as Hypervisor, AUTOSAR Classic, and Adaptive AUTOSAR, comparing their roles in modern vehicle design. Key topics include the scale of vehicle architectures and the standard frameworks supporting both vehicle architecture and the IoT landscape.

Chapter 4, *Vehicle Diagnostics*, introduces key diagnostic protocols in modern automotive systems: **Unified Diagnostic Services** (**UDS**) and **Diagnostic over Internet Protocol** (**DoIP**), integrated with AUTOSAR. We explore UDS for versatile vehicle diagnostics and firmware updates across communication platforms and delve into DoIP for high-speed diagnostic communication over networks, crucial for predictive maintenance. We discuss the diagnostic communication flow and components for remote diagnostics in AUTOSAR-based systems, emphasizing advanced service management for enhanced flexibility and scalability. These protocols ensure efficient, reliable, and secure vehicle diagnostics in today's connected automotive landscape.

Chapter 5, *Next Wave of Vehicle Diagnostics*, covers the evolving landscape of vehicle diagnostics to meet the demands of modern vehicles, including IoT applications. UDS has limitations in adapting to dynamic software-defined vehicles, prompting the need for a more flexible protocol. Enter **Service-Oriented Vehicle Diagnostics** (**SOVD**), the next generation of diagnostic protocols tailored for modern vehicles. This chapter provides insights into SOVD, including a demonstration and comparison with UDS. Key topics covered include the necessity beyond UDS, an in-depth look at SOVD, and a demonstration of its application.

Chapter 6, Exploring Secure Development Processes for Automotive IoT, explores how automotive IoT brings new cybersecurity threats and as such there is a need for cybersecurity and for establishing secure software development processes. This chapter discusses security processes and software development methodologies including ISO/SAE 21434, ASPICE for Cybersecurity, the NIST Cybersecurity Framework, ISO 27001, OWASP, and DevSecOps. Additionally, specific cybersecurity activities in the secure software development life cycle are presented.

Chapter 7, Establishing a Secure Software Development Platform, shows how to establish a secure software development platform to help develop secure software for automotive IoT. This chapter gives step-by-step practical guidance on how to establish such a platform and explains the benefits of using this platform approach. Furthermore, several different application security testing approaches are described, as well as how to handle vulnerability management and how to automate security testing.

Chapter 8, Securing the Software Supply Chain, discusses the risks in the software supply chain, due to the plethora of software for automotive IoT use cases provided through it, and presents several practical suggestions on how to address the risks. For example, topics on **Cybersecurity Interface Agreement for Development (CIAD)**, vendor security assessments, open-source software, and **Software Bill of Material (SBOM)** will be covered.

Chapter 9, System Design of an Automotive IoT Application, details the end-to-end system design of remote vehicle diagnostics use case. It explores the critical balance of desirability, feasibility, and viability in system design, emphasizing a user-centric approach. It provides a comprehensive overview of system components, from telematics gateways to cloud platforms, detailing the technologies and design considerations involved.

Chapter 10, Developing an Automotive IoT Application, explores the software design and development process of automotive IoT applications. It covers cloud backend deployment models, service models, and IoT application architecture. The chapter details software components for both cloud and vehicle telematics gateways, emphasizing the importance of remote diagnostics and predictive maintenance. It also discusses the development process for cloud and embedded software, highlighting key differences and considerations.

Chapter 11, Deploying and Maintaining an Automotive IoT Application, delves into the deployment and maintenance of automotive IoT applications, emphasizing the DevSecOps life cycle. The chapter details activities, tools, and interactions throughout the process, highlighting how deployment pipelines are established and managed across all stages. It also covers security integration, coding, building, testing, releasing, deploying, operating, and monitoring, providing a comprehensive guide to ensuring rapid deployment and maintaining high-quality standards in automotive IoT applications.

Chapter 12, *Processes and Practices*, explores processes and practices in automotive IoT software development. It covers Automotive SPICE®, functional safety (ISO 26262), and other key processes such as DFMEA and 5 Why Root Cause Analysis. It emphasizes the importance of processes in achieving high-quality software and provides insights into their practical application. The chapter also discusses the challenges and benefits of adopting these processes, highlighting their role in ensuring safety, reliability, and continuous improvement in automotive software engineering.

Chapter 13, *Embedded Automotive IoT Development*, explores embedded development for automotive IoT applications. It covers essential electrical engineering concepts, device drivers, memory management, and **key performance indicators** (**KPIs**). The chapter also delves into automotive operating systems, hypervisors, and the software development ecosystem, emphasizing the importance of collaboration and supplier management in the automotive industry.

Chapter 14, *Final Thoughts*, offers the authors' perspectives and insights based on their experiences. It then recaps the book to show you how everything is connected.

Conventions used

There are a number of text conventions used throughout this book.

`Code in text`: Indicates code words in text, database table names, folder names, filenames, file extensions, pathnames, dummy URLs, user input, and Twitter handles. Here is an example: "The client stores the connection configuration for each available resource in a set of `.rdp` files."

A block of code is set as follows:

```
#include <iostream>
#include <cpprest/http_listener.h>
#include <cpprest/json.h>

using namespace web;
using namespace web::http;
using namespace web::http::experimental::listener;

const utility::string_t base_url = U("http://localhost:8080");
```

Bold: Indicates a new term, an important word, or words that you see onscreen. For instance, words in menus or dialog boxes appear in **bold**. Here is an example: "You will then see the **Sign in to your account** popup."

> **Tips or important notes**
> Appear like this.

Get in touch

Feedback from our readers is always welcome.

General feedback: If you have questions about any aspect of this book, email us at `customercare@packtpub.com` and mention the book title in the subject of your message.

Errata: Although we have taken every care to ensure the accuracy of our content, mistakes do happen. If you have found a mistake in this book, we would be grateful if you would report this to us. Please visit `www.packtpub.com/support/errata` and fill in the form.

Piracy: If you come across any illegal copies of our works in any form on the internet, we would be grateful if you would provide us with the location address or website name. Please contact us at `copyright@packt.com` with a link to the material.

If you are interested in becoming an author: If there is a topic that you have expertise in and you are interested in either writing or contributing to a book, please visit `authors.packtpub.com`.

Share Your Thoughts

Once you've read *Building Secure Automotive IoT Applications*, we'd love to hear your thoughts! Scan the QR code below to go straight to the Amazon review page for this book and share your feedback.

https://packt.link/r/1-835-46550-1

Your review is important to us and the tech community and will help us make sure we're delivering excellent quality content.

Free Benefits with Your Book

This book comes with free benefits to support your learning. Activate them now for instant access (see the "*How to Unlock*" section for instructions).

Here's a quick overview of what you can instantly unlock with your purchase:

PDF and ePub Copies	Next-Gen Web-Based Reader

- Access a DRM-free PDF copy of this book to read anywhere, on any device.
- Use a DRM-free ePub version with your favorite e-reader.

- **Multi-device progress sync**: Pick up where you left off, on any device.
- **Highlighting and notetaking**: Capture ideas and turn reading into lasting knowledge.
- **Bookmarking**: Save and revisit key sections whenever you need them.
- **Dark mode**: Reduce eye strain by switching to dark or sepia themes

How to Unlock

Scan the QR code (or go to `packtpub.com/unlock`). Search for this book by name, confirm the edition, and then follow the steps on the page.

Note: Keep your invoice handy. Purchases made directly from Packt don't require one

Part 1: Introduction to Automotive IoT

This part introduces the reader to automotive IoT by first giving a background on the current automotive landscape and trends, followed by explaining the changes happening in various technologies used in the automotive industry, and then finally doing a deep dive into a couple of automotive IoT use cases that will be referenced throughout the book.

This part has the following chapters:

- *Chapter 1, Automotive Technology Trends*
- *Chapter 2, Introducing Automotive IoT Use Cases*

Automotive Technology Trends

The automotive industry is drastically changing. With technology advancements in other industries, including software development methodologies and frameworks, network connectivity, **Internet of Things** (**IoT**), and cloud infrastructure, the automotive industry is evolving to deploy novel solutions that make use of these new technology advancements. To utilize these new technology advancements, **automotive IoT applications** are now being developed and deployed in the automotive industry.

To set the stage and better understand the context, we will first review the current automotive industry trends. We will then give an introduction to automotive IoT and describe the overall ecosystem. We will also provide some example use cases for automotive IoT that explain the end-to-end communication flow.

This chapter will help prospective and existing automotive IoT engineers and managers to better understand the underlying automotive technology trends that are driving automotive IoT development.

In this chapter, we are going to cover the following main topics:

- Overview of current automotive trends
- Introduction to automotive IoT

Overview of current automotive trends

Welcome to the wonderful world of automotive! Writing a section on automotive trends is always challenging since trends are continuously evolving. Thus, this section becomes a snapshot of the current trends at the time of this writing. Of course, we recognize that something that may be a trend in this snapshot of today may have lost steam in some time and that there may be a new trend, buzzword, or hot topic when you are reading this book.

For example, an excerpt of automotive trends from a book in the 1980s may have sounded something like the following:

"There is a new in-vehicle communication protocol called Controller Area Network (CAN) that will revolutionize communication between Electronic Control Units (ECUs). CAN offers several benefits including reduced complexity of in-vehicle networks, improved scalability, and reduced weight and cost of the wiring. Using CAN allows ECUs to communicate with low latency and high reliability, making it possible to develop new and more advanced features. For example, using CAN it is possible to increase the vehicle performance and fuel efficiency by improved interaction between the engine control unit and the transmission control unit. Another example is the improved safety and performance of the Anti-lock Braking System (ABS) thanks to better communication between the ABS, the engine control unit, and the traction control system. Besides internal vehicle technology advancements, we are also going to see new cars equipped with Compact Disc (CD) players to enhance the user experience, allowing drivers and passengers to listen to their favorite music on the go. For example, the latest CD player system includes a CD changer that allows for up to 6 CDs to be preloaded, and the driver can select which CD to play using buttons on the dashboard."

The only thing missing from this excerpt is the trend of launching a new line of cars that can transform into robots (although these cars were from a toy line called Transformers).

Fast forward to 2024, and the automotive trends have drastically changed. There are several noteworthy trends that we would like to discuss and it would not be possible to include all of them, so we will focus on the most important trends and those that will have a major impact in the automotive industry, causing disruption to the status quo. While technical improvements to vehicle systems are continuously being developed, equally important are the advancements made to improve the user experience and growing beyond the fundamental transportation use case of vehicles to provide end-to-end mobility and entertainment solutions.

We will cover the following automotive trends in the rest of this section:

- CASE
- SDV and SOA
- Mobile apps and the cloud
- Modern software development
- Standards and regulations

CASE

The term **Connected, Autonomous, Shared, Electric (CASE)** appeared a few years ago to describe the future of mobility and has been a main driver for leading several of the current automotive trends.

Connected

Most vehicles today contain several connectivity interfaces allowing the vehicles to communicate with external entities. In recent years we have seen the addition of support for wireless communication such as Wi-Fi, Bluetooth, **vehicle-to-everything** (**V2X**), 5G, **Ultrawide Band** (**UWB**), and so on. The **Connected** trend allows vehicles to interact with various services on the internet, smart devices, and other vehicles. As such, connectivity plays an integral role in supporting a large number of use cases including software updates over the air, remote diagnostics, remote keyless entry, multimedia playback, safety warnings, mobile apps, and more.

Autonomous

SAE J3016 [1] defines six levels of automation from 0 to 5 for **Autonomous** vehicles. While features to support drivers with some level of autonomy (Levels 1-2) have been around for decades, such as adaptive cruise control or lane keep assist, more advanced features are continuously being developed. For example, there are many vehicles under development supporting Levels 3 and 4, where the vehicle can take control of steering, braking, and acceleration, allowing the driver to be engaged in other activities. For Level 3, the driver may be asked to take over control in situations where the vehicle is unable to make certain decisions but in Level 4 the vehicle is fully capable of self-driving within a certain area (however, the driver is still able to manually take control of the vehicle if needed). Level 5 is a fully autonomous vehicle with no pedal or steering wheel (that is, the driver is not able to take manual control), and not restricted to any specific areas.

Shared

The concept of **Shared** mobility allows for new use cases and services to be provided for vehicles. Advancements in connectivity, software development, cloud, and mobile apps have paved the way for a plethora of new shared mobility services. An example is ride-hailing services such as Uber and Lyft, where a user can request a ride from a driver using a smartphone app. It allows users who do not own a car or who are visiting a city to easily get around. Another example is car-sharing services, where a user can use a smartphone app to rent a car by the hour or day. It is a convenient and cost-efficient way to get around since a user can typically find a nearby parked car participating in the car-sharing program.

Electric

The **Electric** trend has given birth to **Electric Vehicles** (**EVs**), which have gained enormous traction in the past few years. EVs provide many benefits over traditional internal combustion engine vehicles, including no harmful emissions during driving and improved user experience due to electric motors providing a smoother and more comfortable ride. As such, many governments are setting ambitious targets for EVs. For example, the USA has the goal of EV sales making up 50% of automobile sales by 2030 [2], the EU wants 55% of new cars to be EV by 2030, and 100% by 2035 [3], and Japan has set a target of EV sales being 100% of automobile sales by 2035 [4].

SDV and SOA

More recently, new terms such as **Software-Defined Vehicle (SDV)** and **Service-Oriented Architecture (SOA)** have overtaken CASE as the new trends. Let us understand them more in detail.

SDV

SDV has gained popularity in recent years as the main new trend in automotive technology. The term however also brings some ambiguity with it since its concept is based on the more generic word **software**. It is important to note that software for vehicles has been around for many decades, starting with simple ECU software to perform functions such as engine control and fuel injection. For example, in the 1970s, ECU software was small and often written in assembly language. Fast-forward to the 2020s, and ECU software is extremely different: much larger codebases, often using the C and C++ programming languages, more advanced functionality, and greater complexity. Using a generic term such as software may lead one to ask what the term SDV actually entails, since software for vehicles has been around for more than 50 years.

One interpretation is that it is not only software running on ECUs controlling simple vehicle functionality. SDVs promote a shift from relying on hardware to relying on software for its functionality. That is, the value comes from removing the dependence on hardware when designing the software, and shifting to a **software-first approach** where the software architecture and functionality are designed first and then the specific required hardware is defined. One additional consideration is that the hardware specifications contain room in terms of performance for improving the software and adding new functionality in the future. In particular, connectivity is a prerequisite to allow for **Over-The-Air (OTA)** software updates and configuration changes to the vehicle functionality over time. In other words, an SDV has been designed with a **software-centric architecture**, where software controls vehicle functionality, which allows for flexibility and the customization of functionality.

SDVs provide several benefits for the automotive industry as a whole. For example, one benefit is improved fuel economy and reduced emissions as the engine software can be tweaked continuously with minor software patches for optimal performance. Another benefit is increased safety as SDVs can use more complex software utilizing the interaction between different ECUs or software modules to provide more advanced safety features such as collision mitigation systems or driver assist systems. Moreover, since SDVs are designed with a software-centric architecture, it is possible to provide new functionality and services after the vehicle has been produced. For example, a new solution that was not considered during development can be designed and developed afterward due to the software-centric architecture nature of SDVs. Thus, auto manufacturers can equip SDVs with new functionality post-production such as autonomous driving functionality and advanced infotainment features, which could lead to improved user experience and increased value of the SDV.

SOA

SOA is another relatively new term in the automotive industry used to describe next-generation automotive software development. Previous generations of automotive software and in-network communication were designed and built using static concepts where, in general, the software and the signal-based communication approach were fixed. This allowed for robust, high-reliability, and low-latency functions and communication. The drawback was that it typically limited the functionality to more simple features. With the development of more advanced automotive software that requires more dynamic, modular, and scalable solutions, the concept of SOA is becoming commonplace.

In SOA, an automotive software application is designed to provide a collection of independent services. Each service provides a specific function, such as providing the engine speed, getting the steering wheel angle, controlling the brakes, adjusting the air conditioner, and so on. These services use a common set of protocols and **Application Programming Interfaces** (**APIs**) to communicate. This modular approach is very flexible as it allows for services to easily be updated or replaced without interfering with the rest of the system.

SOA provides several benefits, including modularity and flexibility as mentioned previously, where services can be designed, developed, and tested independently as well as updated or replaced easily on vehicles in the field without causing disruption to the existing functionality. Moreover, SOA allows for new functions to be developed and added after production using a common set of protocols or APIs. This enables scalability as new functionality developed for new vehicles can also be added to older vehicles.

With more automotive organizations applying SOA to their automotive software development, this trend will overall allow for more efficient, reliable and scalable automotive solutions to be developed and deployed.

Mobile apps and the cloud

One major scope change is that the focus for software development in the automotive industry is not only on the embedded systems in the vehicle itself. The scope is increasing and expanding to also include developing and maintaining software for mobile apps and cloud solutions to support the end-to-end ecosystem for mobility services.

Mobile apps

Besides embedded software, most **Original Equipment Manufacturers** (**OEMs**) offer a large number of **mobile apps** to their users. These mobile apps provide several types of features. For example, some apps provide an extension to access vehicle features remotely, that is, apps that can be used to remotely lock/unlock the car, open the trunk, or remotely start the engine.

Some apps access various data from the vehicle to provide the user with relevant information, such as driving range and fuel consumption. Some apps contain functions to provide full functionality for a certain use case. For example, for EV charging, apps can be used to check the current battery status, enable and monitor the charge level, and finalize the transaction and pay for the charging.

Other apps provide mobility services. For example, with these apps it is possible to rent a vehicle for 15 minutes or pay per hour.

The cloud

There are two main phases to consider in the context of the **cloud**: **development** and **operations**.

During development, it is important to note that more development is occurring in the cloud. Various code repositories, build environments, toolchains, and test environments are managed in the cloud. This allows for scalability and performance. To meet increasing demands as more projects are onboarded, new build environments, development toolchains, and test environments can easily be spun up, or cloud instances running in environments with improved performance can be used.

Another crucial aspect is that testing on virtual platforms allows for breaking away from the dependence on hardware. That is, it is possible to perform various functional testing and security testing on virtual platforms before hardware is available. This allows organizations to shift left and perform testing months earlier, becoming able to find and fix issues earlier and thus reducing costs.

Furthermore, automotive organizations typically offer a large number of cloud-based services during operations. The use cases provided by mobile apps typically have a corresponding backend services component hosted in the cloud. Examples of use cases include OTA updates, data analytics, and predictive diagnostics and maintenance.

Modern software development

With the development of more advanced vehicle features and mobility services, there is increased demand from users for new functionality and improved user experience. Similar to other industries such as the smartphone sector, users expect constant new features and updates. As such, there is a need to change the traditional automotive software development approaches to more agile approaches.

To support the ever-changing software requirements, automotive organizations have started to transform into software companies. In some cases, separate software development houses are established or acquired (such as Woven [5] or CARIAD [6]). In other instances, an internal software development team is consolidated from various ECU domains into one consolidated software development division (as with Bosch [7]). These software-focused departments typically follow more agile development approaches based on **Development, Security, and Operations** (**DevSecOps**). DevSecOps is an approach to establish a company culture of automation and built-in security. It requires close collaboration between development teams, operations teams, and security teams.

Furthermore, as software development approaches are changing, the **software composition** is also drastically changing. In the past, for many smaller ECUs, it was common that software was developed at tier 1 suppliers. As systems became more complex, software codebases grew and comprised software from third-party suppliers, tier 1s, tier 2s, and OEMs, as well as open-source software components.

End-to-end applications also span across vehicles, the cloud, and mobile apps, and as such require development using non-embedded programming languages and frameworks that are considered "new" to the automotive industry.

Thus, the deployment of modern software development approaches includes both changes in technologies, such as changes to software composition and the usage of new programming languages and frameworks, and changes in development methodologies and organizational structure.

Standards and regulations

Another important trend to bring up regarding the automotive industry is that of **standards** and **regulations**. Due to its safety-critical nature, the automotive industry is heavily regulated. With the introduction of more advanced functionality controlled by software coupled with more communication interfaces, which increases exposure and widens the attack surface, several cybersecurity standards and regulations have been introduced lately.

Figure 1.1 provides an overview of the relevant standards to help automotive organizations work in a more structured way to develop safer, more reliable and secure systems.

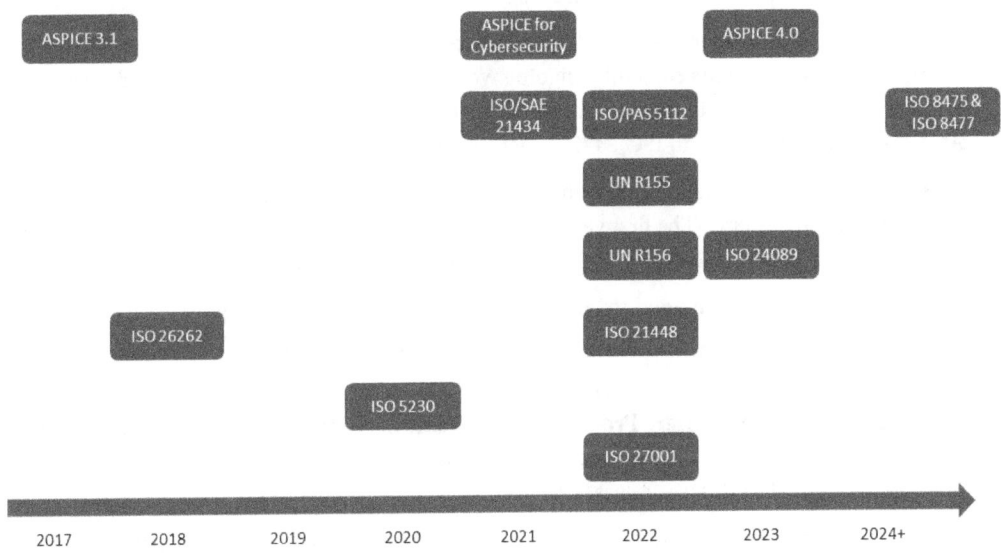

Figure 1.1 – Overview of relevant standards and regulations

Please note this is not an exhaustive list but rather provides an overview of standards and regulations relevant to current trends:

- **UN R155**: This is a regulation for **Cybersecurity Management System** (**CSMS**) focusing on establishing proper security processes in the organization and performing cybersecurity activities during the product development lifecycle. It became mandatory for type approval in UNECE member countries for new vehicle types in 2022 and for all vehicle types in 2024.

- **UN R156**: This is a regulation for **Software Updates Management System** (**SUMS**). It provides requirements for establishing a secure update communication channel to vehicles, ensuring updates are stored securely on the backend and verified on the vehicle side before performing updates. It became mandatory for type approval in UNECE member countries for new vehicle types in 2022 and for all vehicle types in 2024.

- **ISO/SAE 21434**: This is a cybersecurity engineering standard for the automotive industry released in 2021. It provides numerous requirements for security on the organizational and project levels for the entire product lifecycle. It can serve as a backbone to fulfill the requirements of CSMS for UN R155.

- **ISO 24089**: This is a software update standard for the automotive industry released in 2023. It can help automotive organizations fulfill requirements for establishing SUMS in UN R156.

- **ISO/PAS 5112**: This provides guidelines for performing audits of organizational requirements for ISO/SAE 21434 and was released in 2022.

- **ISO 8475 and ISO 8477**: To provide additional guidance on activities and requirements defined in ISO/SAE 21434, there is currently ongoing work on two new cybersecurity standards: ISO 8475, which provides guidance on **Targeted Attack Feasibility** (**TAF**) and **Cybersecurity Assurance Level** (**CAL**), and ISO 8477, which covers verification and validation activities.

- **ISO 26262**: This is the safety engineering standard for the automotive industry and applies to all safety-related systems. The first version was originally published in 2013 and an updated version was published in 2018. There is currently work ongoing on a new version.

- **ISO 21448**: This was developed to cover safety-related systems that require external input, ensuring that **Safety Of The Intended Function** (**SOTIF**) is achieved. The current version was released in 2022.

- **ASPICE**: **Automotive Software Process Improvement Capability dEtermination** (**ASPICE**) 3.1 was released in 2017 and provides a framework and a maturity model for automotive development. ASPICE 4.0 was released in 2023.

- **ASPICE for Cybersecurity**: ASPICE 3.1 was updated to include several new specific cybersecurity activities and released as ASPICE for Cybersecurity in 2021.

- **ISO 5230**: With increased usage of **Open-Source Software** (**OSS**), Open Chain published the ISO 5230 standard in 2020 to provide guidance on the management of OSS and the software supply chain.

- **ISO 27001**: Going beyond automotive development, automotive organizations also need to consider information security and establish an **Information Security Management System** (**ISMS**). ISO 27001 is not automotive-specific, but provides general requirements on how to manage information security considering people, processes, and technology. The current version of ISO 27001 was released in 2022.

After providing this overview of current automotive trends to help set a baseline for you, we will now continue with an introduction to automotive IoT.

Introduction to automotive IoT

Based on the aforementioned trends, we see new automotive use cases and solutions emerging that span across the entire ecosystem of connected vehicles to backend cloud solutions, user devices, and mobile apps. This section only aims to give a brief introduction to automotive IoT, which is described in more detail in *Chapter 2, Introducing Automotive IoT Use Cases*.

Automotive IoT

In this book, we will refer to these use cases as **automotive IoT** use cases. Before going further in this book, we will outline a definition of automotive IoT as it is a relatively new term and may have different interpretations depending on your background.

> **Automotive IoT definition**
> Automotive IoT is defined as use cases and solutions with specific purposes for the automotive industry that encompass vehicles and backend solutions and/or other relevant smart devices such as mobile devices.

An overview of the automotive IoT ecosystem is depicted in *Figure 1.2*.

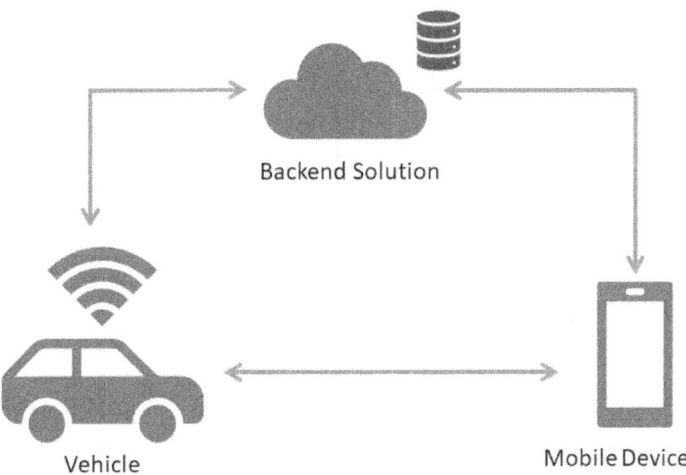

Figure 1.2 – Overview of the automotive IoT ecosystem

The automotive IoT ecosystem comprises three main entities: **Vehicle**, **Backend Solution**, and **Mobile Device**. The vehicle contains some communication module that provides several means of communication with external entities. Commonly supported interfaces include cellular communication such as 4G and 5G, as well as short-range communication such as Wi-Fi, Bluetooth, UWB, and **Near Field Communication** (**NFC**). The backend solution is typically hosted in the cloud and provides various functionality, for example, to send commands to and receive data from the vehicle. The backend solution also processes the collected data and offers various features to the user's mobile device. The mobile device contains the relevant app that allows the user to access features either directly on the vehicle using short-range wireless communication such as Wi-Fi and Bluetooth, or indirectly by communicating through the backend solution to the vehicle.

Automotive IoT use case examples

Using the automotive IoT definition given previously, the following are examples of automotive IoT use cases:

- **Phone as a key**
- **Remote diagnostics**
- **Vehicle management**

As shown in *Figure 1.2*, these use cases encompass two or three entities in the ecosystem:

- **Vehicle**, **Backend Solution**, and **Mobile Device** (mobile app) for the **phone as a key** use case
- **Vehicle** and **Backend Solution** for the **remote diagnostics** and **vehicle management** use cases

These three use cases are briefly described as follows.

Phone as a key

The phone as a key use case allows users to install an app on their mobile device that then serves as a key for their car. This means that users can use their phone to unlock the car door to get into the vehicle and start the engine.

Remote diagnostics

The remote diagnostics use case allows an OEM to remotely collect diagnostics data periodically as well as specific log data about certain incidents, such as component failures. The collected data is then processed in the OEM backend to help with **predictive diagnostics**. This allows the system to give indications and foresee potential component failures, providing warnings to the driver to bring their vehicle in for a checkup or component replacement before the component fails.

Vehicle management

The vehicle management use case gives companies an overview of their fleet of vehicles to better understand the vehicle usage and maintenance required. For example, the following data and conditions may be monitored and managed: vehicle location, fuel consumption, maintenance appointment schedules, driver behavior and safety, and optimized route planning and vehicle usage. The vehicle management system could also be used to ensure compliance with regulations by offering evidence based on the data collected from the vehicles.

Data management for automotive IoT use cases

It is important to note that these use cases typically handle various types of data. For example, user identities, credentials for authentication, and relevant vehicle, backend, and mobile app data are handled in the phone as a key use case.

Likewise, log data and diagnostics data are collected from the vehicles for the remote diagnostics and vehicle management use cases. Moreover, credentials for authentication and relevant vehicle data and backend data are also handled in these use cases.

Summary

This chapter presented an introduction to the current automotive trends. Common terminology and concepts were described, including CASE, SDV, SOA, mobile apps and the cloud, modern software development approaches, and relevant standards and regulations. Moreover, we gave an introduction to the automotive IoT ecosystem and presented several automotive IoT use cases. These topics allow you to establish the solid foundation needed to consume the following chapters in the book. The journey will equip you with skills for understanding and navigating the complex landscape of automotive IoT, from evolving vehicle architectures to secure development practices and holistic application development.

In *Chapter 2, Introducing Automotive IoT Use Cases*, we will provide several examples of automotive IoT use cases in more detail.

References

- [1] SAE J3016_201806 Standard, https://www.sae.org/standards/content/j3016_201806/
- [2], White House Fact Sheet on Affordable Electric Vehicles, https://www.whitehouse.gov/briefing-room/statements-releases/2023/04/17/fact-sheet-biden-harris-administration-announces-new-private-and-public-sector-investments-for-affordable-electric-vehicles/
- [3], European Parliament Press Release on Fit for 55 Initiative, https://www.europarl.europa.eu/news/en/press-room/20230210IPR74715/fit-for-55-zero-co2-emissions-for-new-cars-and-vans-in-2035
- [4], U.S. Department of Commerce Market Intelligence on Japan's Transition to Electric Vehicles, https://www.trade.gov/market-intelligence/japan-transition-electric-vehicles
- [5], Toyota Woven Platform, https://woven.toyota/en/
- [6], Volkswagen Cariad Technology, https://cariad.technology/
- [7], Bosch Car Software and Electronics Stories, https://www.bosch.com/stories/car-software-electronics/

Get This Book's PDF Version and Exclusive Extras

Scan the QR code (or go to `packtpub.com/unlock`). Search for this book by name, confirm the edition, and then follow the steps on the page.

Note: Keep your invoice handy. Purchases made directly from Packt don't require one.

2
Introducing Automotive IoT Use Cases

This chapter introduces automotive IoT use cases. The integration of the **Internet of Things** (**IoT**) into automotive industry marks a revolutionary step forward, enhancing vehicle functionality and driver experience through a myriad of connected applications. These advancements not only improve safety but also transform vehicles into highly personalized environments tailored to individual preferences and needs. From predictive maintenance and optimized fleet management to advanced driver-assistance systems and enhanced infotainment, the applications of IoT in vehicles are reshaping the automotive industry. This chapter explores the multifaceted use cases of IoT within the automotive sector, highlighting how this technology is steering us toward a more connected and efficient future.

In this chapter, we're going to cover the following main topics:

- Enhanced driver experience and safety
- Optimized fleet management
- Connected mobility revolution

The IoT is enhancing the range of applications within vehicles, offering significant benefits to the automotive industry. By embedding sensors, processors, and connectivity throughout vehicles, the automotive industry unlocks a treasure trove of data and capabilities. Let us explore a range of use cases that are reshaping the automotive industry for both consumers and manufacturers.

Enhanced driver experience and safety

Enhanced driver experience and safety are revolutionized by automotive IoT through features such as connected car services for remote access, **Advanced Driver-Assistance Systems (ADASs)** acting as a co-pilot, personalized in-car experiences for maximum comfort, and the convenience of using your phone as a key.

Connected car services

IoT allows vehicles to connect to the cloud, enabling features such as real-time traffic updates, navigation with pinpointed accuracy, and remote diagnostics. An example is a car that reroutes you around unexpected congestion or alerts you to potential mechanical issues before they become a major problem. The following figure shows how the user can configure and remotely access the connected car via a mobile phone.

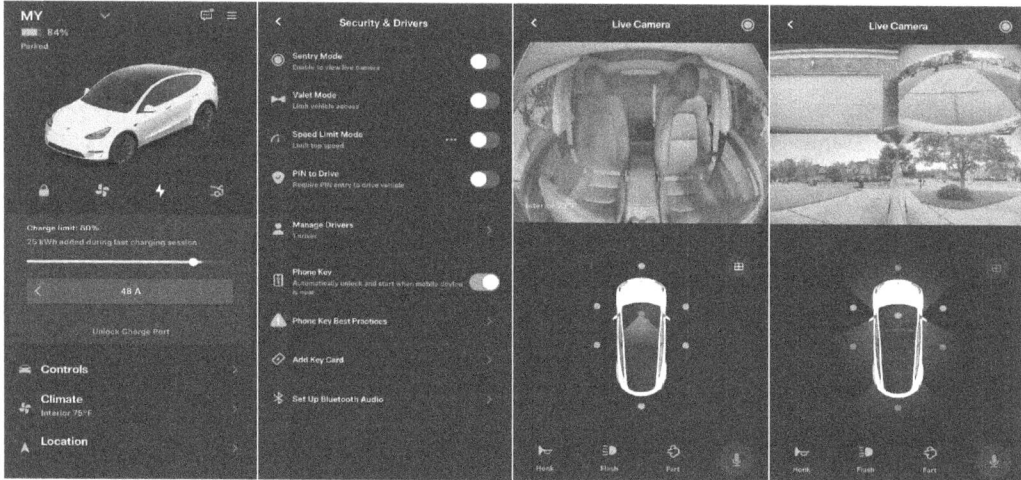

Figure 2.1 – Connected car end user mobile app facilitates remote access (Tesla Model Y)

Connected car services are like superpowers for a regular car, turning it into a high-tech companion. Cars these days have their own built-in cellular connection, just like a smartphone. This allows them to connect to the internet and a wide range of services. Through a smartphone app connected to the car, we can unlock a variety of features remotely. This could include locking or unlocking the doors, checking the fuel level or battery life (for electric vehicles), and even starting the car to preheat or cool the cabin before we get in. We can also remotely check the surroundings of the car for safety by leveraging the built-in car cameras.

Advanced driver-assistance systems

A vehicle with an ADAS is equipped with sensors such as LiDAR, radar, and a camera that work together to provide features such as automatic emergency braking, lane departure warning, and blind-spot detection. These systems leverage real-time data to enhance safety and reduce accidents, as shown in the following diagram, in which a vehicle detects an obstruction in its path and not only takes a corrective action but also relays the information to other connected vehicles that are heading in that direction to slow down.

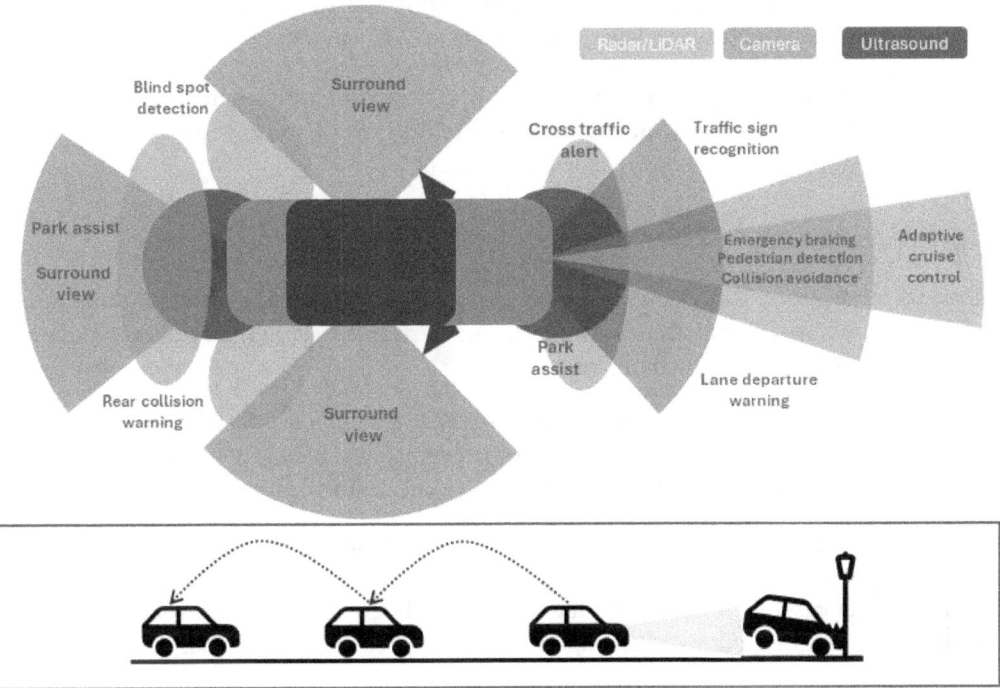

Figure 2.2 – ADAS enhanced with IoT

An ADAS greatly benefits from IoT technology, by connecting vehicles to a network of sensors and devices, and it can be enhanced in several ways:

- It improves situational awareness as IoT sensors can provide the ADAS with a broader range of data about the vehicle's surroundings. This can include information from other vehicles, weather conditions, and road infrastructure. With this data, the ADAS can make more informed decisions and react to potential hazards more effectively.

- IoT allows for real-time data exchange between vehicles and the cloud. This enables the ADAS to receive updates on traffic conditions, accidents, and road closures. This can be crucial for features such as adaptive cruise control and lane departure warning.

- IoT sensors can be used to monitor driver behavior, such as whether they are drowsy or distracted. If the system detects a potential problem, it can issue a warning or even take corrective action, such as pulling the car over to a safe location.

Here are some examples of ADAS functionalities that leverage IoT:

- IoT can provide **Automatic Emergency Braking** (**AEB**) with data on nearby vehicles and pedestrians, allowing it to react more quickly and effectively to potential collisions
- **Lane Departure Warning** (**LDW**) systems can use real-time traffic data from IoT to account for congestion and adjust lane departure warnings accordingly

Overall, ADASs and IoT are a powerful combination that can significantly improve road safety and the driving experience. As both technologies continue to develop, we can expect to see even more innovative ADAS features emerge that leverage the power of IoT.

Personalized in-car experience

IoT personalizes the in-car experience. Biometric sensors can adjust cabin temperature and music preferences based on driver identity. Connected infotainment systems offer a wider range of entertainment options and seamless integration with smartphones.

Connected infotainment systems become a seamless extension of your smartphone, having your favorite music streaming service, audiobooks, or podcasts readily available on the car's dashboard. The system integrates seamlessly with the smartphone for safe, hands-free communication.

Personalization goes beyond just temperature and music. IoT connects your car to a wider ecosystem of services. For instance, the navigation system integrated with your calendar could learn your frequent destinations and automatically set them up before you get into the car. It automatically suggests alternate routes based on real-time traffic data, helping you avoid delays on your usual commute.

The level of personalization can be tailored to one's preferences. Imagine having different profiles for various drivers. Parents could set specific temperature and music preferences for their children, while also having their own personalized profiles.

This personalized in-car experience with IoT is all about creating a comfortable, convenient, and even safer driving experience. It's like having a co-pilot who anticipates your needs, adjusts the environment to your liking, and keeps you connected and informed throughout your journey. Imagine a future where your car transforms from a mode of transportation to a personalized sanctuary on wheels.

Phone as a key

The **phone as a key** application offers users the capability to remotely manage various vehicle functions, including starting/stopping the engine, locking/unlocking doors, manipulating the trunk, and adjusting windows. Additionally, it facilitates easy key sharing over the phone, proving particularly useful in scenarios such as fleet management, shared vehicles, and rental cars. Phone as a key systems use Bluetooth/NFC for advanced features and stronger security with encryption and remote access control. Passive keys rely on radio signals, offering convenience, but these are more vulnerable to relay attacks. Phones offer dynamic security updates, enhancing overall safety.

It's important to note that the quality and quantity of features supported by the phone as a key application can vary between different **Original Equipment Manufacturers** (**OEMs**). For instance, Ford provides a dedicated application called *FordPass* [2], supporting remote features such as starting/stopping, locking/unlocking doors, and checking the vehicle's fuel/charge status. Tesla's mobile app provides more comprehensive functionality where we can remotely monitor the vehicle's internal and external environments for safety, schedule charging, and set charge limits.

Phone as a key systems will perform functions such as starting/stopping the engine, locking/unlocking doors, and so on. This functionality can be implemented directly through the local system using the following features:

- **Near Field Communication** (**NFC**): NFC is a close-range wireless technology used in phones. With NFC, users can unlock and control their vehicles by tapping their smartphones near an NFC reader on the car. A unique aspect of NFC is that it works even if the phone battery is dead, making it super convenient. This means users can still use their phones as keys for keyless entry and vehicle control, even if the phone's battery is dead. NFC's simple and battery-independent feature boosts the overall ease and dependability of keyless entry systems.

- **Bluetooth Low Energy** (**BLE**): BLE is a wireless communication technology specifically designed for short-range data exchange between devices while consuming minimal power. In the context of the phone as a key application, BLE is employed to enable the phone to remotely control various vehicle functions. This includes actions such as starting or stopping the engine, locking or unlocking doors, opening the trunk, and adjusting windows. BLE's energy-efficient design makes it well suited for such applications, allowing users to manage their vehicles conveniently from their smartphones without significant power consumption.

- **Ultra-Wide Band** (**UWB**): UWB is another wireless technology used in phones for better accuracy and security. Unlike BLE, which focuses on low-power and short-range communication, UWB stands out for its super-precise location information. When added to the phone as a key application, UWB makes keyless entry safer by pinpointing exactly how close your phone is to the vehicle. This accuracy lowers the risk of unauthorized access, giving a more secure keyless experience than traditional methods.

NFC, BLE, and UWB are wireless technologies, and each has its own benefits. BLE is good at saving energy and works well with many devices, such as wearables and IoT gadgets. NFC is great for short-distance communication, and it's simple to use without needing a lot of battery power. This makes it perfect for things such as keyless entry systems. UWB is super precise, especially in knowing where things are. It makes keyless entry more accurate and secure. When choosing between BLE, NFC, and UWB, it depends on what you need, such as saving energy, keeping things simple, or having extra precise security. The following diagram illustrates the phone connecting with and without the internet to the vehicle providing remote and local access.

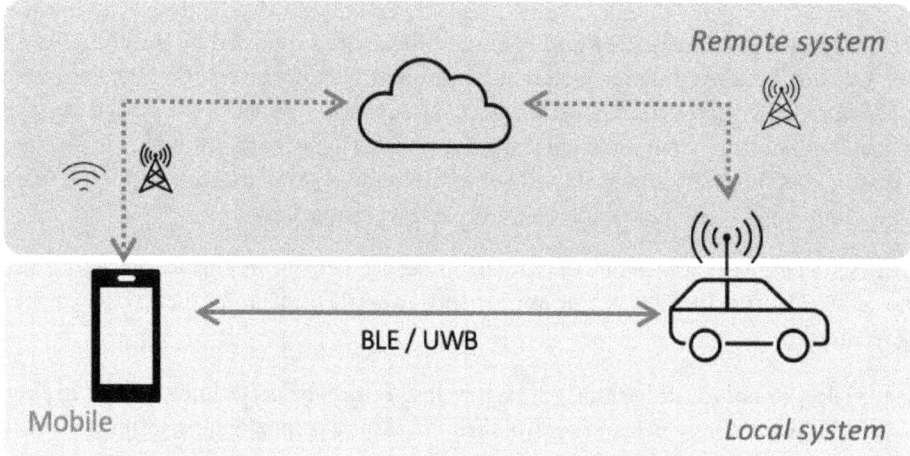

Figure 2.3 – Phone as a key provides remote and local access

We can use the phone as a key in two ways.

- The first way does not need the internet and is called a local system. It works with NFC, BLE, or UWB.
- The second way, called a remote system needs an internet connection.

We'll talk more about local and remote systems next.

Phone as a key – local system

In this scenario, the phone initiates a direct secure connection with the vehicle using BLE/UWB/NFC. Once the secure connection is established, the phone and vehicle exchange commands/responses such as "door lock" or "door unlock" based on user actions. The local system has limitations as it only works when the phone comes into range of the vehicle within a limited distance, such as a few meters.

Figure 2.3 shows the high-level view of the phone as a key with local and remote systems. In the local system case, the phone communicates with the vehicle through NFC, BLE, or UWB. In the remote system case, communication between the phone and the vehicle occurs via the cloud using cellular networks.

Phone as a key – remote system

In this scenario, the internet is used for communicating between the phone and the vehicle, which gets rid of distance limits. However, both the vehicle and the phone need to be connected to the internet. As you can see in *Figure 2.4*, the vehicle requires access to a cellular network for internet connectivity.

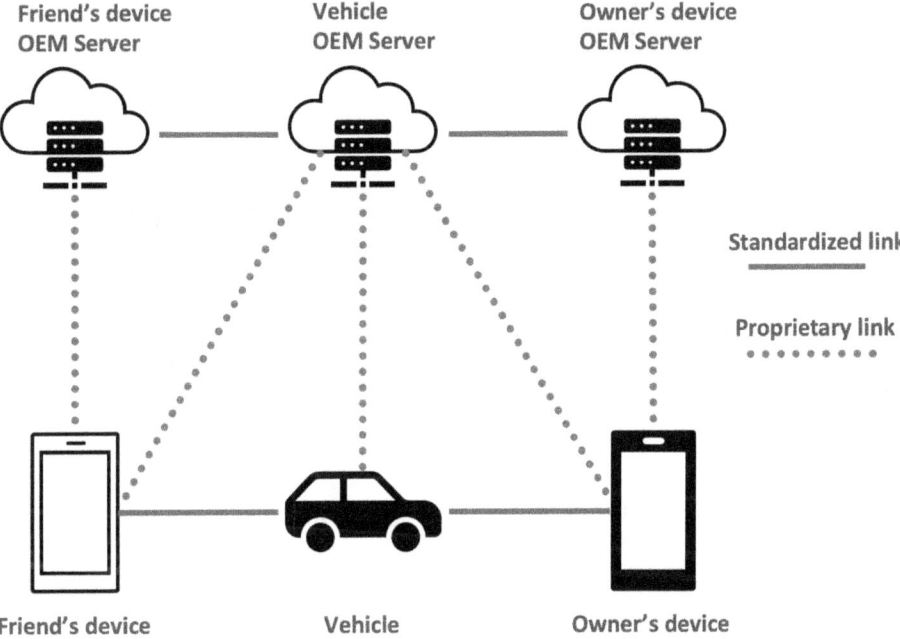

Figure 2.4 – Sharing vehicle keys via mobile phone [1]

The remote system approach lets a person control (including locking/unlocking doors and opening the trunk) a vehicle remotely. The cloud helps with communication, acting as a secure hub to manage access. Cloud servers check whether users are allowed, and store rules such as location-based access, allowing only certain phones, and revocation policies. The cloud also makes updates and keeps an eye on access, making things more secure with geolocation.

Optimized fleet management

Optimized fleet management leverages real-time vehicle tracking and telematics to not only monitor location and fuel efficiency but also utilize predictive maintenance and driver performance monitoring to create a safer, more cost-effective transportation system.

Real-time vehicle tracking and telematics

Fleet managers gain real-time visibility into vehicle location, fuel consumption, and driver behavior. This data allows for optimized route planning, improved fuel efficiency, and targeted maintenance schedules, leading to significant cost savings.

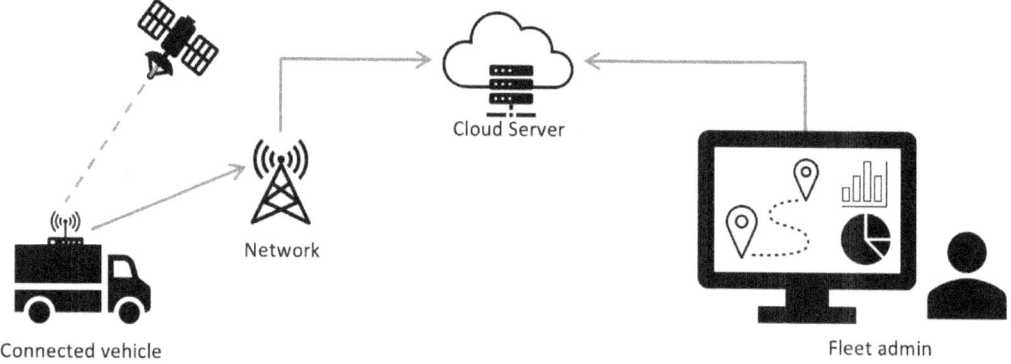

Figure 2.5 – Connected vehicle sending information to the cloud server

Figure 2.5 shows how a connected vehicle derives its location using GPS satellites and then relays its location and other information to the cloud server via the cellular network. The fleet admin receives all the information on their dashboard.

Driver performance monitoring

Fleet managers can monitor driver behavior for metrics such as speeding, harsh braking, and idle time. This data can be used to implement driver coaching programs, promoting safer and more fuel-efficient driving habits. In addition to promoting safer driving habits, the data collected through driver performance monitoring can also be utilized by insurance companies for risk assessment. By analyzing metrics such as speeding, harsh braking, and idle time, insurers can gain insights into the behavior of individual drivers and adjust premiums accordingly. Drivers who demonstrate safer driving habits as indicated by the monitoring system may qualify for lower insurance premiums, incentivizing them to maintain good driving practices.

Predictive maintenance

IoT sensors constantly monitor vehicle health, enabling predictive maintenance. By analyzing data on engine performance, tire pressure, and other parameters, potential problems can be identified and addressed before they lead to costly breakdowns. For example, sensors monitoring engine health analyze real-time data, predicting when critical parts are likely to fail. This allows for timely maintenance, reducing the risk of unexpected breakdowns and improving vehicle reliability. Another example includes **Tire Pressure Monitoring Systems (TPMSs)**, which are valuable for remote diagnostics by providing real-time alerts on tire pressure issues, enhancing safety, fuel efficiency, and reducing tire wear. Integrated into remote diagnostics, TPMS data aids in preventive maintenance, operational efficiency, and cost savings by enabling timely interventions for tire issues, optimizing maintenance schedules, and improving fleet management through data analytics. The following diagram illustrates a use case where predictive analytics running on the vehicle or in the cloud notify or alert the user to initiate a remote preventive maintenance/diagnostics session.

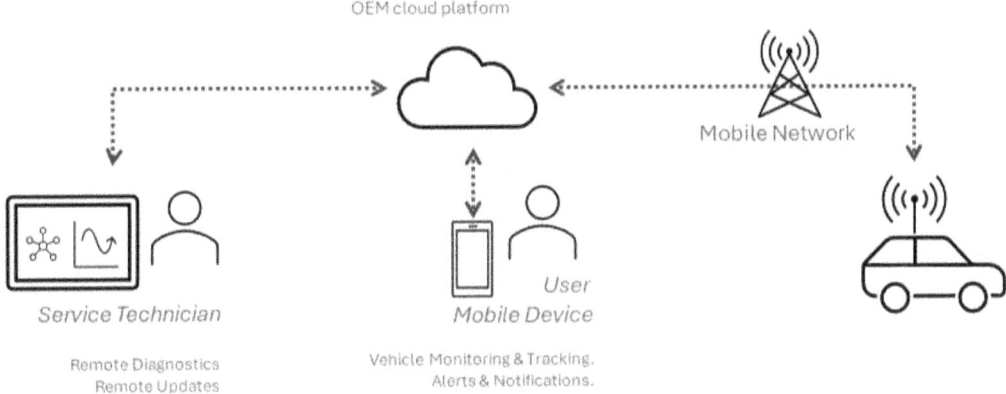

Figure 2.6 – Predictive maintenance/remote diagnostic

Predictive maintenance benefits vehicle owners by alerting them to the vehicle's health status, reducing trips to dealers, and lowering overall repair costs. Software updates can remotely fix some issues, a trend that becomes more common with software-defined vehicles. Software-defined vehicles use software for flexible control of driving dynamics and features, moving away from traditional hardware control. Dealers benefit by knowing the vehicle's status, identifying malfunctions, scheduling maintenance, predicting the need for roadside assistance, and ordering parts based on advanced diagnostic information.

Predictive maintenance relies on vehicle diagnostic data analytics and remote diagnostics. Remote diagnostics is a process that allows the monitoring and assessment of a system, equipment, or vehicle's performance, condition, and potential issues from a distant location, often facilitated through connected technologies such as Wi-Fi, cellular networks such as 4G and 5G, and data transmission. Drivers can receive real-time support if anything goes wrong while driving the vehicle. For example, if the airbag light turns on, the service center can retrieve the data and determine the best course of immediate action.

Figure 2.6 illustrates the high-level view of predictive maintenance/remote diagnostics. Users can receive location updates and status alerts for their vehicles through their phones, while technicians can remotely access diagnostic information by connecting to the vehicle via the cloud.

Some key challenges in predictive maintenance include data privacy and the need to safeguard user data, especially with increased connectivity and remote diagnostics, and the maintenance of cloud systems, ensuring the reliability and security of cloud-enabled functionalities.

Connected mobility revolution

In the connected mobility revolution, IoT is taking us beyond driving, with smart parking solutions, vehicles talking to everything (V2X communication), and even connected supply chains and manufacturing seamlessly working together for a more efficient transportation ecosystem.

Smart parking solutions

As shown in *Figure 2.7*, IoT occupancy sensors can detect vacant parking spaces in real time, guiding drivers to available spots and reducing congestion caused by circling for parking. Imagine a world where parking frustration becomes a thing of the past.

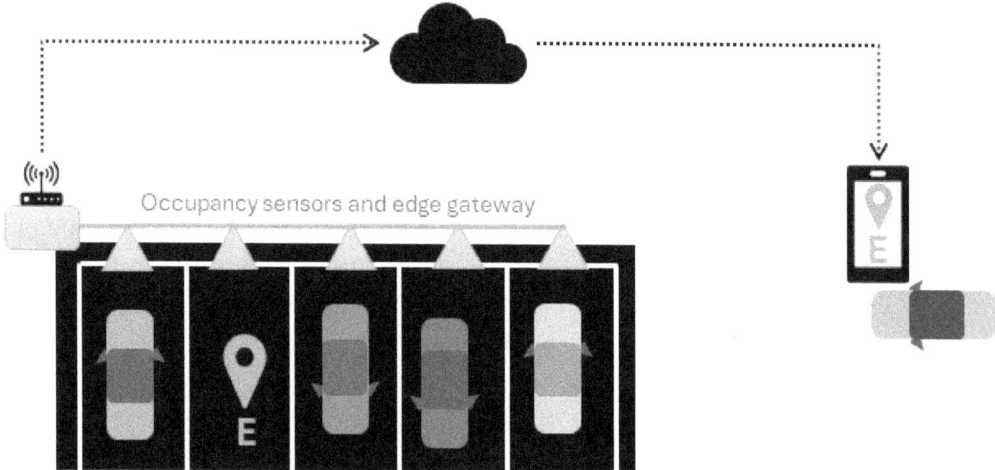

Figure 2.7 – Smart parking solution

The system is composed of sensors, cameras, and apps that streamline the parking experience. Sensors (usually underground) or cameras detect whether a parking spot is occupied, and this data is then sent to the cloud via an edge gateway. This data is then made available to the mobile app that shows the driver's available parking in real time. These systems improve the traffic flow by reducing the time drivers spend searching for parking. They can also help cities gather data on parking usage patterns, allowing for better management of parking spaces.

Vehicle-to-Everything (V2X) communication

V2X heralds a transformative era in transportation, where vehicles seamlessly interact with each other and with various elements of the urban landscape, from infrastructure to pedestrians and cyclists. By leveraging **Dedicated Short-Range Communication** (**DSRC**) [3] technology or **Cellular V2X** (**C-V2X**) [4][5], V2X facilitates a dynamic exchange of real-time data, revolutionizing road safety, traffic management, and urban mobility.

Vehicles can communicate with each other and with roadside infrastructure, enabling features such as collision avoidance and automated traffic management. As you can see in *Figure 2.8*, V2X paves the way for safer, more efficient, and potentially autonomous transportation systems.

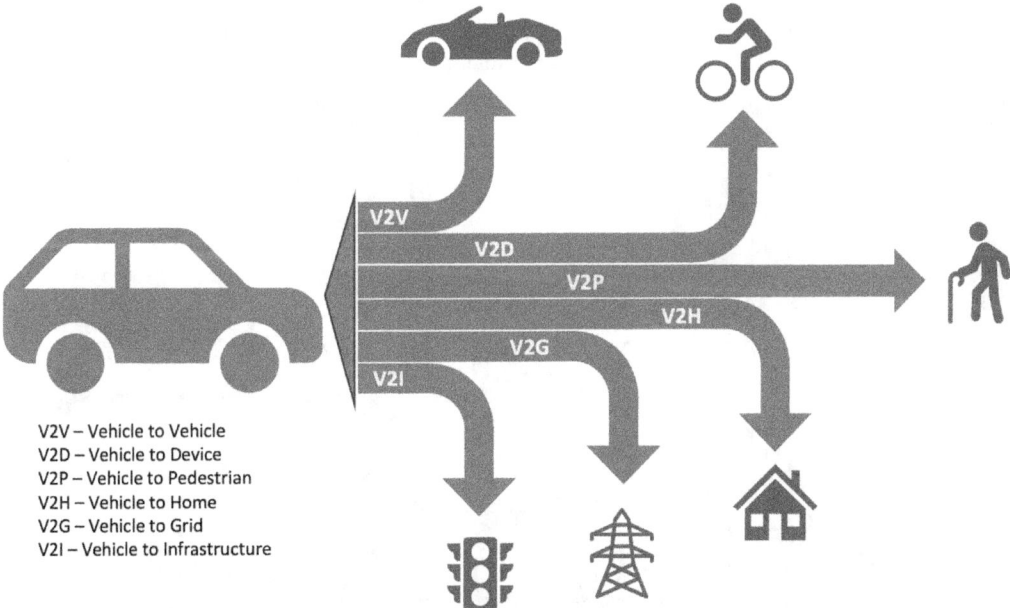

V2V – Vehicle to Vehicle
V2D – Vehicle to Device
V2P – Vehicle to Pedestrian
V2H – Vehicle to Home
V2G – Vehicle to Grid
V2I – Vehicle to Infrastructure

Figure 2.8 –V2X communication model

V2X empowers vehicles to anticipate and mitigate potential hazards through proactive communication. Whether it's alerting drivers to nearby vehicles (V2V), communicating with pedestrians (V2P), or integrating with infrastructure (V2I), this system creates a comprehensive safety net. For instance, vehicles can receive signals from traffic lights, preemptively adjusting speed for upcoming green lights or warning drivers of approaching emergency vehicles. This proactive approach not only reduces the risk of accidents but also enhances overall road safety. Let's look at the V2X model:

- **V2V**: Alerts one vehicle to the presence of another; can talk using DSRC technology
- **V2D**: Vehicles communicate with cyclists' V2D devices, and vice versa
- **V2P**: Car communication with pedestrians with approaching alerts, and vice versa
- **V2H/V2G**: Vehicles/fleets will act as supplement power supplies to the home/grid
- **V2I**: Alerts vehicles to traffic lights, traffic congestion, and road conditions

In a smart city ecosystem enabled by V2X communication, traffic management becomes more agile and responsive. As you can see in *Figure 2.9*, emergency vehicles approach, traffic lights intelligently prioritize their passage, seamlessly switching to green to facilitate their swift movement. Drivers receive real-time alerts about road conditions, hazards, or construction work directly on their dashboards, enabling them to navigate congestion and delays more effectively. Moreover, smart navigation systems leverage real-time traffic data to optimize routes, suggesting the optimal speed for a smooth, frustration-free journey.

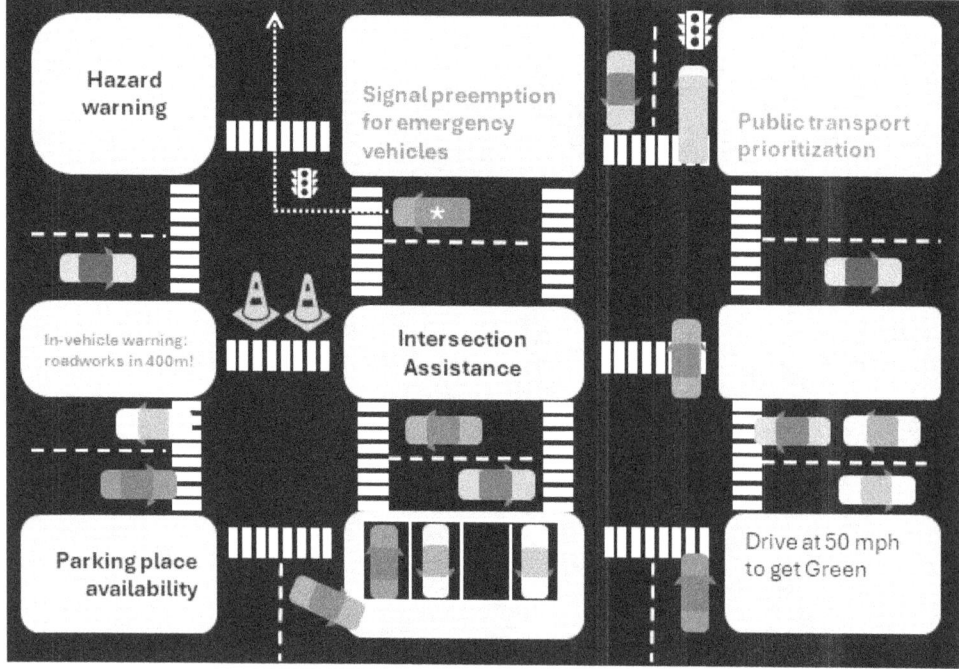

Figure 2.9 – Smart city use cases

V2X doesn't just benefit individual drivers; it transforms the entire urban mobility landscape. Gone are the days of circling endlessly for parking; in-vehicle displays guide drivers straight to available spaces, reducing congestion and emissions. Furthermore, public transportation systems gain priority access to navigate city congestion, ensuring timely arrivals and departures for commuters. By seamlessly integrating vehicles with city infrastructure, V2X paves the way for a more efficient, sustainable, and interconnected urban transportation network.

In summary, V2X communication represents a paradigm shift in transportation, fostering safer roads, efficient traffic management, and enhanced urban mobility. By harnessing the power of real-time data exchange, cities can embrace a future where transportation is not just about getting from point A to point B but about doing so seamlessly, safely, and sustainably.

Connected supply chain and manufacturing

In the automotive industry, the integration of IoT technology revolutionizes the manufacturing process, creating a seamlessly connected ecosystem that extends from the supply chain to the production floor. By leveraging IoT devices and sensors, automotive manufacturers gain unprecedented visibility and control over every aspect of their operations, driving efficiency, reliability, and innovation. IoT-enabled tracking systems meticulously monitor the movement of parts and materials at every stage of the supply chain, from sourcing to assembly. Real-time data analytics provide insights into inventory levels, shipment status, and potential bottlenecks, empowering manufacturers to optimize logistics, reduce lead times, and enhance overall supply chain agility.

On the production floor, IoT sensors embedded in machinery and equipment transform traditional manufacturing into a smart, data-driven operation. These sensors continuously monitor the health and performance of critical assets, detecting anomalies and predicting maintenance needs before issues arise. Predictive maintenance not only minimizes unplanned downtime but also maximizes equipment uptime, optimizing productivity and reducing operational costs. Furthermore, real-time insights from production-line sensors enable manufacturers to fine-tune workflows, identify process inefficiencies, and implement continuous improvements for enhanced quality and throughput. The connected supply chain and manufacturing paradigm foster collaboration across the automotive ecosystem, facilitating seamless communication and coordination between suppliers, manufacturers, and distributors. Through IoT-enabled platforms, stakeholders can share real-time data, collaborate on demand forecasting, and proactively address supply chain disruptions. This collaborative approach promotes agility, resilience, and innovation, enabling automotive companies to adapt rapidly to changing market dynamics and customer demands.

In essence, connected supply chain and manufacturing initiatives empower automotive manufacturers to embrace a new era of efficiency, agility, and competitiveness. By harnessing the power of IoT technology, companies can optimize their operations, drive sustainable growth, and deliver superior products that meet the evolving needs of the modern automotive market.

Summary

In this chapter, we explored various IoT integrations in the automotive industry that significantly enhance vehicle functionality and driver safety through connected car services, ADASs, and personalized in-car experiences. IoT facilitates innovations such as remote vehicle management via smartphones, real-time traffic updates, remote diagnostics, and predictive maintenance, improving both convenience and efficiency. Additionally, IoT extends its benefits to optimized fleet management and revolutionary V2X communication, which improves traffic management and urban mobility.

In the next chapter, we will delve into the evolution of vehicle architectures and frameworks, examining how they establish a foundational platform to support a variety of automotive IoT use cases.

References

- [1]: `https://carconnectivity.org/wp-content/uploads/2022/11/CCC_Digital_Key_Whitepaper_Approved.pdf`
- [2]: `https://www.ford.com/support/category/fordpass/`
- [3]: `https://en.wikipedia.org/wiki/Dedicated_short-range_communications`
- [4]: `https://en.wikipedia.org/wiki/Cellular_V2X`
- [5]: `https://auto-talks.com/technology/c-v2x-technology/`

Get This Book's PDF Version and Exclusive Extras

Scan the QR code (or go to `packtpub.com/unlock`). Search for this book by name, confirm the edition, and then follow the steps on the page.

Note: Keep your invoice handy. Purchases made directly from Packt don't require one.

Part 2: Vehicle Architectures

This part primarily focuses on the evolution of vehicle architecture, its associated technologies, and vehicle diagnostics as well. It also explores the technological advancements that enable IoT applications within vehicles, including middleware software, tools, and technologies, illustrated with relevant examples.

This part has the following chapters:

- *Chapter 3, Vehicle Architecture and Framework*
- *Chapter 4, Vehicle Diagnostics*
- *Chapter 5, Next Wave of Vehicle Diagnostics*

3

Vehicle Architecture and Frameworks

Over the past 25+ years, vehicle architecture has experienced substantial transformations. Initially, it relied on a distributed system, with each **electronic control unit** (**ECU**) dedicated to a single function, often without significant software involvement. However, a noticeable shift has occurred toward a more integrated approach, where a single ECU now handles multiple software functions. Furthermore, vehicle design has witnessed a transition from mechanical components to electronic components. As a result, the number of ECUs has exponentially increased over time. In this chapter, we will explore vehicle architectures, hypervisor, and available frameworks such as **Classic Automotive Open System Architecture** (**AUTOSAR**), and Adaptive AUTOSAR. We will also discuss support for the tools required for AUTOSAR configurations. Finally, we will compare Classic AUTOSAR with Adaptive AUTOSAR.

In this chapter, we're going to cover the following main topics:

- Scale of vehicle architectures
- Standard frameworks to support vehicle architecture and IoT

The scale of vehicle architecture

Vehicle architecture has evolved over the years, transitioning from a distributed architecture to a more centralized one. This shift aims to support new technologies and features, including connected vehicles, software-defined vehicles, 5G, and IoT use cases such as efficient remote diagnostics and Phone as a key (which was introduced in *Chapter 2, Introducing Automotive IoT Use Cases*). This evolution also optimizes the hardware and expands the software scope, which, in turn, reduces the product's overall cost and improves the upgradability and maintainability aspects of the product. For example, remote diagnostics can identify issues without the need to perform a physical inspection, while OTA updates allow for seamless software enhancements.

Distributed architecture

Distributed architecture is widely employed in traditional vehicles and is where each ECU is responsible for one function, such as audio. These ECUs communicate over the CAN. As illustrated in *Figure 3.1*, vehicle communication becomes complex due to the wiring and the number of channels inherent in the distributed architecture, leading to an overall increase in the weight of the vehicle. Vehicle users often need to physically visit service centers for issue resolution in vehicles due to no/limited remote diagnostics, OTA update capabilities, and the need for hands-on inspection and professional intervention.

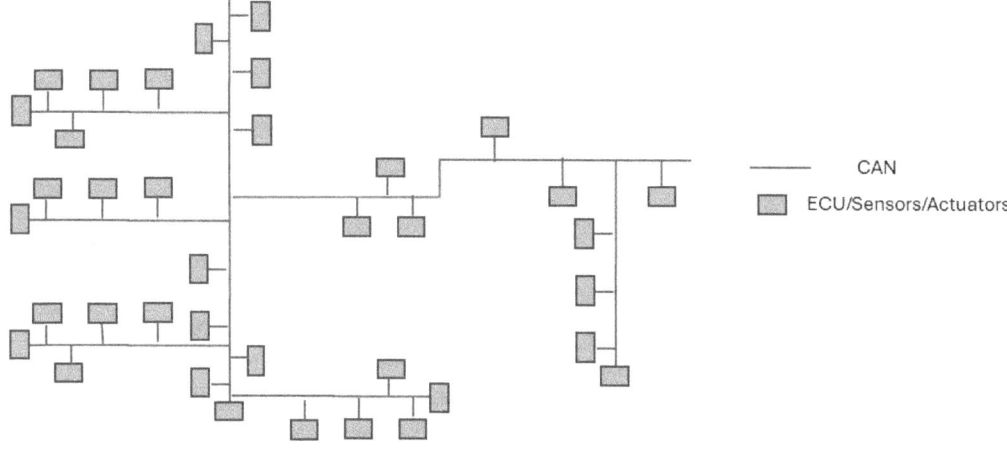

Figure 3.1 – Distributed vehicle architecture

Some of the shortcomings of distributed architecture are addressed as part of the centralized zonal domain architecture with the introduction of the domain controller. We'll cover this in more detail in the next section.

Centralized zonal domain architecture

A domain controller refers to a centralized computing unit that manages and controls specific functions within a vehicle's electronic architecture, consolidating various functionalities for improved efficiency and integration.

In this architecture, several related functionalities are grouped and assigned dedicated ECUs, such as the **autonomy domain controller (ADC)**, **cockpit domain controller (CDC)**, and **vehicle domain controller (VDC)**, as shown in *Figure 3.2*. GM, Ford, Stellantis, and other OEMs are already using the current centralized zonal domain architecture.

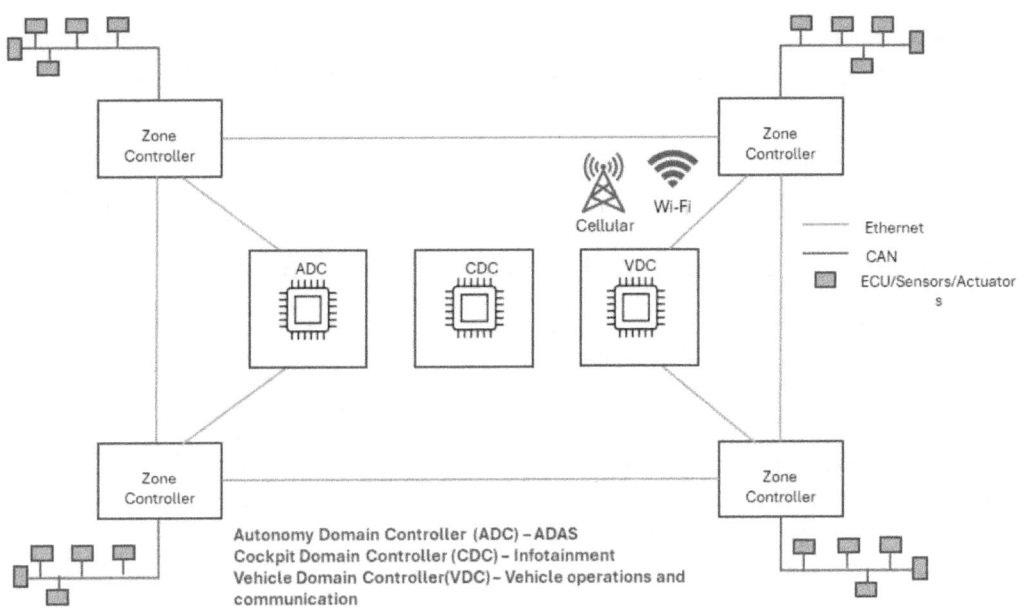

Figure 3.2 – Centralized zonal domain architecture

ADC could involve processing the sensor data, decision-making, and how the vehicle is controlled. For example, autonomous braking requires sensor data to be processed so braking can occur autonomously based on the surrounding environment and driving conditions.

CDC supports infotainment features such as supporting media (audio/video), navigation, projecting phone screens, and providing applications such as Pandora, Google/Apple Maps, and others.

VDC mainly supports communication to the external world via Ethernet, CAN, Wi-Fi, a cellular networks such as 4G and 5G. Other ECUs in the vehicle communicate with external world-like devices on IoT via VDC.

Each domain controller contains one or more **systems on a chip** (**SoCs**). Each SoC contains multiple cores, and each core operates at multiple **gigahertz** (**GHz**) speeds. These SoCs are capable of performing complex tasks in a short period. An example of such a SoC is the Qualcomm Snapdragon processor. More details about this SoC can be found in reference [1].

The zone controller shown in *Figure 3.2* is an **input/output** (**I/O**) extender that supports various communication protocols, such as CAN, LIN, and Ethernet. The zone controller communicates with the domain controller via Ethernet and with legacy ECUs through protocols such as CAN and LIN. Serving as a gateway, it converts messages between legacy ECUs and the domain controller, performing pre-processing for seamless communication.

This architecture not only minimizes vehicle wiring but also enhances communication and enables new features, such as OTA updates and remote diagnostics, through the implementation of a domain controller.

In the next wave of architecture, ADC, CDC, and VDC are combined into a single ECU called the central computer, which consists of multiple domains. We'll explore this architecture in the next section.

A central computer with multiple domain-specific SoCs

In this architecture, the ECU incorporates multiple SoCs, with each dedicated SoC assigned to a specific domain controller, such as ADC, CDC, or VDC, as shown in *Figure 3.3*. These individual domain controllers are consolidated into a single ECU known as the central computer, resulting in cost savings and less space being required in the vehicle. A successful implementation requires efficient inter-SoC communication, effective power and thermal management, robust security, and isolation measures, as well as support for development and debugging tools to integrate multiple domain controller functionalities into the central computer.

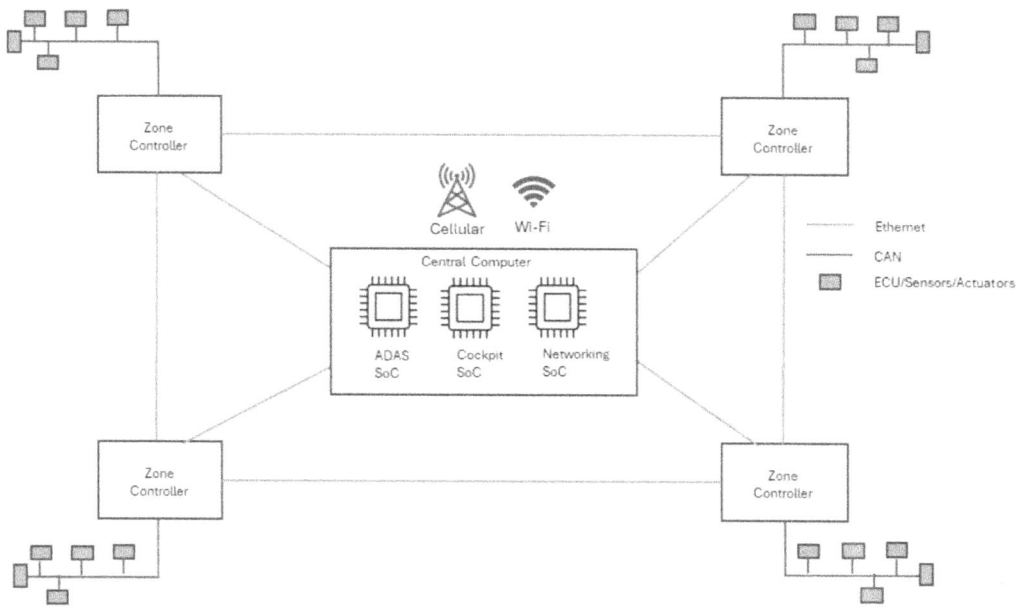

Figure 3.3 – A central computer with multiple domain-specific SoCs

In the next section, we will explore the possibility of further streamlining and reducing costs by replacing multiple SoCs with a single powerful SoC, eliminating the need for inter-SoC communication.

A central computer with a single SoC

In this architecture, a single SoC handles ADC, CDC, and VDC functions using virtualization, as shown in *Figure 3.4*. Virtualization refers to abstracting computing resources, such as CPU time, memory, and I/O, to create virtual instances within a vehicle's electronic architecture. This technology enables multiple functions or applications to be consolidated onto a single hardware platform, often a powerful SoC.

The primary benefits of virtualization in automotive systems include optimizing resource utilization, enhancing flexibility in managing various software components, facilitating efficient integration of diverse functionalities, and supporting easier software updates and upgrades. Overall, virtualization contributes to a streamlined vehicle architecture, enabling greater adaptability and scalability in response to evolving automotive technologies and requirements.

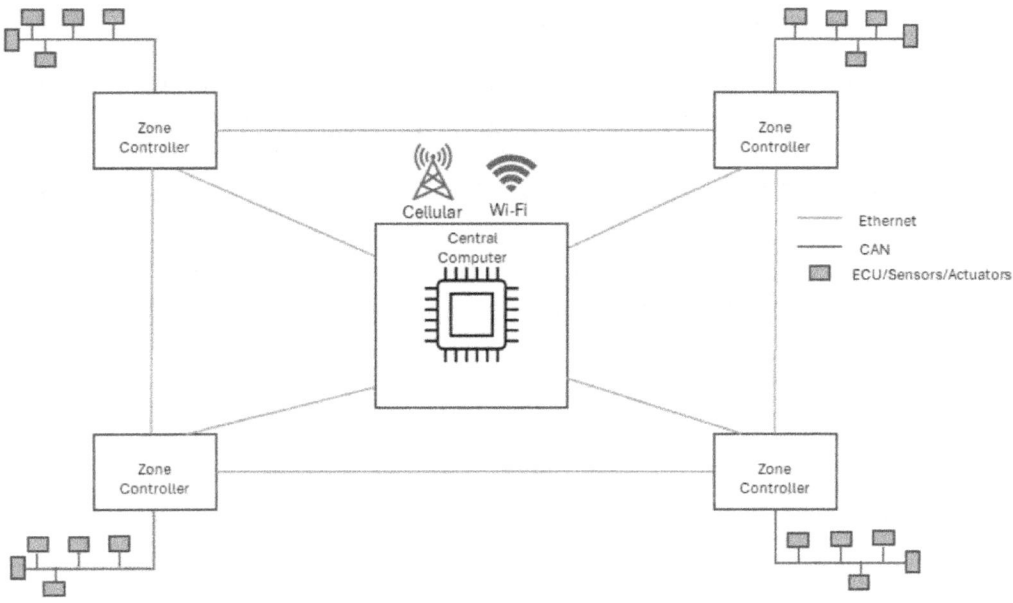

Figure 3.4 – A central computer with a single SoC (fully virtualized)

In automotive systems, a hypervisor enables virtualization by managing and allocating hardware resources to multiple functions or applications on a single SoC. For more information about hypervisors, please see the SoC Hypervisor section. Overcoming challenges with hypervisor use in automotive systems involves implementing advanced real-time scheduling algorithms to meet stringent requirements, optimizing hypervisor resource management for reduced overhead, and employing robust safety and security measures, such as stringent secure boot and frequent SW updates via OTA, to mitigate potential vulnerabilities and ensure the overall integrity of the system. Secure Boot is a security feature that ensures that only authenticated and trusted software is allowed to run during a device's startup process.

In the next section, we'll explore standard automotive frameworks that are available for implementing IoT and automotive use cases in both the domain controller and central computer architectures.

Standard frameworks to support vehicle architecture and IoT

This section covers the internal architecture of the ADC/CDC/VDC ECUs which comprise microcontrollers and SoCs. The variation in the number of microcontrollers and SoCs is based on functionalities, features supported, and the required number of inputs and outputs. Before delving into the standard framework, we'll provide a high-level overview of the domain controller to understand where these frameworks are applied.

A high-level overview of the domain controller

Figure 3.5 shows a domain controller that encompasses both a microcontroller and a SoC. Microcontrollers are capable of handling small, time-critical applications such as power management and communication stacks such as CAN. On the other hand, SoCs are capable of handling resource-intensive applications such as navigation, media, OTA, and remote diagnostics, all of which require more processing time and memory.

The interaction between the microcontroller and SoC is facilitated through communication mediums such as **Serial Peripheral Interface (SPI)**, **Universal Asynchronous Receiver/Transmitter (UART)**, and Ethernet. One or more of these communication mediums will be utilized, depending on the design and complexity of the product:

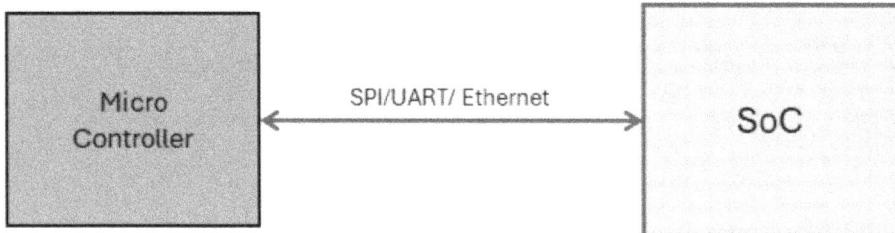

Figure 3.5 – Domain controller view

SPI is a synchronous serial communication protocol that's frequently employed for data transfer between SoC and one or more peripheral devices, such as a microcontroller. It utilizes four signal lines, namely **Master In Slave Out (MISO)**, **Master Out Slave In (MOSI)**, **Serial Clock (SCLK)**, and **Slave Select/Chip Select (SS/CS)**, allowing for efficient and simultaneous bidirectional communication. The Electrical engineering section in *Chapter 13*, *Embedded Automative IoT Development*, provides detailed insights into the operation of SPI, including an illustrative example.

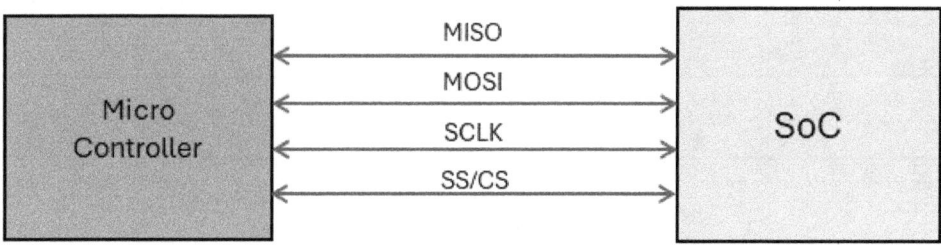

Figure 3.6 – Microcontroller and SoC via SPI

Figure 3.6 illustrates the interactions between the microcontroller and the SoC through SPI. Based on the design, one functions as a master while the other operates as a slave node in the communication setup.

UART can be used in place of SPI based on the required speed and hardware support. UART typically operates at lower speeds than SPI. SPI can achieve higher data transfer rates, making it faster for applications that require rapid communication.

UART is a serial communication protocol that's commonly used for transmitting and receiving data between devices. It operates asynchronously, meaning data is sent without a shared clock signal, and instead, start and stop bits define the data frames. UART typically uses two wires: one for **transmitting data (TX)** and one for **receiving data (RX)**.

In the next section, let's explore details about hypervisors well-suited for SoC-based platforms and understand the details.

SoC hypervisor

SoC contains ample processing power and memory, enabling it to execute complex activities swiftly, typically within seconds. An SoC with multiple cores (4, 8, 16, or more) enables parallel code execution, with each core operating concurrently to boost computational efficiency. SoC supports a rich set of features in terms of both software and hardware. For instance, Qualcomm Snapdragon chipsets offer diverse capabilities such as audio processing, video rendering, navigation, virtualization, and high-performance computing. *Figure 3.7* provides a high-level overview of the software architecture of the SoC. Adaptive AUTOSAR can be hosted on one or more **virtual machines (VMs)**.

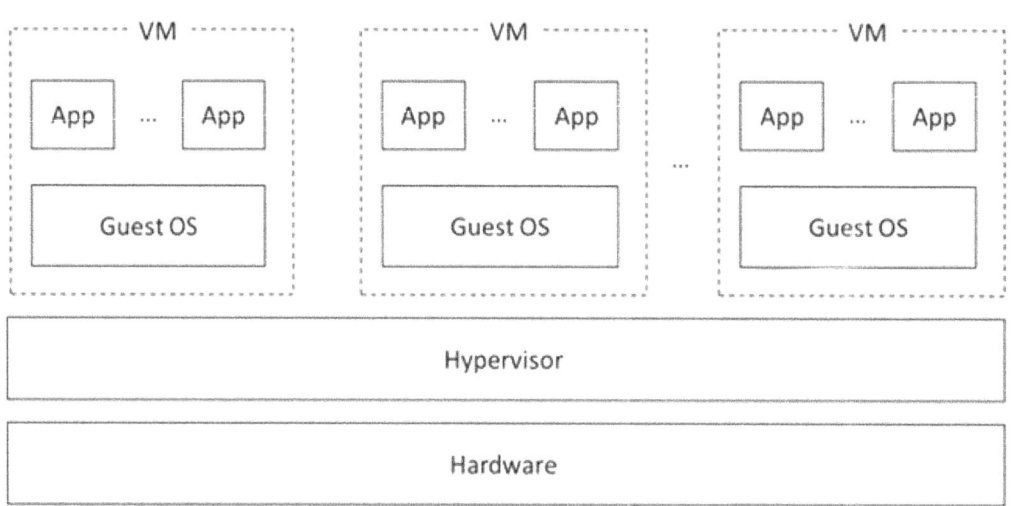

Figure 3.7 – SoC SW view (type 1 hypervisor)

A hypervisor serves as a software or hardware layer that enables virtualization, allowing multiple OSs to run on a single physical hardware. It efficiently manages and allocates resources, ensuring each **virtual machine** (**VM**) operates independently, thereby enhancing efficiency and isolation. A VM is a software-based simulation of a computer system, enabling the execution of applications in a controlled environment and optimizing resource utilization on constrained hardware. This effective hardware utilization minimizes product costs.

There are two types of hypervisors available. Type 1 hypervisors interact directly with hardware without OS involvement, making them preferred in embedded automotive systems for superior performance and real-time capabilities. *Figure 3.7* shows the type 1 hypervisor view. In contrast, type 2 hypervisors, such as Oracle VirtualBox, operate atop an OS and find use in development and testing environments. Further details on Type 1 and Type 2 hypervisors are covered in the Hypervisor section of *Chapter 13, Embedded Automotive IoT Development*.

Apps are software applications that implement business logic for features such as software updates, navigation, remote diagnostics, and phone as a key. They interact with underlying middleware and low-level drivers to execute end-to-end use cases.

Guest OSs are OSs that run over the hypervisor layer. Examples include QNX, Automotive Grade Linux, Android, and embedded **real-time operating systems** (**RTOSs**). RTOSs are designed to support time-critical activities in the embedded system.

An embedded system is a computer system that's designed to perform a dedicated function within a larger device, often with real-time processing requirements, and is tightly integrated into the device it controls. An example of this is the ECU in a vehicle.

The following section explores standard frameworks, including Classic **Automotive Open System Architecture** (**AUTOSAR**), which is used on microcontrollers, and Adaptive AUTOSAR for SoC-based/hypervisor-based solutions.

Classic AUTOSAR

The Classic AUTOSAR software framework is extensively used on microcontrollers. It serves as a standardized automotive software architecture that's designed to establish an open and standardized platform for managing functions within vehicles. *Figure 3.8* provides a high-level view of the Classic AUTOSAR software.

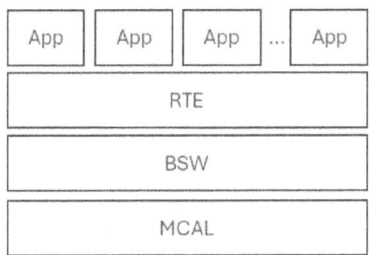

Figure 3.8 – Classic AUTOSAR

Let's take a closer look at what's involved:

- **Microcontroller Abstraction Layer** (**MCAL**) includes drivers and a hardware interface layer for direct communication with hardware.
- **Basic Software** (**BSW**) acts as a middleware component with communication stacks such as CAN, LIN, Ethernet/DoIP, as well as **operating system** (**OS**), diagnostic, power management, memory, and other middleware components.
- **Real-Time Engine** (**RTE**) serves as an abstraction layer, connecting the application layer to the BSW layer.

The application layer contains business logic to meet specific requirements. For insights into the internals of Classic AUTOSAR, refer to the official documentation [2].

Implementing IoT-related use cases on a low-end microcontroller is infrequent due to limited processing power and memory. However, it is feasible on high-end microcontrollers with multiple cores and high clock speeds where clock speeds greater than a few hundred **megahertz** (**MHz**) and multiple **megabytes** (**MB**) of memory are supported. Classic AUTOSAR supports DoIP components for wired/wireless network communication, facilitating the implementation of remote diagnostics use cases.

Classic AUTOSAR is well-suited for microcontrollers due to its design for resource-constrained environments. It efficiently manages the limited processing power and memory that's typical in microcontrollers, making it suitable for automotive control systems with real-time requirements.

The next section provides a concise overview of AUTOSAR components and their interconnected roles in enabling remote diagnostics.

Classic AUTOSAR components overview

Classic AUTOSAR contains several components, and the list is growing to support new features. *Figure 3.9* shows Classic AUTOSAR, including its components, which are required for remote diagnostics and basic functionalities.

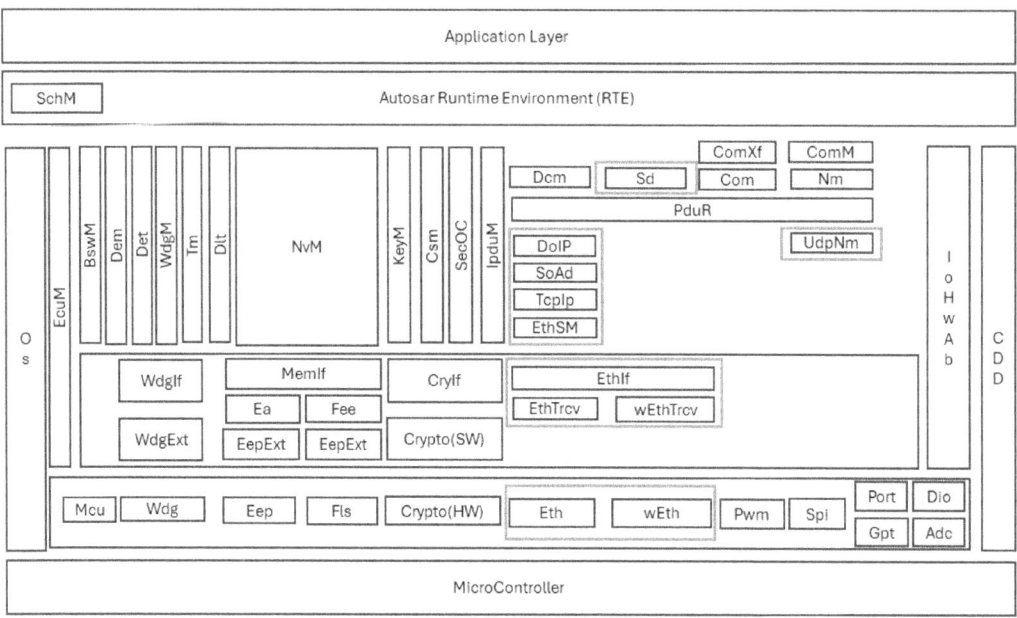

Figure 3.9 – Classic AUTOSAR with Ethernet/DoIP components

Supporting remote diagnostics in the Classic AUTOSAR environment involves a complex interplay between various software components designed for efficient network communication, diagnostics, and vehicle management. Here are some of the key components that are required to implement **Diagnostics over Internet Protocol (DoIP)**:

- **Ethernet Transceiver (EthTrcv)**: Manages the physical layer for Ethernet communication, including signal transmission and reception.

- **Ethernet Driver (Eth)**: Interfaces with the Ethernet hardware to send and receive Ethernet frames at the **data link layer** (**DLL**).

- **Ethernet Switch (EthSwt)**: Enables routing and switching of Ethernet frames within the vehicle to support complex networking scenarios.

- **Ethernet Interface (EthIf)**: Provides a standardized interface for upper-layer protocols to access Ethernet communication services.
- **Ethernet State Manager (EthSM)**: Manages the state of the Ethernet controller, coordinating initialization, communication readiness, and error handling.
- **TCP/IP Stack (TcpIp)**: Implements the TCP/IP protocol suite, enabling IP-based communication over Ethernet.
- **Service Discovery (Sd)**: Facilitates the discovery of network services, essential for locating remote diagnostic services.
- **Socket Adapter (SoAd)**: Acts as an intermediary between the TcpIp stack and upper-layer protocols, managing socket connections.
- **DoIP**: Supports diagnostics communication over IP networks, crucial for remote diagnostics. Let's explore more details of DoIP in the next chapter.
- **Diagnostic Communication Manager (DCM)**: Coordinates diagnostic communication, handling diagnostic session control, and request processing.
- **Non-Volatile Memory Manager (NVMM)**: Manages read, write, and management operations on **non-volatile memory (NVM)**, crucial for storing diagnostic information.
- **Communication Manager (ComM)**: Manages network communication states, including wake-up and sleep transitions, to optimize communication resources.
- **BSWM (BswM)**: Coordinates the interaction and operational modes of various **Basic Software (BSW)** modules based on vehicle state.
- **Network Management (NM)**: Manages the vehicle's network nodes, including node availability and network organization.
- **Security Module (SecOC or CSM)**: Ensures the security of diagnostic communication, providing authentication and encryption services to protect against unauthorized access.
- **Runtime Environment (RTE)**: Provides the middleware that facilitates communication between application **software components (SWCs)** and the AUTOSAR BSW.
- **Application**: High-level SWCs that implement the actual diagnostic logic and other functionalities.

Software development with Classic AUTOSAR requires extensive configuration and code generation. Many vendors offer tools to simplify and enhance the efficiency of this configuration process. In the next section, we'll explore the development tools used in Classic AUTOSAR.

Classic AUTOSAR development tools

There are multiple tools available, mainly from AUTOSAR solution providers such as Vector, **Elektrobit** (**EB**), ETAS, KPIT, and Mentor Graphics, among others. For example, Vector offers the DaVinci Configurator and DaVinci Developer tools for generating configurations for the BSW, diagnostics, and applications.

Figure 3.10 shows a high-level view of a Classic AUTOSAR tool, including types of input and output files.

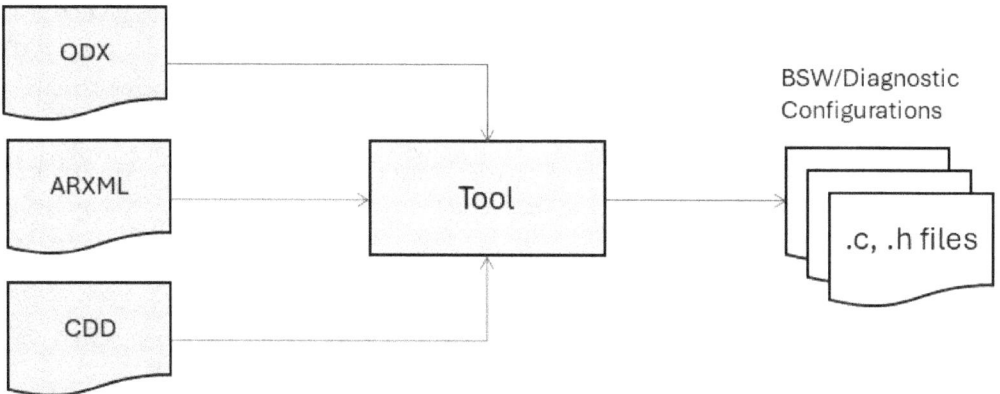

Figure 3.10 – Development tools for Classic AUTOSAR

Additionally, chip manufacturers provide tools to generate **Microcontroller Abstraction Layer** (**MCAL**) configurations, which consist of low-level software and drivers, using their proprietary and well-known tools, such as EB tresos.

The Vector DaVinci Configurator supports the generation of MCAL and BSW configurations. Each chip manufacturer may also offer their own tools to facilitate the creation of MCAL configurations. It's important to note that the Vector DaVinci Configurator is specifically designed for generating MCAL and BSW configurations. For generating code for application SWCs, Vector provides another tool called DaVinci Developer. For more details, please refer to Vector's official website [3].

An AUTOSAR tool accepts various formats of inputs, and *Figure 3.9* shows some of the commonly used formats:

- **AUTOSAR XML (ARXML)**: This is utilized within the AUTOSAR framework to describe the architecture, configuration, and components of automotive software systems.
- **CANdela Diagnostic Data (CDD)**: This format, developed by Vector, specifies diagnostic information and behavior. It is used with only Vector's diagnostic tools.
- **Open Diagnostic data eXchange (ODX)**: ODX standardizes the exchange of diagnostic data. It details vehicle diagnostic procedures and information for a wide range of diagnostic tools.

ODX aims to keep diagnostic data from OEMs separate from tool manufacturer software, enabling OEMs to switch tool vendors without altering their diagnostic data.

Depending on the product, tools, and solutions used, it may be necessary to utilize some or all of these formats. For instance, in a project utilizing the AUTOSAR framework, ARXML files alone can encapsulate both AUTOSAR configurations and diagnostic information. OEMs may choose to use one, two, or all these formats based on the overall system design.

Classic AUTOSAR's architecture and protocol constraints make it less suitable for handling advanced/complex applications such as remote diagnostics. Its design is optimized for microcontrollers with limited resources, such as memory, processing power, and support for Ethernet/Wi-Fi technologies, which are essential for implementing remote diagnostic functionalities. The good news is the shortcomings of Classic AUTOSAR are addressed in Adaptive AUTOSAR. In the next section, let's explore more about the internals of Adaptive AUTOSAR.

Adaptive AUTOSAR

The **Adaptive AUTOSAR** architecture is based on **Service Oriented Architecture** (**SOA**). SOA is a model where services are accessible over a network, residing on the same or different ECUs. These self-contained units operate independently, with clients (service users) detached from the services. The runtime coupling between services and clients is dynamic, allowing applications to consume or offer services. Essentially, SOA enables applications to openly present their functionalities as network-accessible services, fostering dynamic and adaptive interactions between applications. *Figure 3.11* provides a high-level view of Adaptive AUTOSAR.

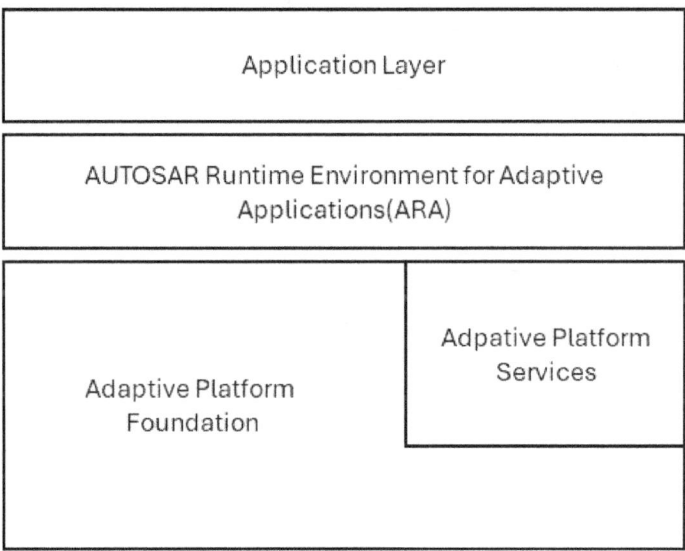

Figure 3.11 – Overview of Adaptive AUTOSAR

The application layer comprises applications that can be dynamically started or stopped at runtime and independently downloaded through a software update, eliminating the need for a complete software reflash.

The **AUTOSAR Runtime Environment for Adaptive Applications (ARA)** facilitates dynamic linking between applications and functional clusters. ARA is comprised of application interfaces provided by functional clusters. Within ARA, software is organized into functional clusters, each representing a grouping of requirements based on specific aspects they address. Simply put, a functional cluster is a collection of requirements related to a particular aspect. The main purpose of a functional cluster is to provide functionalities in the form of services to the application, such as SW download services and diagnostics.

The Adaptive Platform Foundation offers fundamental functionalities and services such as logging, communication within and outside the ECU, memory services and persistency, time synchronization, and cryptography. On the other hand, Adaptive Platform Services provide advanced offerings, such as diagnostics, software updates, security management, and remote diagnostic support. Adaptive AUTOSAR is built upon a microservice architecture that enables the dynamic deployment and management of software components. This approach involves developing applications as a suite of loosely coupled, independently deployable services, each focused on specific business functionalities. Such modularity is particularly advantageous for complex applications and services that require high data processing rates and extensive communication capabilities.

In the Adaptive AUTOSAR framework, support for remote diagnostics, including the communication stack DoIP, relies on specialized components designed for high-level communication and data processing tasks. Unlike Classic AUTOSAR, which follows a more static approach tailored for embedded systems, Adaptive AUTOSAR embraces a dynamic Service-Oriented Architecture (SOA), making it well-suited for handling complex applications.

Figure 3.12 illustrates Adaptive AUTOSAR, including its components, which are required for remote diagnostics and basic functionalities.

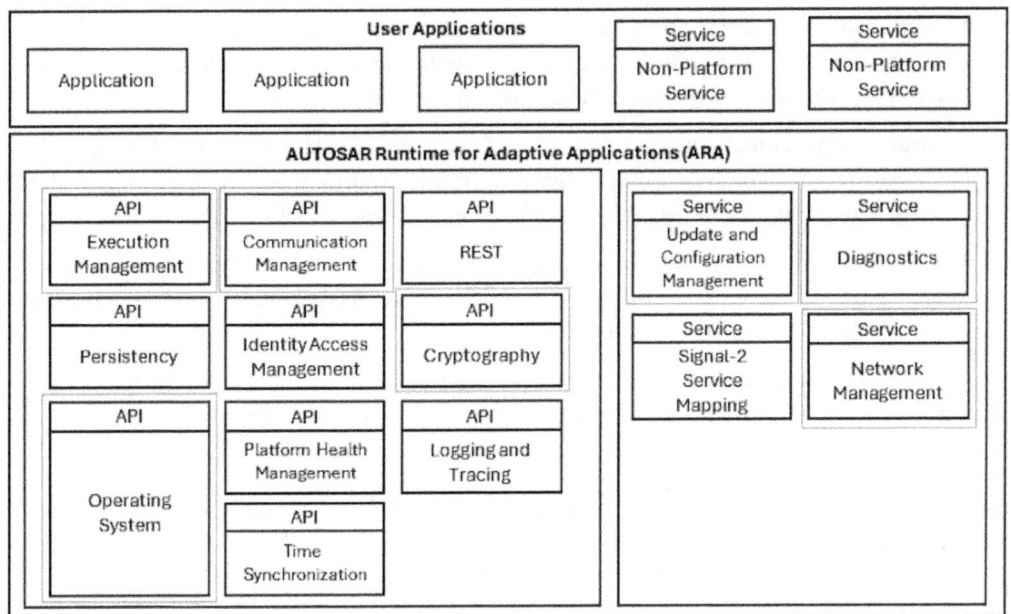

Figure 3.12 – Adaptive AUTOSAR

Here are some of the key components of Adaptive AUTOSAR:

- **Diagnostic Manager**: Manages the diagnostic communication, handling requests from external diagnostic tools and routing them to appropriate applications or services within the vehicle. It interprets diagnostic messages and manages the diagnostic logic and data exchange.

- **Communication Management**: Manages the establishment and control of network connections, including IP-based communications for remote diagnostics. This involves handling the network state, bandwidth allocation, and prioritization of diagnostic messages.

- **Scalable service-Oriented MiddlewarE over IP (Some/IP) and SD**: SOME/IP is a middleware protocol used for service-oriented communication over IP networks in automotive applications. SD allows services to be discovered and announced over the network, crucial for identifying and accessing diagnostic services remotely.

- **Security Module**: Ensures the security and integrity of diagnostic communications. This includes authentication, authorization, and encryption to protect against unauthorized access and ensure data privacy.

- **Persistent Storage Management**: Manages read, write, and storage operations for diagnostic data on non-volatile memory. This is crucial for storing fault codes, configuration data, and other diagnostic information that must be retained across vehicle restarts.

- **Execution Management**: Controls the execution environment for adaptive applications, including starting, stopping, and monitoring the health of applications. For diagnostics, it ensures that diagnostic services are readily available and responsive when needed.

- **Network Management**: Similar to its role in Classic AUTOSAR but adapted for the high-bandwidth and dynamic nature of Adaptive AUTOSAR networks. It oversees the configuration, control, and monitoring of network elements to support diagnostics over IP.

- **Log and Trace Management**: Provides logging and tracing capabilities for diagnostic activities, enabling the recording and analysis of diagnostic sessions. This is essential for debugging and monitoring the health of the vehicle systems.

- **Update and Configuration Management**: Manages the deployment of software updates and configuration changes, which can be initiated remotely as part of diagnostic procedures. This includes handling software updates, parameter changes, and calibration updates.

These components work together in the Adaptive AUTOSAR platform to enable robust, secure, and efficient remote diagnostic capabilities, leveraging IP-based communication and service-oriented architectures to support advanced automotive applications.

The Adaptive AUTOSAR tool is a key player in simplifying and improving the development work for Adaptive AUTOSAR. In the next section, we will explore the Adaptive AUTOSAR tool to learn about the input and output files used/generated by this tool.

Adaptive AUTOSAR development tools

Multiple vendors support Adaptive AUTOSAR development tools for adding/modifying configurations, new code, logic, and so on. Typically, these tools accept diagnostic files such as ODX, CDD, **Diagnostic Extract Template** (**DTEXT**), and ARXML configurations. Also, some tools accept JSON files, depending on the vendor tool being used. These tools also support various views for editing configurations, adding new code, creating models, and so on. *Figure 3.13* shows the tools' input and output details.

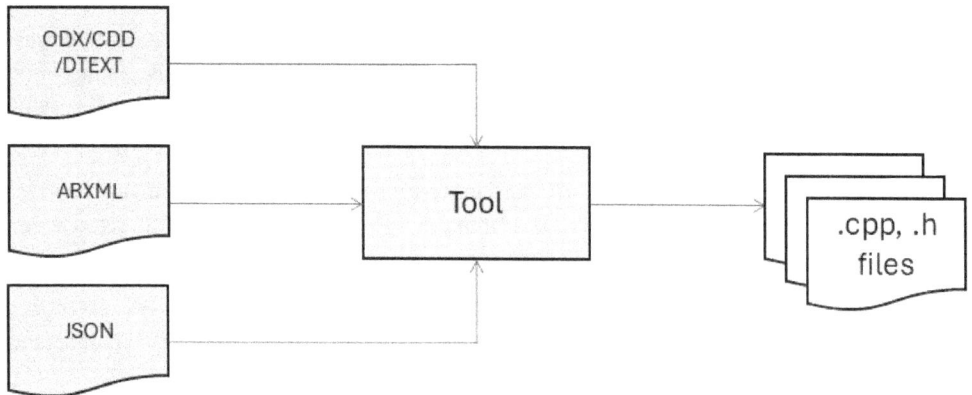

Figure 3.13 – Adaptive AUTOSAR tool

Vendors such as Vector, EB, ETAS, and KPIT support development tools for Adaptive AUTOSAR.

For example, Vector supports a tool called DaVinci Adaptive AUTOSAR, which supports ARXML, CDD, and JSON files as input. They also support multiple views for adding new models, creating/updating ARXML files, and generating `.cpp` and `.h` files.

For more details about DaVinci Adaptive AUTOSAR, please refer to reference [4].

In the next section, let's go over the Classic versus Adaptive AUTOSAR details by comparing the key characteristics.

Classic vs Adaptive AUTOSAR

Table 3.1 offers a concise comparison between Classic AUTOSAR, which is suited for traditional automotive systems, and Adaptive AUTOSAR, which is designed for high-performance computing.

Aspect	Classic AUTOSAR	Adaptive AUTOSAR
Architecture	Layered, deeply integrated	Modular, service-oriented
Design Goal	Stability, predictability	Flexibility, adaptability
Computing Environment	Resource-constrained embedded systems	High-performance computing platforms
Programming Languages	C	C++
System Behavior	Static, deterministic	Dynamic
Timing	Real time, with fixed timing schedules	More flexible; can adapt to varying computational loads
Use Cases	Traditional automotive functions (for example, engine control)	Advanced functions (for example, autonomous driving and advanced driver-assistance systems (ADASs))
Remote Diagnostics Support	Limited, primarily through external mechanisms	Directly supported, including advanced features such as DOIP
Development	Focus on reliability and safety	Emphasizes rapid development and deployment of new services

Table 3.1 – Classic versus Adaptive AUTOSAR

With this, we have come to the end of the chapter.

Summary

Vehicle architectures have transitioned from distributed to centralized systems, embracing connected IoT features. Current trends favor centralized zonal architectures, with the future moving toward a single-processor central computer. Standard frameworks, such as Classic AUTOSAR and Adaptive AUTOSAR, along with development tools and hypervisor technology, support these advancements.

Now that we've discussed the scale of vehicle architecture, gone through the high-level details of Classic AUTOSAR, understood the basics of the Adaptive AUTOSAR framework, and explored hypervisors, we will cover other essential components that enable remote diagnostics. In the next chapter, we will delve into diagnostic protocols and their support in Classic and Adaptive AUTOSAR for implementing remote diagnostics on the vehicle side.

References

- [1]: https://www.qualcomm.com/products/automotive/cockpit
- [2]: https://www.autosar.org/standards/classic-platform
- [3]: https://www.vector.com/us/en/products/products-a-z/software/#c6985
- [4]: https://www.vector.com/us/en/products/products-a-z/software/davinci-developer-adaptive/

Get This Book's PDF Version and Exclusive Extras

Scan the QR code (or go to `packtpub.com/unlock`). Search for this book by name, confirm the edition, and then follow the steps on the page.

Note: Keep your invoice handy. Purchases made directly from Packt don't require one.

4
Vehicle Diagnostics

This chapter delves into crucial diagnostic protocols in modern automotive systems, focusing on **Unified Diagnostic Services** (**UDS**), **Diagnostic over Internet Protocol** (**DoIP**), and their integration with **Automotive Open System Architecture** (**AUTOSAR**). We begin with UDS, a protocol for vehicle diagnostics and firmware updates that's adaptable to multiple communication platforms. Then, we examine DoIP, which enables high-speed diagnostic communication over Ethernet and wireless networks, vital for predictive maintenance and remote diagnostics. The chapter also covers the diagnostic communication flow and essential components of remote diagnostics in AUTOSAR-based systems, highlighting advanced diagnostic service management for enhanced flexibility and scalability. These protocols and frameworks together ensure efficient, reliable, and secure vehicle diagnostics in the connected automotive landscape.

In this chapter, we're going to cover the following main topics:

- UDS
- DoIP
- Diagnostic communication workflow in Classic AUTOSAR
- Diagnostic service management in Adaptive AUTOSAR

UDS

UDS is a communication protocol widely used in the automotive industry for diagnostics, firmware updates, routine testing, and more. UDS communication is conducted in a client-server relationship, with the client being a tester tool and the server being a vehicle's **electronic control unit** (**ECU**).

UDS enables the following use cases between the tester and the vehicle ECU:

- Reading/clearing **diagnostic trouble codes** (**DTCs**) for troubleshooting vehicle issues
- Extracting parameter data values such as temperature, state of charge, **vehicle identification number** (**VIN**), and so on
- Initiating diagnostic sessions to, for example, test safety-critical features
- Modifying ECU behavior via resets, firmware flashing, and settings modification

Initially, UDS was designed for CAN-based communication. Later, it was adapted for **Local Interconnect Network** (**LIN**), Ethernet, and wireless communication platforms. UDS has been widely used for vehicle diagnostics for a long time. There is further scope to continue to use it for future diagnostic use cases as well.

In the next section, let's explore the UDS message structure and the types of messages supported.

UDS message structure

At a high level, a UDS message contains a **Service Identifier** (**SID**), an optional sub-function byte, request data parameters, and padding bytes, as shown in *Figure 4.1*.

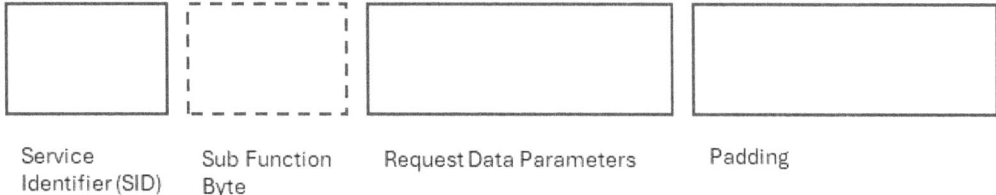

Figure 4.1 – The UDS message format

The SID specifies the type of service being requested or serviced. *Table 4.1* shows the list of supported services in UDS.

SID	Description
0x10	**DiagnosticSessionControl**: Manages diagnostic sessions (default/programming/extended/supplier), which can alter the ECU's behavior to provide different levels of access to diagnostic functions
0x11	**ECUReset**: Resets the ECU in various specified modes
0x14	**ClearDiagnosticInformation**: Clears diagnostic information (e.g., DTCs) from the ECU
0x19	**ReadDTCInformation**: Reads diagnostic trouble codes and associated information from the ECU

SID	Description
0x22	**ReadDataByIdentifier**: Reads data (e.g., sensor data, configuration settings) from the ECU identified by a data identifier
0x23	**ReadMemoryByAddress**: Reads a block of data from a specific memory address in the ECU
0x24	**ReadScalingDataByIdentifier**: Reads the scaling information for a specific data identifier
0x27	**SecurityAccess**: Requests access to protected services, requiring a security key exchange
0x28	**CommunicationControl**: Controls the ECU's communication capabilities (e.g., enabling or disabling certain communication types)
0x2A	**ReadDataByPeriodicIdentifier**: Reads specified data at a defined rate
0x2C	**DynamicallyDefineDataIdentifier**: Allows for the creation of custom data identifiers for specific datasets
0x2E	**WriteDataByIdentifier**: Writes data to the ECU identified by a data identifier
0x2F	**InputOutputControlByIdentifier**: Controls the behavior of the ECU's inputs and outputs
0x31	**RoutineControl**: Manages the execution of diagnostic routines within the ECU
0x34	**RequestDownload**: Initiates a data download to the ECU (e.g., for flashing new firmware)
0x35	**RequestUpload**: Initiates a data upload from the ECU
0x36	**TransferData**: Transfers data to or from the ECU as part of a download or upload process
0x37	**RequestTransferExit**: Signals the end of a data transfer session
0x3D	**WriteMemoryByAddress**: Writes data to a specific memory address in the ECU
0x3E	**TesterPresent**: Indicates to the ECU that the diagnostic tool is still present to keep the session active
0x83	**AccessTimingParameter**: Manages timing parameters for diagnostic access
0x84	**SecuredDataTransmission**: Provides a secure channel for data transmission
0x85	**ControlDTCSetting**: Controls the setting of DTCs
0x86	**ResponseOnEvent**: Configures the ECU to automatically send certain responses upon predefined events

Table 4.1 – SID types

The sub-function byte is used in some UDS request frames. For example, in the case of *DiagnosticSessionControl*, this will be used to indicate the diagnostic session type, such as default mode, programming mode, extended mode, or supplier mode.

In many UDS request services, a variety of request data parameters are employed to offer additional configuration for a request, beyond the SID and the optional sub-function byte. For example, consider Figure 4.2.

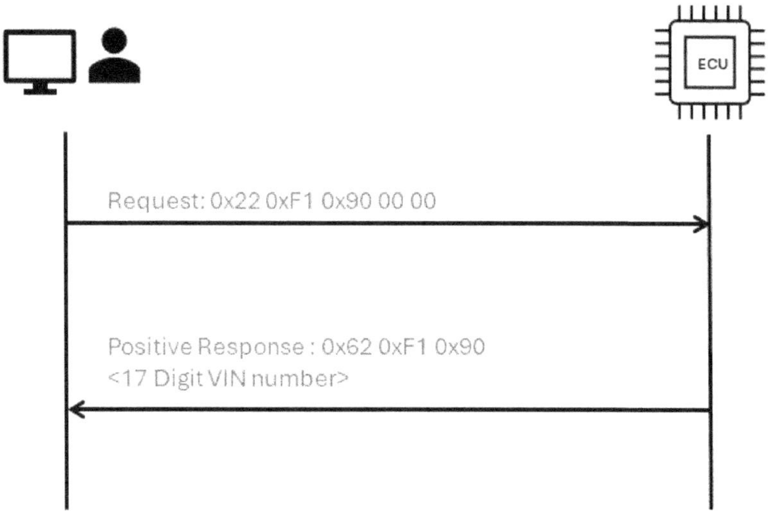

Figure 4.2 – A UDS message positive response

In Figure 4.2, the service **0x22** (ReadDataByIdentifier) includes a 2-byte **data identifier** (**DID**) value. For instance, 0xF190 is a DID that indicates a request to read the VIN from the vehicle's ECU.

When an ECU provides a positive response to a UDS request sent by the tester, the structure of the response frame mirrors that of the request frame. For instance, a positive reply to a service 0x22 request will include the response SID 0x62 (which is 0x22 plus 0x40) followed by the 2-byte DID, and then the actual data payload corresponding to the requested DID. Typically, the format of a positive UDS response message varies depending on the specific service involved. Consider the following examples:

The request message for the VIN is *0x22 0xF1 0x90 00 00*....

A positive response message for the VIN would be *0x62 0xF1 0x90 <17 Digit VIN Number>*.

In some cases, an ECU may provide a negative response to a UDS request.

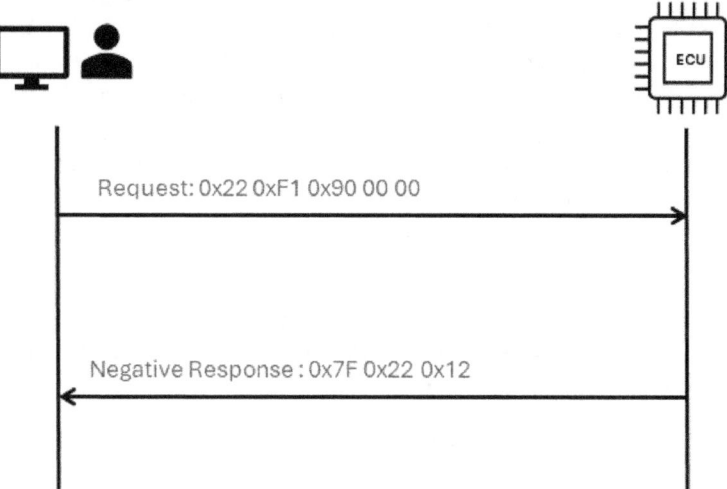

Figure 4.3 – A UDS message negative response

For example, in *Figure 4.3*, if the DID is not supported by the ECU, then a negative response message would be *0x7F 0x22 0x12*. Here, *0x7F* indicates a negative response, *0x22* represents the SID number, and the third byte (*0x12*) signifies the error code indicating that the condition is not correct.

For more details on UDS, please refer to references [1].

Next up, we'll explore the DoIP, mainly used for vehicle diagnostics over Ethernet, and later extended to support wireless diagnostics. We chose DoIP because it's a standard way to diagnose vehicles over both Ethernet and wireless communication in the automotive industry.

DoIP

The automotive industry is going through a major transition due to **autonomous driving** (**AD**), **software-defined vehicles** (**SDVs**), and connected vehicles. AD refers to vehicles that can operate without human intervention. These vehicles use advanced technologies such as sensors, cameras, radar, and **artificial intelligence** (**AI**) to navigate and make decisions on the road. An SDV is a concept where much of the vehicle's functionality is controlled and defined by software rather than traditional hardware. This approach allows for greater flexibility and adaptability as features and functions can be updated or modified through software updates, similar to how apps are updated on a smartphone. Connected vehicles are equipped with communication technologies that allow them to share data with other vehicles, infrastructure, and the cloud. This connectivity enhances safety, efficiency, and convenience. Example enhancements include vehicles communicating with each other to avoid collisions or accessing real-time traffic information for optimized routing.

Advancements such as AD, SDVs, and connected vehicles demand large amounts of information exchange between the ECU and the external world. The ECU is a critical component in vehicles that is responsible for controlling a specific function. Today's vehicles may contain 100 ECUs or more, controlling functions that range from the essential (such as engine and power-steering control) to comfort (such as power windows, seats, and **heating, ventilation, and air conditioning** (**HVAC**), infotainment that is responsible for entertainment (audio/video), communication and navigation systems, and security and access (such as door locks and keyless entry). ECUs also control passive safety features such as airbags, and even basic active safety features such as automatic emergency braking. Each ECU typically contains a dedicated chip that runs its own software or firmware and requires power and data connections to operate.

The need for large amounts of information exchange necessitates high-speed communication in today's vehicles. Ethernet, **wireless local area network** (**WLAN**) using Wi-Fi, and 4G/5G cellular networks are popular choices to support higher data transfer.

Ethernet ranges from 100 **megabits per second (Mbps)** to 1 **gigabit per second (Gbps)** or more, Wi-Fi can reach speeds of around 3.5 Gbps, 4G cellular networks offer speeds between 10 and 100 Mbps, and 5G can exceed 1 Gbps.

DoIP is a standard communication protocol used in the automotive industry for diagnostics and vehicle maintenance over **Internet Protocol** (**IP**). IP can function over Ethernet, Wi-Fi, and cellular networks. DoIP enables diagnostic communication between a vehicle's ECUs and diagnostic tools over Ethernet, Wi-Fi, and cellular networks. DoIP over Wi-Fi and cellular networks enables remote diagnostics.

In this book, we use DoIP to implement predictive maintenance use cases. Predictive maintenance is realized using remote diagnostics.

Let's take a closer look at DoIP by examining how it connects with the broader framework of network communication, known as **Open Systems Interconnection** (**OSI**). This will help us understand the role of DoIP in networks, its message format, and how DoIP messages flow, which is covered in the next section.

Figure 4.4 shows the mapping of the OSI layer to DoIP, along with the frame format at each layer.

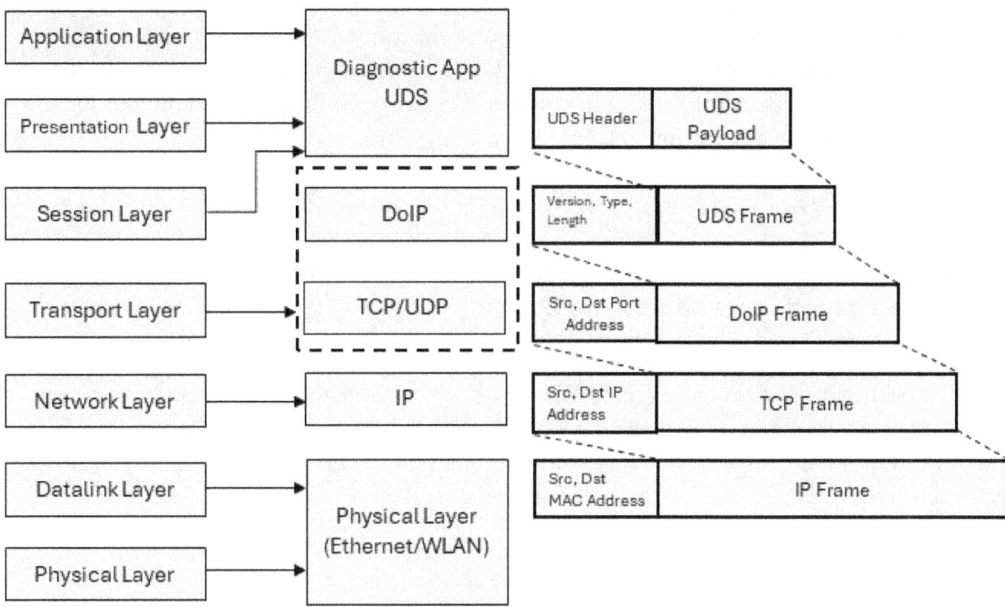

Figure 4.4 – OSI-to-DoIP mapping with packet information

Let's break down the communication/protocols used in DoIP:

- **UDS**: UDS is a diagnostic communication protocol utilized at the application layer for communication between a diagnostic tester and the ECUs in a vehicle.
- **DoIP layer**: DoIP serves as a transport protocol at the transport layer for diagnostic communication between ECUs and the tester. The tester is a tool/application on a PC/remote PC.
- **Transmission Control Protocol (TCP)/User Datagram Protocol (UDP) layer**: This facilitates the transmission of UDS messages over IP networks such as Ethernet and WLAN. TCP and UDP are Operating at the transport layer, these are protocols that DoIP can use to transmit diagnostic messages. TCP offers reliable, connection-oriented communication that's crucial to data integrity, while UDP provides faster, connectionless communication that's suitable for real-time data transfer.
- **IP layer**: IP, operating at the network layer, is responsible for addressing and routing data packets. DoIP, being an IP-based protocol, utilizes IP for addressing and transmitting diagnostic messages.
- **Physical layer (Ethernet/WLAN)**: Ethernet and WLAN serve as physical-layer technologies, providing the medium for data transmission. DoIP can use either Ethernet or WLAN as the physical layer for communication between ECUs in a vehicle and the tester.

In summary, UDS defines how diagnostic information is structured and exchanged at the application layer, DoIP acts as the transport protocol at the transport layer for transmitting UDS messages over IP networks, TCP/UDP ensure reliable communication at the transport layer, IP handles addressing and routing at the network layer, and Ethernet/WLAN serve as the physical layer. Together, these components create a comprehensive and standardized framework for diagnostic communication in modern vehicles, ensuring efficient and reliable vehicle diagnostics.

DoIP message format

DoIP messages refer to the data packets exchanged between ECUs for diagnostic and maintenance purposes. These messages facilitate communication between diagnostic tools and ECUs over networks such as Ethernet, Wi-Fi, and cellular networks.

This section primarily covers details of the DoIP message format, including information about each field such as the length and purpose of the message. *Figure 4.5* shows the DoIP message format from the physical layer to the transport layer. It provides comprehensive information on the structure and transmission across different layers of the communication system.

The Ethernet frame in *Figure 4.5* contains **cyclic redundancy check** (**CRC**) and is mainly used to check the integrity of messages received over the vehicle network.

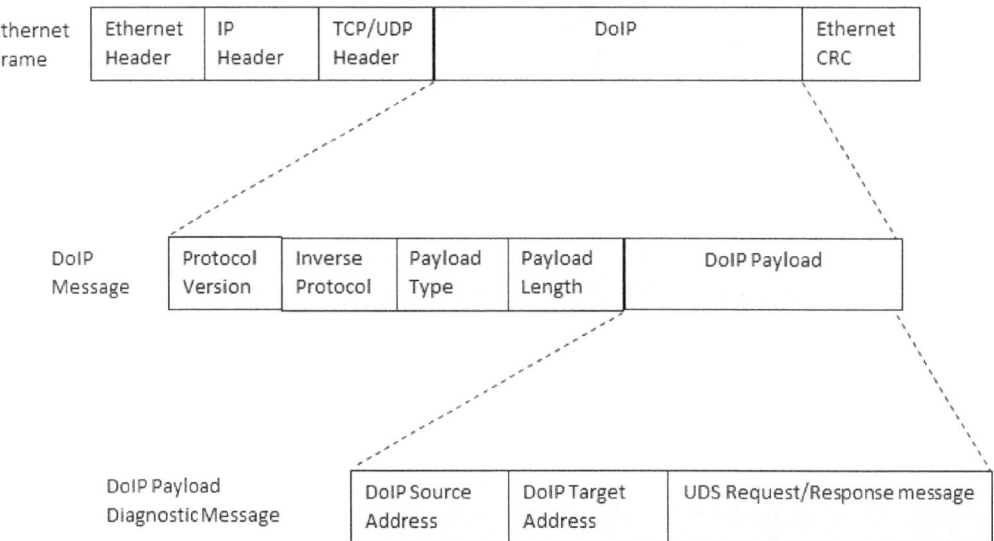

Figure 4.5 – The DoIP message format

Next, we will delve into more details about the DoIP message format, exploring elements such as field names, lengths, and descriptions. Each field is comprehensively explained in a tabular format for a clearer understanding. *Table 4.2* contains details on DoIP frame fields and descriptions.

Field Name	Length	Description
Protocol version	1 byte	Specifies the protocol version of DoIP packets: 0x00: Reserved 0x01: ISO/DIS 13400-2:2010 0x02: DoIP ISO 13400-2:2012 0x03-0xFE: Reserved by ISO 13400 0xFF: Default value for vehicle identification request messages
Inverse Protocol	1 byte	The bit-wise inverse of the protocol version. This acts as a verification pattern, ensuring proper formatting of received DoIP messages.
Payload type	2 bytes	Guides the interpretation of data following the generic DoIP header, covering gateway commands, diagnostic messages, and other information types.
Payload length	4 bytes	Payload length excluding the DoIP header. Length will vary based on the payload type. Maximum size is limited by the underlying transport/physical layer. For example: Ethernet and Wi-Fi support ~1500 bytes and cellular networks support more than 1500 bytes.
Payload message	Based on the payload length value	Message specific to the payload type. Message length is mentioned in the previous field.

Table 4.2 – DoIP message format details

In DoIP, the terms *sender node* and *receiver node* refer to the entities involved in the communication process within a vehicle network.

The sender node is typically an ECU or a device within the vehicle's communication system that initiates the transmission of DoIP messages. The sender node will support the payload types listed in *Table 4.4*. It is responsible for generating and sending diagnostic messages containing information related to the vehicle's health, status, or other relevant data.

The receiver node is another ECU or device in the vehicle network that is capable of receiving and processing DoIP messages sent by the sender node. Its role is to accept incoming DoIP messages, extract the diagnostic information, and respond as needed. This could involve acknowledging receipt, processing the diagnostic data, or triggering specific actions within the vehicle. The receiving node will receive the payload types listed in *Table 4.3*.

Type Value	Payload Type Name	Connection Kind	Purpose
0x0000	Generic DoIP header negative acknowledgment	UDP/TCP	This is used to respond when an invalid DoIP header is received. The negative acknowledgment (NACK) code will specify the exact reason as part of the response message.
0x0001	Vehicle identification request message	UDP	This is used to request information from all participants in the network.
0x0002	Vehicle identification request message with EID	UDP	This is used to request information from the participant ECU in the network with a matching EID.
0x0003	Vehicle identification request message with VIN	UDP	This is used to request information from the participant ECU in the network with a matching 17-digit VIN.
0x0004	Vehicle announcement message/vehicle identification response message	UDP	This is used to receive/send vehicle announcement information and node information on the network. It enables other nodes on the network to be aware of the availability of other nodes and their supported functions.
0x0005	Routing activation message	TCP	This enables the activation or deactivation of a single diagnostic message path to handle different protocols (such as UDS and on-board diagnostics (OBD)) and allows treating individual testers differently.
0x0008	Alive Check response	TCP	This is used to manage various tester connections. If the Source Address field in the received Alive Check response matches the registered Source Address field of the socket connection where the response was received, the DoIP module will take no action.
0x4001	DoIP entity status request	UDP	This is used to determine the node type (gateway or node), the maximum number of allowed connections, the count of active connections, and other relevant information.

Type Value	Payload Type Name	Connection Kind	Purpose
0x4003	Diagnostic power mode information request	UDP	This inquires about the power mode status of the DoIP node, indicating whether it is in a Ready or Not Ready state and so forth.
0x8001	Diagnostic message	TCP	This is for diagnostic request messages.

Table 4.3 – DoIP payload types received by a DoIP entity

The DoIP sender node will support the payload types in *Table 4.4*.

Type Value	Payload Type Name	Connection Kind	Purpose
0x0000	Generic DoIP header negative acknowledgment	UDP/TCP	When the system receives an invalid DoIP header, it utilizes this mechanism to respond. The NACK code is employed to specify the exact reason as part of the response message.
0x0004	Vehicle announcement message/vehicle identification response message	UDP	This is used to receive/send vehicle announcement information and node information on the network. It enables other nodes on the network to be aware of the availability of other nodes and their supported functions.
0x0006	Routing activation response	TCP	This confirms route activation.
0x0007	Alive Check request	TCP	This sends an Alive Check request to the other end of the connection.
0x4002	DoIP entity status response	UDP	This sends information about the node type, maximum number of connections, and other relevant details.
0x4004	Diagnostic power mode information response	UDP	This sends the power mode status, indicating whether the node is ready or not.
0x8002	Diagnostic message positive acknowledgment	TCP	This is a diagnostic acknowledgment message.
0x8003	Diagnostic message negative acknowledgment	TCP	This is a diagnostic negative acknowledgment message with a NACK code.

Table 4.4 – DoIP payload types transmitted by a DoIP entity

At this point, you have learned about the DoIP message format in detail, so now, let's understand how these messages are transmitted in the vehicle network.

DoIP example message flow

Figures 4.6 and *4.7* illustrate the typical DoIP message flow with a diagnostic request and response. DoIP uses UDP for quick setup due to its speed and efficiency, transitioning to TCP for ongoing, reliable diagnostic communication that ensures data integrity. *Figure 4.6* depicts the message flow between the DoIP node and the tester. The DoIP node represents the ECU that supports the DoIP protocol, while the tester is a tool/diagnostic application on a remote machine that also supports the DoIP protocol.

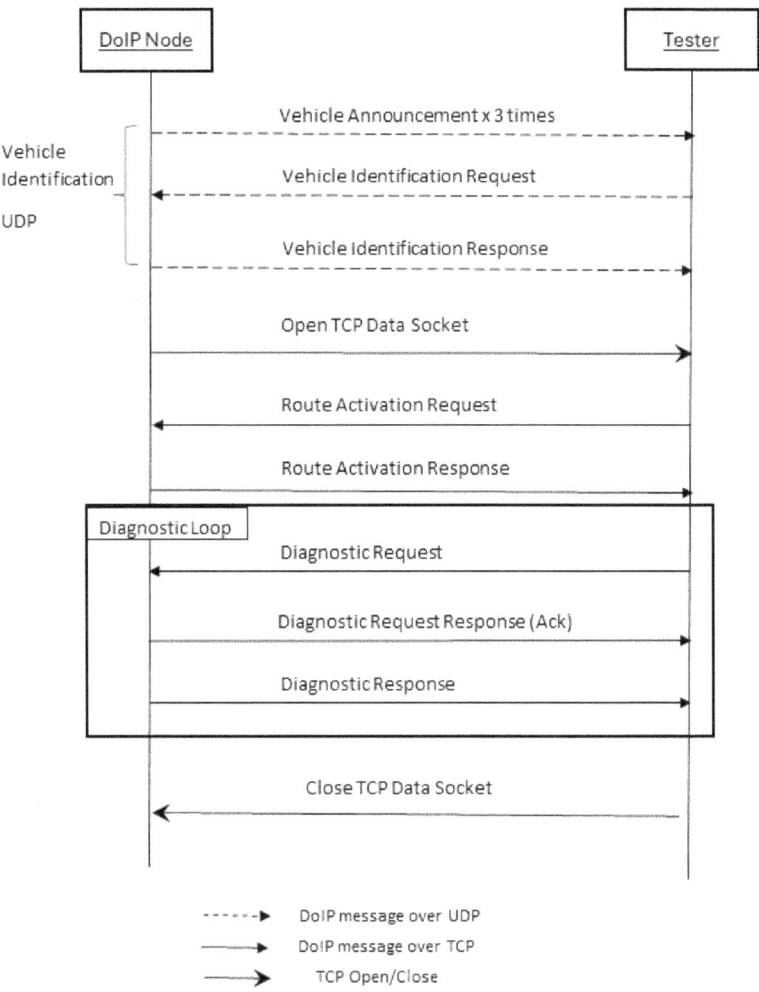

Figure 4.6 – A typical diagnostic message flow between the tester and the DoIP node

Figure 4.7 provides an example of the message flow between the DoIP gateway, the non-DoIP node, and the tester. The DoIP gateway supports gateway functionality, communicating with the external world, such as the tester, and connecting with other ECUs in the vehicle that do not support DoIP, called **non-DoIP nodes**. These ECUs support other vehicle communication protocols such as **CAN** and **LIN**.

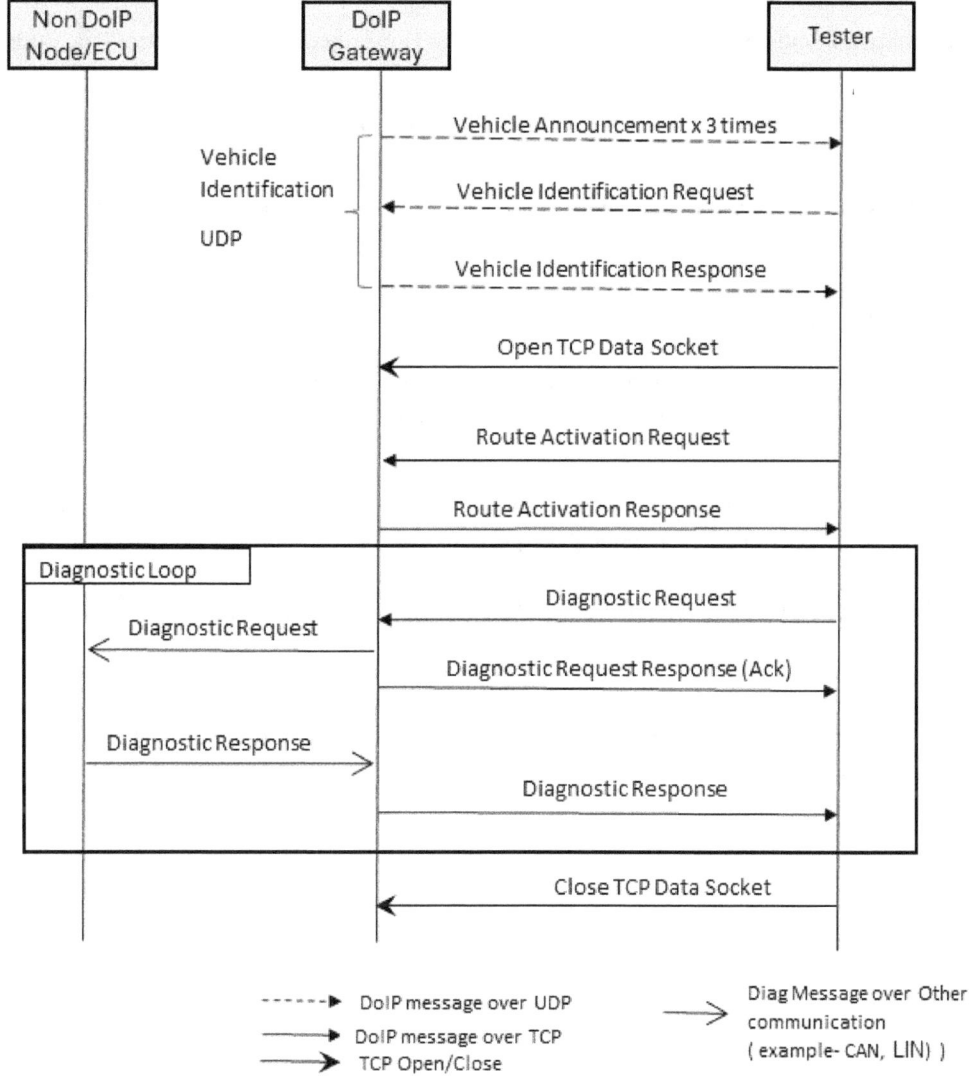

Figure 4.7 – A typical diagnostic message flow between the tester and the DoIP gateway

At this point, you have learned about DoIP, the mapping of DoIP layers to OSI layers, and the details of the message format and the message flow between DoIP, non-DoIP nodes, the DoIP gateway, and testers on the remote machine. DoIP is widely used for remote diagnostics over IP networks, which is key to implementing predictive maintenance applications/use cases. DoIP is also used to support **over-the-air** (**OTA**) updates, and DoIP packets carry UDS messages.

Some key challenges in DoIP include the following:

- **Security implementation**: Addressing challenges in implementing robust security measures for vehicle communication
- **Integration complexity**: Managing the integration of older and newer systems to ensure seamless compatibility
- **User adoption**: Educating users about new features and ensuring the smooth adoption of IoT applications in vehicles
- **System standardization**: Establishing standards for consistent and interoperable communication protocols in vehicles

Overcoming challenges in DoIP requires making communication secure, handling system integration, assisting users in adapting, and establishing standard protocols for smooth operation. For more details on DoIP, please refer to references [2].

The next section covers the high-level diagnostic communication flow in Classic AUTOSAR, including DoIP and the other key essential components required.

Diagnostic communication workflow in Classic AUTOSAR

Figure 4.8 shows the typical flow for diagnostic communication. Diagnostic requests flow from the Ethernet driver to **Diagnostic Communication Manager** (**DCM**). DCM will process the request and delegate it to the application if any additional processing is required. In the case of DoIP, initial communication occurs over UDP and will later switch to TCP.

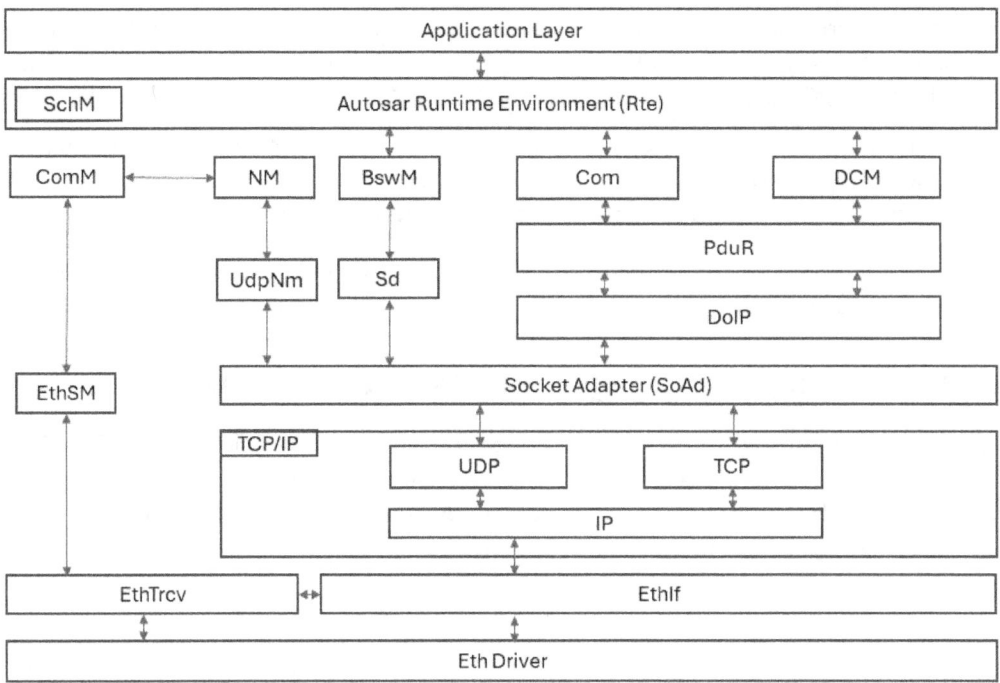

Figure 4.8 – Classic AUTOSAR with DoIP

The AUTOSAR components shown in *Figure 4.8* can be categorized into the following elements that are essential for remote diagnostics:

- **Network setup**: **EthTrcv**, **Eth**, and **EthSwt** establish the physical and data link layer infrastructure, with EthIf providing a uniform interface for upper layers.

- **Communication management**: **EthSM** oversees the operational state of the Ethernet communication, ensuring readiness for data transmission.

- **IP networking**: **TCP/IP** enables the establishment of IP-based communication, crucial for diagnostics over Ethernet.

- **Service location and connection management**: **Sd** discovers diagnostic services, while **SoAd** manages the socket connections needed for protocol-specific communication, including DoIP for diagnostics.

- **Diagnostics processing**: DoIP facilitates the transmission of diagnostic messages over IP. **DCM** manages the diagnostic logic, handling requests and responses between external diagnostic tools and vehicle ECUs.

- **System coordination**: **NVM** stores diagnostic session data persistently, **ComM** optimizes communication channels, **BswM** orchestrates software module interactions, and **NM** ensures network integrity and management.
- **Software integration**: **RTE** acts as a conduit, enabling seamless communication between the application layer (diagnostic services) and lower-level software modules, ensuring that diagnostics can be performed remotely with high efficiency and reliability.

In the next section, let's gain an insight into diagnostic service management, a key enabler of diagnostic functionality in Adaptive AUTOSAR.

Diagnostic service management in Adaptive AUTOSAR

The diagnostic service management response handling essentially mirrors the functionality of the DCM BSW module of the AUTOSAR Classic platform. It is responsible for the processing and dispatching of diagnostic services. *Figure 4.9* illustrates the processing steps and functional blocks of the **Diagnostic Manager** (**DM**)'s diagnostic service management part.

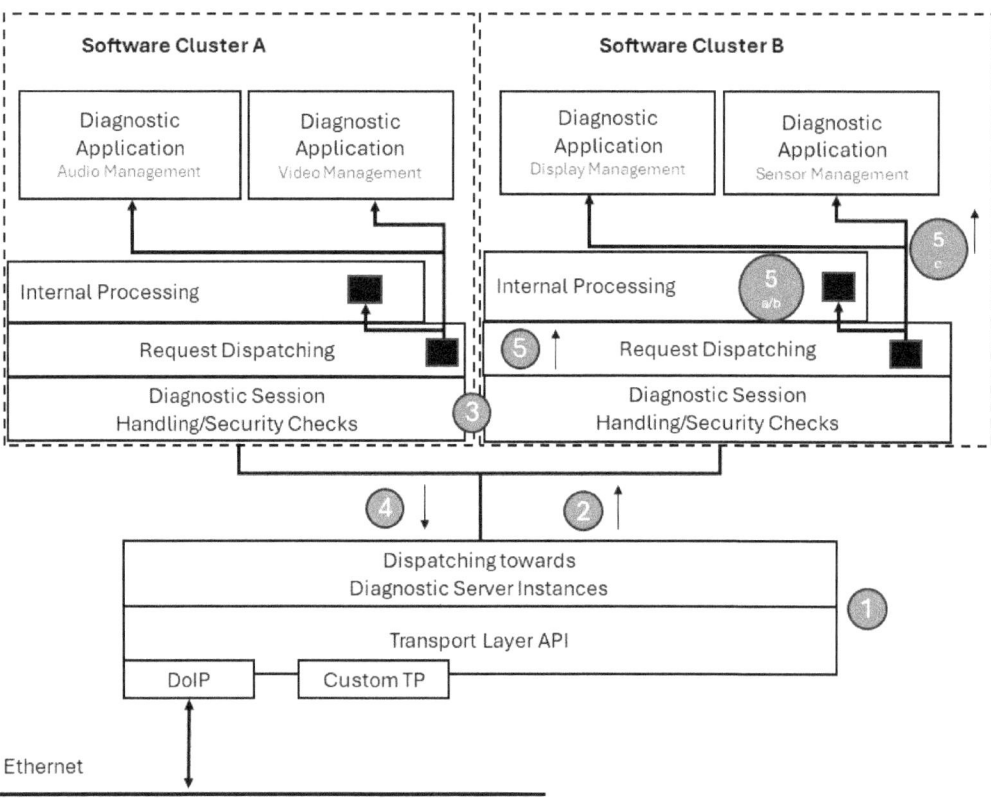

Figure 4.9 – Diagnostic service management

Diagnostic service management is responsible for the following:

1. Receiving UDS diagnostic request messages from the network layer and extracting UDS information, independent of the transport layer
2. Dispatching a request to the diagnostic server based on the target address and type (physical/functional)
3. Checking whether a request is permitted in the current session and security context
4. If not permitted, sending a negative UDS response back through the network layer
5. If permitted, processing a request based on the DM's configuration and request type and dispatch, as follows:

    A. Internally within the DM's diagnostic service function

    B. Internally within the DM's event memory management function

    C. Delegate to an external Adaptive Application for processing – for example, audio/video/display/sensor management diagnostic application

A software cluster in Adaptive AUTOSAR is the following:

- A deployable group of one or more Adaptive Applications
- The smallest unit for independent deployment in the Adaptive AUTOSAR environment
- Managed by the Adaptive Platform for life cycle processes such as deployment and execution
- Enables communication within and across clusters for complex functionalities

In the next section, let's connect back to the remote diagnostics use case mentioned in *Chapter 2, Introducing Automotive IoT Use Cases*, and also understand DoIP, UDS, Adaptive AUTOSAR, and Classic AUTOSAR use.

Reflecting on the application of remote diagnostics

Figure 4.10 shows how the integration of Classic AUTOSAR and Adaptive AUTOSAR frameworks occurs on the vehicle side. These frameworks streamline the implementation of DoIP and UDS, thus enabling remote diagnostics. The availability of various off-the-shelf components from vendors such as Vector, EB, ETAS, KPIT, and others accelerates the integration of remote diagnostics within vehicles.

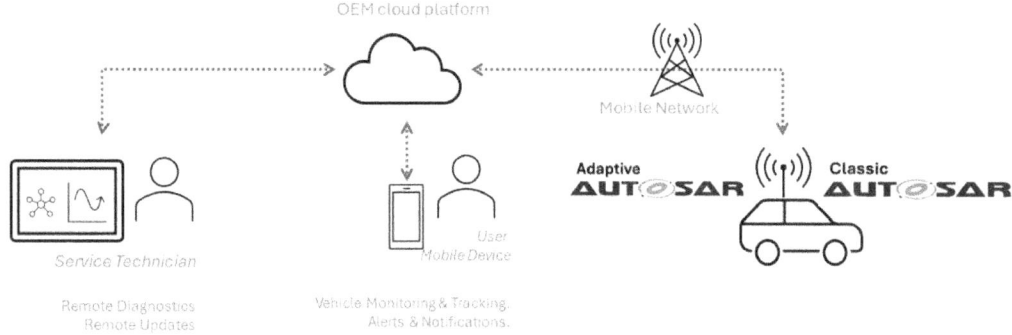

Figure 4.10 – A remote diagnostics use case

With this, we have come to the end of the chapter.

Summary

UDS is widely used for vehicle diagnostics and is heavily utilized in CAN-based communication. Later, DoIP was introduced to support diagnostics over IP, and it also carries UDS messages. Both Classic and Adaptive AUTOSAR platforms support UDS/DoIP-based diagnostics. Many vendors in the market provide off-the-shelf components related to UDS and DoIP-based diagnostics as part of the AUTOSAR deliverables, which expedite the implementation of diagnostic features in vehicle ECUs.

The next chapter covers the next wave of vehicle diagnostics, focusing on the details of **service-oriented vehicle diagnostics** (**SOVD**), which will be key in supporting remote diagnostics in the forthcoming wave of vehicle architectures.

References

1. [1]: https://www.iso.org/standard/72439.html
2. [2]: https://piembsystech.com/doip-protocol/?expand_article=1

5

Next Wave of Vehicle Diagnostics

In this chapter, we will get an understanding of the next generation of diagnostics, better suited to advancements in modern vehicles, including the **Internet of Things (IoT)** field. **Unified Diagnostic Services (UDS)** is not well-suited to modern vehicle architecture, as it is static in nature and inflexible for software-defined vehicles, which tend to change more often. There is a need for a dynamic diagnostic protocol to meet the demands of modern technologies. This chapter covers **Service-Oriented Vehicle Diagnostics (SOVD)**, the next wave of diagnostic protocols for modern vehicles. It also provides a reference SOVD demo, as well as a comparison between UDS and SOVD.

In this chapter, we're mainly going to cover the following topics:

- Needs beyond UDS
- SOVD
- SOVD example, demo, and details

Technical requirements

The following are the basic requirements needed to understand/implement SOVD:

- Programming language skills such as C++/Java/Python/JavaScript
- Knowledge of the basics of **Hypertext Transfer Protocol (HTTP)** and **Representational State Transfer (REST)** methods
- Knowledge of the basics of **JavaScript Object Notation (JSON)** or **eXtensible Markup Language (XML)**
- Basic knowledge of client-server, UDS, and SOVD protocols

Needs beyond UDS

The rationale for exploring options beyond UDS stems from its reliance on a static configuration presented in the **Open Diagnostic data eXchange** (**ODX**) file format. This configuration model proves unsuitable for software-defined vehicles, which require the flexibility to accommodate frequent software changes. Given that high-performance computing systems serve as pivotal facilitators for software-defined vehicles, software is expected to assume increasing responsibilities.

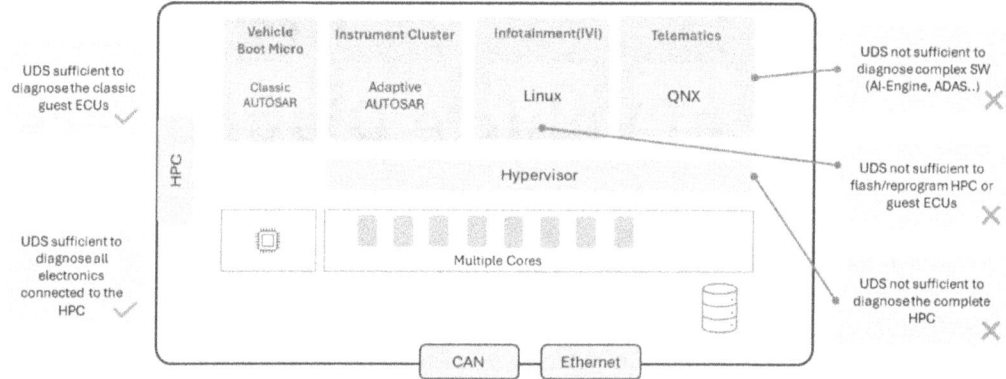

Figure 5.1 – UDS boundaries

In the case of UDS, clients (testers/testing tools) need to have a UDS stack built in to translate UDS messages to human-readable information and vice versa. *Figure 5.1* shows a high-level view of what UDS can support well in the **High-Performance Computing** world and what its limitations are. UDS is adequate in supporting diagnostics on classic **Electronic Control Units**. However, it is not adequate to support diagnostics for whole HPCs or complex applications such as **Advanced Driver Assistance Systems**. Additionally, UDS is not capable of supporting the updating of complete HPCs, that includes multiple guest ECUs/machines.

At the other end of the spectrum, even though HPCs are popular and used in modern vehicle architecture, there is still a need for Classic AUTOSAR to support legacy ECUs in the system – for example, switches (steering wheel controls).

The **Association for Standardization of Automation and Measuring Systems** (**ASAM**) is developing the Open Vehicle Diagnostics specs to overcome some of the limitations of UDS and better support HPCs/software-defined vehicles. ASAM has also developed SOVD to meet the needs of HPCs/software-defined vehicles and future architectures. To learn more about ASAM, please check reference [1]

SOVD not only supports future architectures but also accommodates Classic AUTOSAR ECUs, Adaptive AUTOSAR, and non-AUTOSAR entities by providing uniform APIs. Let's delve into SOVD in the next section.

SOVD

SOVD is designed to provide a straightforward diagnostic interface for both modern software-defined vehicle architectures and legacy architectures, including classic ECUs. Additionally, SOVD ensures uniform access for remote, proximal, and in-vehicle diagnostic scenarios. *Figure 5.2* provides a high-level view of in-vehicle, proximal, and remote diagnostics.

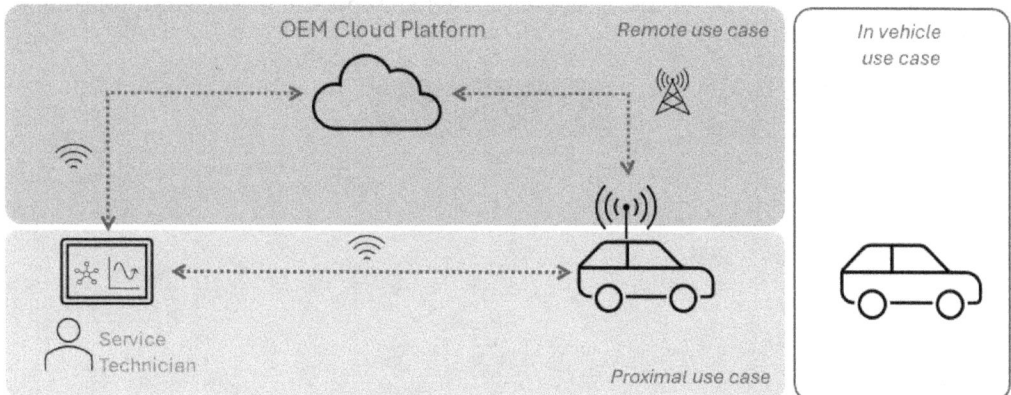

Figure 5.2 – SOVD use cases (in-vehicle, proximal, and remote)

The ECUs can be accessed within the vehicle (**In-Vehicle**), locally, near the vehicle (**Proximity**), or remotely via the cloud (**Remote**). Here is a detailed discussion of the three types of access:

- **In-vehicle access**: This means accessing ECUs from within the vehicle itself, typically for real-time monitoring, error detection, maintenance, and health status checks.
- **Proximity access**: This means accessing ECUs when physically near the vehicle, often required for diagnostics, repairs, or compliance verification, such as in service centers, manufacturing facilities, or during emission checks. The tester will make use of Ethernet or Wi-Fi to connect to the ECU.
- **Remote access**: This means accessing ECUs from a remote location via cloud-based systems or remote communication channels, enabling functionalities such as over-the-air updates, backend evaluation, fleet management, remote assistance, and activation of additional services.

At its core, ASAM SOVD establishes a unified set of specifications and guidelines, facilitating seamless communication between diagnostic tools and ECUs. This communication includes essential functions such as fault diagnosis, parameter adjustments, and comprehensive vehicle testing.

The significance of ASAM SOVD lies in its ability to ensure interoperability across a wide array of diagnostic tools and ECUs. By adhering to this standard, automotive manufacturers and suppliers streamline the diagnostic process, thereby reducing development costs and enhancing efficiency.

Moreover, ASAM SOVD plays a pivotal role in modern vehicle architecture by facilitating the integration of third-party diagnostic tools and software. This standardization fosters innovation and flexibility within automotive systems, ultimately leading to more robust and adaptable vehicle designs.

In essence, ASAM SOVD stands as a foundational element in the evolution of automotive technology, enabling enhanced diagnostics and fostering collaboration across the industry.

Figure 5.3 illustrates multiple connections between a tester and applications utilizing proprietary communication protocols.

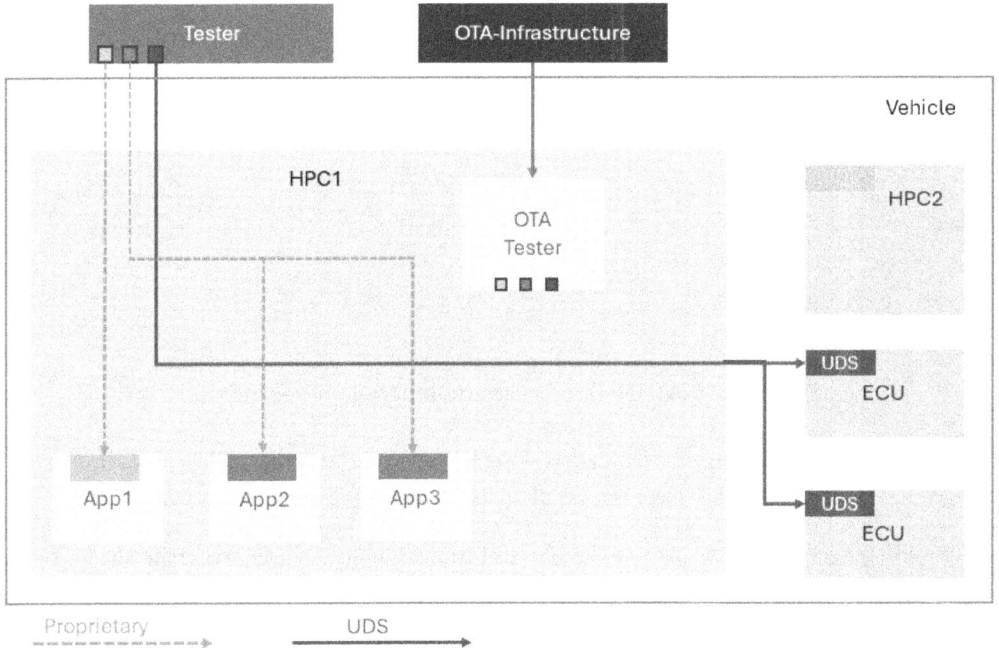

Figure 5.3 – Vehicle without SOVD

In contrast, *Figure 5.4* depicts a single connection between a tester and applications via the SOVD gateway/server.

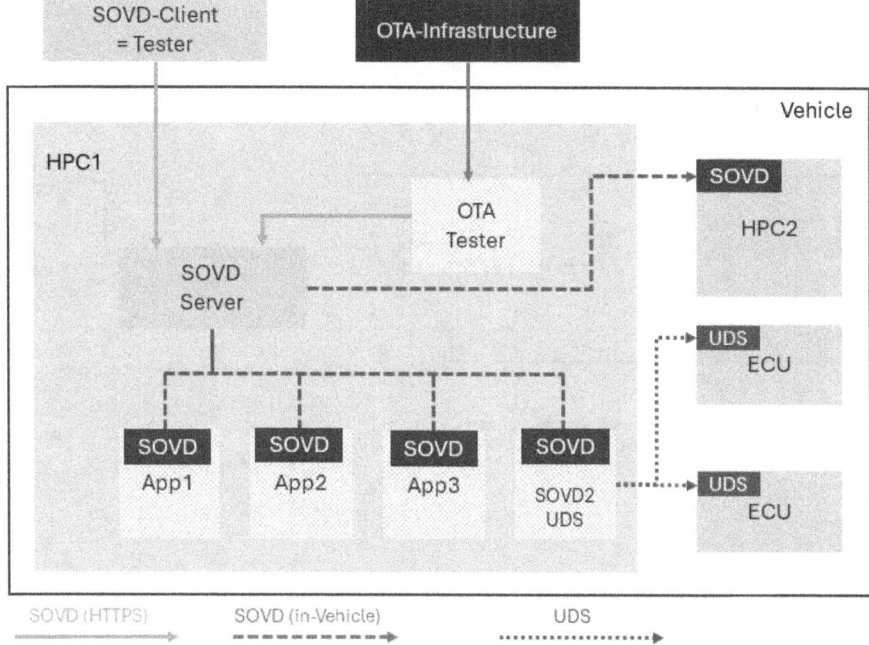

Figure 5.4 – Vehicle with SOVD

In the case of vehicles without SOVD, testers require multiple connections and communication protocols to support proprietary protocols as well as UDS. This significantly increases the complexity of the tester, tools, and vehicle software architecture and makes troubleshooting difficult in the field. However, with vehicles equipped with SOVD, testers can utilize a single protocol regardless of the type of applications they are connecting within the HPCs. Applications should support SOVD interfaces by implementing the GET/POST/PUT/DELETE type of interfaces. Please refer to the SOVD example section to see the sample code required to implement the GET interface.

Figure 5.5 provides a detailed view of SOVD internals, particularly focusing on the vehicle side.

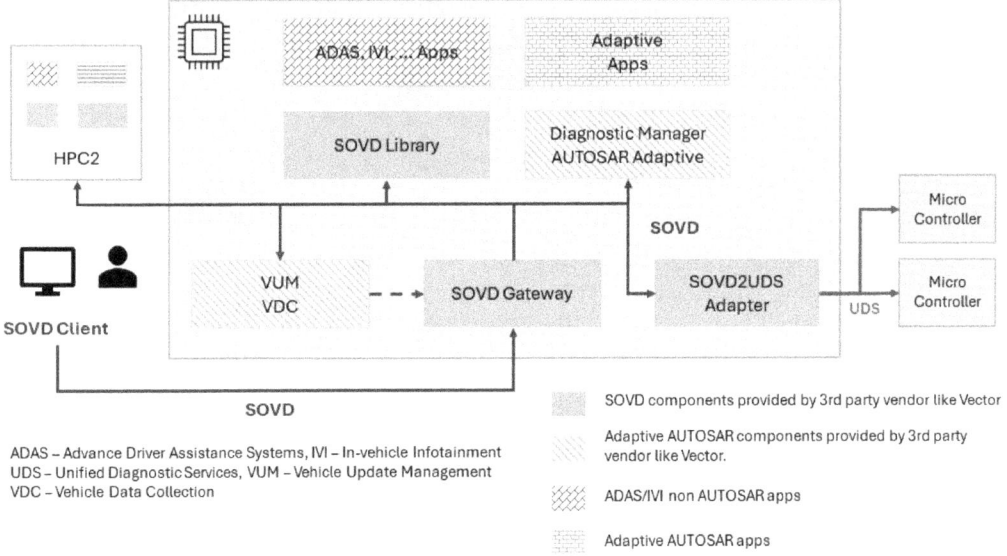

Figure 5.5 – Vehicle with SOVD

To successfully implement SOVD, it is essential to integrate all vehicle components into the SOVD API, including HPC machines, zonal ECUs, sensors, actuators, and other ECUs. This integration ensures a scalable architecture. Here are specific SOVD component details:

- **The SOVD server/gateway**: Centralizes requests and directs them to the appropriate endpoints.
- **The SOVD2UDS adapter**: Links UDS-based microcontrollers, translating diagnostic requests between SOVD and UDS.
- **Diagnostic manager**: Implements SOVD in AUTOSAR adaptive applications using `ara::diag` interfaces, a standardized protocol for diagnostics communication, fault identification/management, session halding, etc. These interfaces also support remote diagnostics.
- **SOVD library**: Offers SOVD-specific functions regardless of the AUTOSAR environment.

SOVD establishes a standardized REST API, which operates independently of the programming language used. This enables the development of new, lightweight diagnostic applications. Using SOVD, diagnostic processes can be written in any modern programming or scripting language, such as Python, to extract and analyze vehicle data. Moreover, the REST API facilitates the creation of enterprise (cloud) applications for larger tasks, such as remote diagnostics or development testing. These applications can run as web applications on smartphones, tablets, and laptops. In the next section, let's delve into a high-level overview of REST.

REST

REST [2] is an architectural style used for designing networked applications based on a client-server model. In this model, the client initiates requests to the server, which processes and responds to those requests. RESTful systems typically use HTTP as the underlying protocol for communication.

REST is characterized by several key principles:

- **Statelessness**: Each request from a client to the server must contain all the necessary information to understand the request. The server doesn't store any client context between requests, making it easier to scale and manage.
- **Uniform interface**: RESTful systems have a uniform interface between components, typically consisting of standard HTTP methods such as GET, POST, PUT, DELETE, and so on. This simplifies communication and promotes interoperability.
- **Resource-based**: Resources are identified by **Uniform Resource Identifiers** (**URIs**), and interactions with resources are performed using standard HTTP methods. Resources can represent any concept or entity, such as users, documents, or products.
- **Representation**: Resources are typically represented in a human-readable format, such as JSON or XML. Clients can request different representations of a resource based on their needs.

Regarding the relationship between SOVD and REST, SOVD could employ RESTful principles for communication between diagnostic tools and vehicle systems. This entails leveraging HTTP methods and URIs to access diagnostic data and perform actions on vehicle components. By embracing RESTful principles, SOVD could offer a standardized and interoperable interface for accessing diagnostic information from vehicles, aligning with the flexibility and simplicity provided by REST. To learn more about REST, please check reference [2].

The next section covers a SOVD example, demo, and other reference details to understand the practical implementation.

SOVD example, demo, and details

Let's look at an example of SOVD in comparison with UDS for better understanding. Also, let's further explore the SOVD demo to gain a comprehensive understanding of how SOVD operates. Additionally, at the end, there is a summary table listing the differences and similarities between SOVD and UDS.

Example of a diagnostic message using UDS and SOVD

Let's examine an example of reading the **Vehicle Identification Number** (**VIN**) using UDS versus SOVD.

Here's an example of reading the VIN via `DID 0xF190`:

```
UDS Request to Target ECU:
0x22 0xF1 0x90
UDS Response from ECU:
0x62 0xF1 0x90 0x56 0x33 0x43 0x54 0x30 0x52 0x56 0x33 0x48 0x31 0x43
0x4C 0x33 0x31 0x32 0x35 0x34
```

The tester tool needs to perform further processing before presenting it to the user:

1. It first needs to map `0xF190` as a VIN.
2. It then converts the hexadecimal to the VIN, `V3CT0RV3H1CL3254`.

In contrast, with SOVD, the tester tool will receive symbolic/readable information, and no additional processing is required. Each component can self-describe its identification data upon request. The `/data/ resource` provides all parameters supported by the respective component. In SOVD, a query string, termed `identData`, enables the filtering of all data by identification data:

```
/data?categories=identData
```

The following example shows how all available identification data can be listed via SOVD:

```
Request:
HTTP GET {base_uri}/components/engine/data?categories=identData
Response:
HTTP OK 200
{
 "items": [
 {
 "id": "vin",
 "name": "Vehicle Identification Number",
 "category": "identData"
 },
 {
 "id": "swversion",
 "name": "Software Version",
 "category": "identData"
 }
 ]
}
```

Once all available identification data is known, users can query the relevant data and their respective values individually.

The following is the request for the readout of the VIN from the engine ECU:

```
Request:
HTTP GET {base_uri}/components/engine/data/vin
Response:
HTTP OK 200
{
  "id": "vin",
  "name": "Vehicle Identification Number",
  "category": "identData",
  "data": "V3CT0RV3H1CL3123"
}
```

This also illustrates that SOVD does not require static configurations such as ODX files. In the next section, we will cover additional reference documentation for SOVD and provide details about the SOVD demo.

Example of an SOVD interface as part of applications on the server side

This example code shows how to implement the GET VIN interface on the server side:

```
#include <iostream>
#include <cpprest/http_listener.h>
#include <cpprest/json.h>

using namespace web;
using namespace web::http;
using namespace web::http::experimental::listener;

const utility::string_t base_url = U("http://localhost:8080");

// Function to handle HTTP GET requests for retrieving VIN
void handle_get_vin(http_request request)
{
    // Simulate retrieving VIN from a database or external source
    utility::string_t mock_vin = U("123456789ABCDEFG");

    // Create a JSON response object
    json::value response;
    response[U("vin")] = json::value::string(mock_vin);
```

```cpp
        // Send the response with status code 200 (OK)
        request.reply(status_codes::OK, response);
}

int main()
{
    // Create an HTTP listener
    http_listener listener(base_url);

    // Bind the handle_get_vin function to the GET method at /vin endpoint
    listener.support(methods::GET, [](http_request request) {
        handle_get_vin(request);
    });

    try
    {
        // Start listening for requests
        listener.open().then([&listener]() {
            std::wcout << "Listening for requests at: " << base_url << std::endl;
        }).wait();

        // Keep the server running until terminated
        std::string line;
        std::getline(std::cin, line);
    }
    catch (const std::exception& e)
    {
        std::cerr << "An error occurred: " << e.what() << std::endl;
    }

    return 0;
}
```

From the preceding snippet:

- Function `handle_get_vin` is responsible for handling HTTP GET requests to retrieve the VIN.
- A mock VIN is provided for simplicity. In the main function, an http_listener object is created and bound to the specified base URL.

- The `handle_get_vin` function is associated with the GET method at the `/vin` endpoint.
- The listener is then started to initiate listening for incoming HTTP requests.
- Upon receiving a GET request to `/vin`, the server responds with a JSON object containing the mock VIN.

SOVD documentation and demo

To explore more details about SOVD, please refer to the videos and documentation available at the Vector website [3]. Vector also provides a console to experience SOVD-based communication [4]. If you need assistance with registration, please refer to the provided registration guide [5].

Figure 5.6 illustrates the Vector SOVD console view during a GET request for the VIN and provides details of its response.

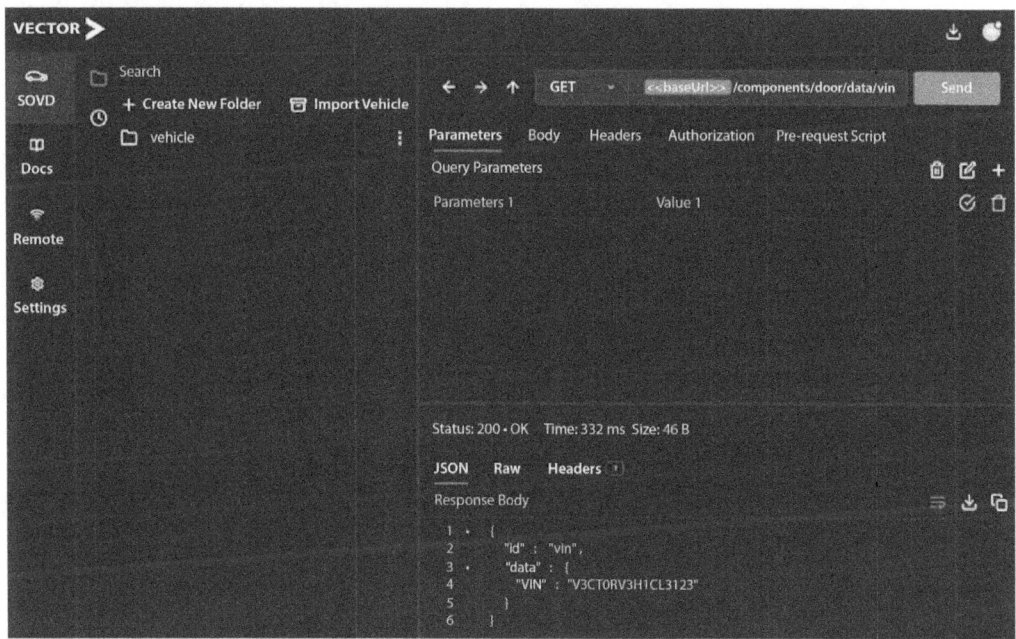

Figure 5.6 – Vector SOVD console

Having explored the details and examples of SOVD messages, as well as the SOVD demo, let's now compare SOVD and UDS based on several key aspects in the next section.

SOVD and UDS comparison

Table 5.1 presents a high-level comparison of UDS and SOVD, outlining some of their key characteristics.

Aspect	UDS	SOVD
Protocol	Standardized protocol used in the automotive industry	Protocol designed for modern software-defined vehicle architectures
Focus	Primarily focuses on traditional diagnostic communication	Focuses on providing a simple diagnostic interface for modern vehicles and legacy architectures
Flexibility	Relies on a static configuration (ODX file format) and is less adaptable to frequent software changes	Designed to accommodate frequent software changes in software-defined vehicles
Compatibility	Compatible with Classic AUTOSAR ECUs and traditional vehicle architectures	Designed to support both modern software-defined vehicle architectures and legacy architectures based on Classic ECUs
Communication	Communication typically occurs over a **controller area network** (**CAN**) or other automotive bus systems	Can utilize various communication protocols, potentially including RESTful principles for remote diagnostics
Features	Offers comprehensive diagnostic capabilities for traditional automotive systems	Provides uniform access for remote, proximity, and in-vehicle diagnostic scenarios, catering to evolving automotive diagnostic needs
Security	Limited built-in security features, often necessitating additional layers of security implementation	Incorporates modern security measures to protect against unauthorized access and data breaches, especially in remote diagnostic scenarios

Table 5.1 – Comparison of UDS versus SOVD

As you have learned so far, SOVD is essential for modern vehicle architecture as it is necessary to support diagnostics for complex applications, including software updates for HPCs.

Summary

This chapter mainly covered what lies beyond UDS and the need for a new diagnostic protocol, by examining UDS limitations and SOVD internals and providing examples. Additionally, it compared UDS and SOVD based on key aspects at the end of the chapter. At this time, you should have a fair understanding of SOVD and its practical implementation and other details.

So far, we have explored the ongoing evolution of vehicle architecture, which encompasses various standard automotive frameworks, such as Classic AUTOSAR, Adaptive AUTOSAR, and hypervisor, supporting guest virtual machines. At the core of modern vehicle architectures lies SOVD, which plays a pivotal role in diagnostics for modern vehicles.

As vehicles become more connected, ensuring the security of remote diagnostics systems is crucial. Cybersecurity measures, such as encryption and secure communication protocols, protect sensitive vehicle data from unauthorized access.

In the next chapter, we will explore cybersecurity in automotive IoT, emphasizing the importance of secure software development processes. The discussion will cover ISO/SAE 21434, ASPICE for cybersecurity, DevSecOps methodologies, and specific activities in the secure software development lifecycle.

References

- [1]: `https://www.asam.net/`
- [2]: `https://developer.mozilla.org/en US/docs/Glossary/REST`
- [3]: `https://www.vector.com/us/en/products/solutions/diagnosticstandards/sovd-service-oriented-vehicle-diagnostics/#c308624`
- [4]: `https://console.sovd.io/`
- [5]: `https://cdn.vector.com/cms/content/know-how/SOVD/doc/SOVD_Console_Registration.pdf`

Get This Book's PDF Version and Exclusive Extras

Scan the QR code (or go to `packtpub.com/unlock`). Search for this book by name, confirm the edition, and then follow the steps on the page.

Note: Keep your invoice handy. Purchases made directly from Packt don't require one.

Part 3: Secure Development for Automotive IoT

Automotive IoT introduces new cybersecurity threats that need to be considered by developers. This part provides the reader with a background on secure development for automotive IoT as well as practical guidance on how to establish a secure development platform and secure the software supply chain.

This part has the following chapters:

- *Chapter 6, Exploring Secure Development Processes for Automotive IoT*
- *Chapter 7, Establishing a Secure Software Development Platform*
- *Chapter 8, Securing the Software Supply Chain*

6
Exploring Secure Development Processes for Automotive IoT

While **automotive Internet of Things (IoT) applications** can bring added benefits to users and auto manufacturers, the deployment of such applications also brings new security considerations.

If these applications do not include proper security solutions or do not follow best practices for secure development, there is a risk of successful cybersecurity attacks, which could have disastrous consequences. Therefore, it is imperative for automotive organizations to understand what those risks are and how to address them appropriately.

To that end, in this chapter, we will first provide an overview of cybersecurity threats applicable to the automotive IoT use cases and explain the need for cybersecurity and why automotive organizations need to establish a secure development process.

We will then review secure development processes and frameworks divided into the vehicle side as well as the non-vehicle side including the backend and mobile device development. For the vehicle side, we will review the **ISO/SAE 21434 Cybersecurity Engineering** [1] and **ASPICE for Cybersecurity** [2] processes to better understand what relevant cybersecurity requirements automotive organizations should consider both on an organizational level as well as on a project level. For the non-vehicle side, we will review the **National Standards Institute of Technology (NIST) Cybersecurity Framework** [3], **ISO 27001** [4], **System and Organization Controls 2 (SOC 2)** [5], and **Open Web Application Security Project (OWASP)** [6] to help better understand what organizations need to consider for cybersecurity for backend solutions and mobile app development.

While some automotive organizations are already embracing **DevSecOps** and **agile development**, we see broader adoption occurring in the industry. As such, we will further elaborate on these topics at the end of this chapter. In this chapter, we will cover the following topics:

- An overview of security threats and the need for security and secure development processes
- ISO 21434 and ASPICE for Cybersecurity
- NIST Cybersecurity Framework, ISO 27001, SOC 2, OWASP
- DevSecOps and agile development

An overview of security threats and the need for security and secure development processes

This section presents an overview of security threats and explains the need for product cybersecurity solutions and why automotive organizations need to establish a secure development process.

New cybersecurity threats

Automotive IoT use cases increase the attack surface and introduce several new security threats to consider. The attack surface is increased as there are new attack vectors that need to be considered. For example, attackers can target the vehicle directly by gaining access to communication interfaces on the vehicle side including Wi-Fi, Bluetooth, and cellular network communication, as well as target weaknesses in the IoT applications on the vehicle side. Attackers can also target weaknesses in communication protocols to the cloud side or exploit vulnerabilities in web apps. Additionally, attackers can find and exploit vulnerabilities in mobile apps.

New security threats include, for example, gaining access to or modifying assets, such as cryptographic credentials, vehicle keys, vehicle data, and log files, used in automotive IoT applications.

Recent reports on automotive cybersecurity [7, 8] show that there are over 200 vulnerabilities reported per year since 2019. Backend server attacks account for 40% of all attacks, which also includes a number of **application programming interface** (**API**)-based attacks. Regarding **open-source software** (**OSS**), a recent report [9] found that 63% of the code bases scanned for the category where *automotive* is included contained high-risk vulnerabilities (classified with a score of 7 or higher).

As new cybersecurity threats emerge, it becomes imperative for automotive organizations to consider appropriate cybersecurity solutions and establish secure software development practices.

Examples of recent attacks

Several reports on attacks on automotive systems have been published in the past. A few examples focusing on IoT types of solutions are described as follows [10].

Multiple weaknesses and vulnerabilities were found that allow an attacker to send remote commands to a vehicle by specifying the **vehicle identification number** (**VIN**) to, for example, unlock, lock, engine start, engine stop, and so on, by abusing the API of **original equipment manufacturer** (**OEM**) backends. It was shown that several OEMs were vulnerable to this type of attack.

There were also weaknesses and vulnerabilities that allowed an attacker to take over user accounts, which included being able to track the location of the associated vehicles, as well as gain access to the users' vehicle information, physical address, phone number, and email address. It was shown that several OEMs and relevant systems managing user information were vulnerable to this type of attack.

These types of recent attacks show that automotive IoT use cases can be targeted by attackers by going after the weakest link, which could be misconfigurations, vulnerabilities on the backend solution, or weak authentication approaches.

Simplified threat model of automotive IoT ecosystem

To provide some practical steps to get started, we will first perform a simplified **threat model** of the overall automotive IoT ecosystem to better understand the involved assets, threats, potential consequences, and the resulting risks. Moreover, we discuss some examples of mitigations to address the identified threats. Threat modeling is a process activity that helps organizations identify threats and assess their severity in order to define appropriate and prioritized security countermeasures.

Automotive IoT use cases generally involve communication between several entities including vehicle, cloud, and mobile device, as well as interactions between various sub-systems in the vehicle from external-facing systems such as **telematics control unit** (**TCU**) or **in-vehicle infotainment** (**IVI**) systems to in-vehicle network **gateways** (**GWs**) and embedded in-vehicle **electronic control units** (**ECUs**) storing relevant automotive IoT app data or providing various vehicle features such as locking/unlocking the doors or starting the engine.

An overview of the automotive IoT ecosystem is depicted in *Figure 6.1*.

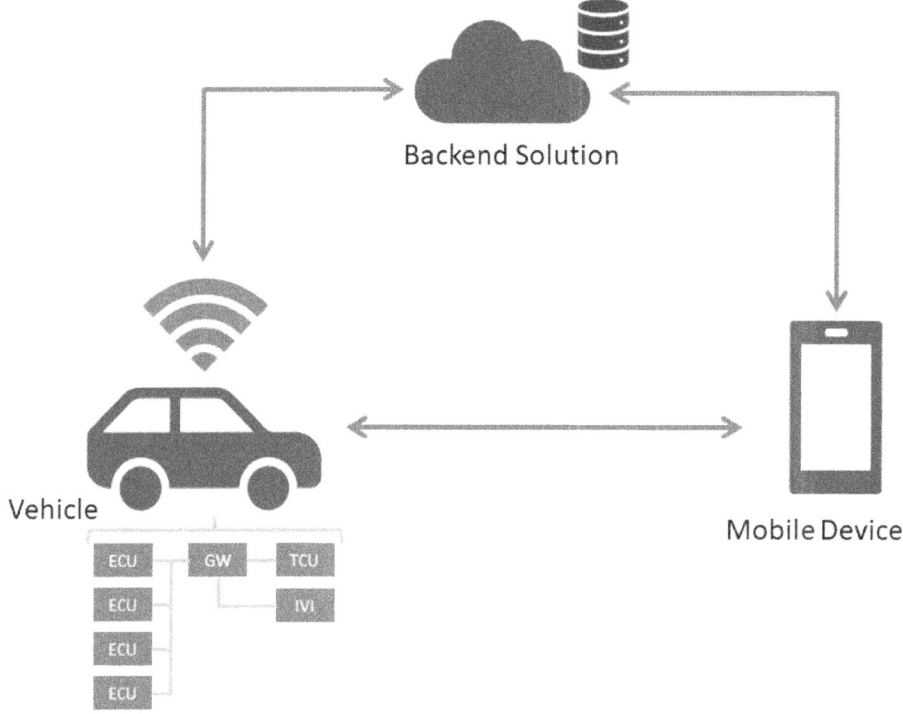

Figure 6.1 – Overview of the automotive IoT ecosystem

As shown in *Figure 6.1*, we have the vehicle (including the **vehicle IoT app**) on the left-hand side, connected to the backend solution (including the **backend IoT app**), depicted in the middle of the figure, as well as the mobile device (including the **mobile IoT app**) shown on the right-hand side. The mobile device is also connected to the backend solution. As mentioned, the vehicle contains an in-vehicle network consisting of TCU, IVI, one or several GWs, and multiple ECUs.

Please note that the example presented in this chapter merely serves to give an overview of the type of assets, threats, consequences, risks, and mitigations that can be considered. Regarding actual automotive IoT products, automotive organizations need to perform specific threat modeling for their automotive IoT applications, use cases, and ecosystems to determine the relevant threats and solutions.

There are numerous approaches to performing threat modeling and risk assessments such as **Threat Analysis and Risk Assessment (TARA)** in ISO/SAE 21434 [1], **NIST Special Publication (SP) 800-30** [11], and **STRIDE** [12]. A general flow of performing threat modeling is depicted in *Figure 6.2*:

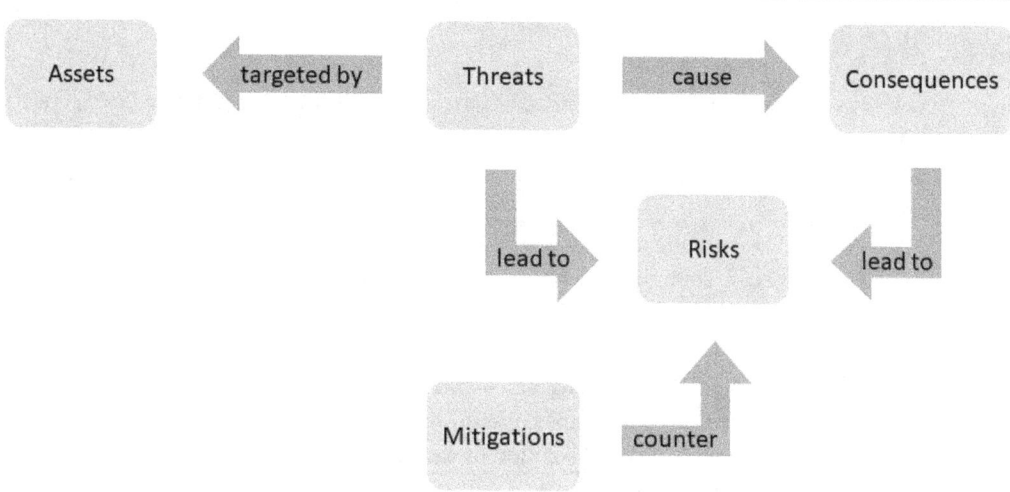

Figure 6.2 – General flow of threat modeling

As shown in *Figure 6.2*, first we identify the assets, which are targeted by threats. These threats cause consequences. Together, the threats and consequences lead to risks. The risks are countered by mitigations. Definitions of these terms based on ISO/SAE 21434, ISO 27001, and ISO 31000 are provided in *Table 6.1*:

Term	Definition
Asset	An object that has value, or contributes to value
Threat	A potential cause of an incident that may result in a breach of information security or the compromise of business operations
Consequence	The outcome of an event, measured against objectives
Risk	The effect of uncertainty on objectives
Mitigation	The action taken to reduce the likelihood of a risk occurring or the consequences of it if it does

Table 6.1 – Definition of terms related to threat modeling

We will follow this general flow in the following subsections as we review the threat model of the automotive IoT ecosystem. Please note that the threat modeling activity would generally go further into identifying assets and analyzing threats against specific systems and subsystems within the different entities, for example, threats against the TCU, IVI, GW, and ECUs in the vehicle relevant to the automotive IoT use case; however, this is left out of this chapter as the purpose is to highlight cybersecurity activities in the development processes and the need for different teams responsible for different entities to collaborate.

Assets

First, the involved assets for the vehicle, backend solution, and mobile device are defined as follows and summarized in *Table 6.2*:

Entity	Asset	Description
Vehicle	Credentials	Used to authenticate to the backend solution and mobile device, to establish secure communication, and to generate/verify digital signatures of data to be sent from the vehicle IoT app or received from the backend solution or mobile device.
	Data	Data collected or processed on the vehicle side used for the IoT use case. Data received from the backend solution or mobile device or data to be sent to the backend solution or mobile device.
	Software	The vehicle IoT app that provides the IoT use case functionality, underlying **operating system (OS)**, communication stack, and so on.
Backend solution	Credentials	Used to authenticate to the vehicle and mobile device, to establish secure communication, and to generate/verify digital signatures of data to be sent from the backend IoT app or received from the vehicle or mobile device.
	Data	Data collected from the vehicle or mobile device and processed on the backend solution used for the IoT use case.
	Software	The backend IoT app that provides the IoT use case functionality, underlying OS/cloud infrastructure, communication stack, and so on.
Mobile device	Credentials	Used to authenticate to the vehicle and backend solution, to establish secure communication, and to generate/verify digital signatures of data to be sent from mobile IoT app or received from the vehicle or backend solution.
	Data	Data collected or processed on the mobile device used for the IoT use case.
	Software	The mobile IoT app that provides the IoT use case functionality, underlying OS, communication stack, and so on.

Table 6.2 – Overview of assets pertaining to the vehicle, backend solution, and mobile device, respectively

Assets have now been identified for the vehicle, backend solution, and mobile device.

Threats

Once we have defined the assets, we proceed with identifying relevant threats against those assets. An overview of these threats is presented in *Table 6.3*.

Entity	Threat description
Vehicle	Unauthorized access to assets through long-range communication interfaces due to misconfiguration or software vulnerabilities.
	Unauthorized access to assets through short-range communication interfaces due to misconfiguration or software vulnerabilities.
	Unauthorized access to assets through physical/local access due to misconfiguration or software vulnerabilities.
Backend solution	Unauthorized access to assets through internet communication due to misconfiguration or software vulnerabilities.
	Denial of service of backend solution through internet communication due to misconfiguration or software vulnerabilities to prevent authorized users from accessing and using backend solution functionality.
	Unauthorized access to assets through physical/local access to server rooms due to misconfiguration, software vulnerabilities, or inadequate physical security.
Mobile device	Unauthorized access to assets through long-range communication interfaces due to misconfiguration or software vulnerabilities.
	Unauthorized access to assets through short-range communication interfaces due to misconfiguration or software vulnerabilities.
	Unauthorized access to assets through physical/local access due to misconfiguration or software vulnerabilities.

Table 6.3 – Overview of threats relevant to the vehicle, backend solution, and mobile device, respectively

At this step, we have now identified threats against the vehicle, backend solution, and mobile device.

Consequences

Based on the previously identified threats, we continue with analyzing the related consequences. That is, we analyze what is the potential impact of a successful attack based on the threats. An overview of these consequences is given in *Table 6.4*.

Entity	Consequence description
Vehicle	Extract sensitive data from one vehicle (e.g., credentials, location data, vehicle data, etc.).
	Modify data to perform arbitrary actions on one vehicle (including locking/unlocking the vehicle, starting the engine, controlling acceleration/steering/braking, performing software updates, etc.).
	Disable vehicle functionality on one vehicle to prevent authorized users from using the vehicle.
Backend solution	Extract sensitive data (including data collected from multiple vehicles) from the backend.
	Modify data to trigger inappropriate behavior on the backend, vehicle, or mobile device based on the processing of the data.
	Take over control of backend functionality to perform arbitrary actions (including triggering actions on vehicles).
	Disable backend functionality to prevent authorized users from using the backend solution.
Mobile device	Extract sensitive data from one mobile device (example, credentials, etc.).
	Modify data to trigger inappropriate behavior on the backend or vehicle based on the processing of the data.

Table 6.4 – Overview of consequences related to the vehicle, backend solution, and mobile device, respectively

We have now provided a few consequences for the vehicle, backend solution, and mobile device, respectively.

Risks

Next, based on the list of threats and consequences, it is possible to calculate the corresponding risks using standardized approaches such as those described in ISO/SAE 21434 [1] and Common Criteria [13]. For example, an organization can map the **attack potential** or **attack feasibility** for different threats to the **damage potential** or **impact** defined for different consequences and calculate the respective risk values.

These risk values can be mapped in a **risk matrix** as shown in *Figure 6.3*.

Impact	Attack Feasibility				
	Very Low	Low	Medium	High	Very High
Very High	Low	Moderate	High	Very High	Critical
High	Low	Moderate	High	High	Very High
Medium	Very Low	Low	Moderate	Moderate	Moderate
Low	Very Low	Low	Low	Low	Low
Very Low	Very Low	Very Low	Very Low	Low	Low

Figure 6.3 – Example risk matrix

The organization would need to define the actual risk values to use when mapping attack feasibility and impact, and each organization may use a slightly different risk matrix.

The organization then defines the **risk treatment options**, for example, based on ISO/SAE 21434, which includes avoiding, reducing, transferring, and accepting the risks.

The organization would also define the threshold for accepting risks, for example, very low and low risks may be accepted as is. In contrast, moderate, high, very high, and critical risks would need to be reduced, transferred, or avoided.

Following a standardized approach to risk management allows an organization to prioritize the risks and make the appropriate risk treatment decisions. Among these risk treatment decisions are mitigations that include suggestions on how to reduce the risks.

Mitigations

A proper threat model would also give suggestions on countermeasures or mitigations focusing on, for example, addressing the very high and critical risks. As this example scenario only serves to give an idea of the type of threats and risks, we will only provide two high-level approaches to mitigations. For the development of actual automotive IoT applications, we recommend that the teams involved perform a proper threat model and identify the relevant threats, consequences, risks, and appropriate mitigations suitable for the actual IoT use cases.

For the purpose of illustrating what type of mitigations organizations can take, we will focus on two categories of mitigations in this chapter: **product security solutions** and **secure development processes**.

Product security solutions overview

Generally, product security solutions are defined to address the threats or risks identified in the threat model or TARA. That is, the need for a certain product security solution or requirement can be traced back to a specific threat or risk. This allows us to achieve an appropriate level of defense for the final product and avoid over-engineering security solutions or applying too lax security controls.

However, in this chapter, we do not perform a full threat model of a specific automotive IoT application from which we can derive appropriate product security requirements. Instead, for the sake of guidance, we provide some general suggestions on product security solutions that can be considered.

For the vehicle, it is common practice to follow a **defense-in-depth** approach, example, use four levels of security: ECU, in-vehicle network communication, **electrical/electronic** (**E/E**) architecture, and external facing communication. Examples of respective security solutions are secure boot for ECU, secure communication using, for example, **Secure Onboard Communication** (**SecOC**) or **Media Access Control Security** (**MACsec**) over in-vehicle network communication, domain/ECU isolation using GWs and E/E architecture design, and lastly, firewalls and secure communication for external facing communication.

For the backend solution, traditional IT security solutions are typically used. Examples include firewalls, **intrusion detection systems** (**IDSs**), authentication and secure communication using **Transport Layer Security** (**TLS**), and **distributed denial-of-service** (**DDoS**) protection using, for example, load balancers.

For mobile devices, common security solutions and approaches include the use of secure storage, isolation between apps, avoiding hardcoded credentials included in the app, and using anti-reverse engineering techniques.

Secure development processes

Secure development processes help organizations develop secure products. There are numerous standards, frameworks, and guiding material that can be considered. In the following sections, we will delve further into automotive-specific standards such as ISO/SAE 21434 and ASPICE for Cybersecurity, which are applicable to vehicle-side development. In addition, we will cover activities for secure development considering more traditional IT environments applicable for the backend solution and mobile device. Thus, in the following sections, we will also discuss the NIST Cybersecurity Framework, ISO 27001, SOC 2, and OWASP.

ISO/SAE 21434 and ASPICE for Cybersecurity

To achieve a certain level of security assurance for automotive IoT products, it is imperative for automotive organizations to follow a standardized and rigorous development process.

In this section, we will review two standardized development processes, namely ISO/SAE 21434 [1] and ASPICE for Cybersecurity [2], and focus on the cybersecurity requirements.

ISO/SAE 21434 Overview

ISO/SAE 21434 was released in 2021 and gives a comprehensive set of cybersecurity engineering requirements for the development of automotive systems. The standard contains 15 clauses providing both **organizational-level** and **project-level** requirements. The first four clauses are more general in nature, defining the scope, normative references and terms, definitions and abbreviated terms, and general considerations, and therefore omitted from further discussion in this book.

An overview of the remaining clauses 5 through 15 is depicted in *Figure 6.4*.

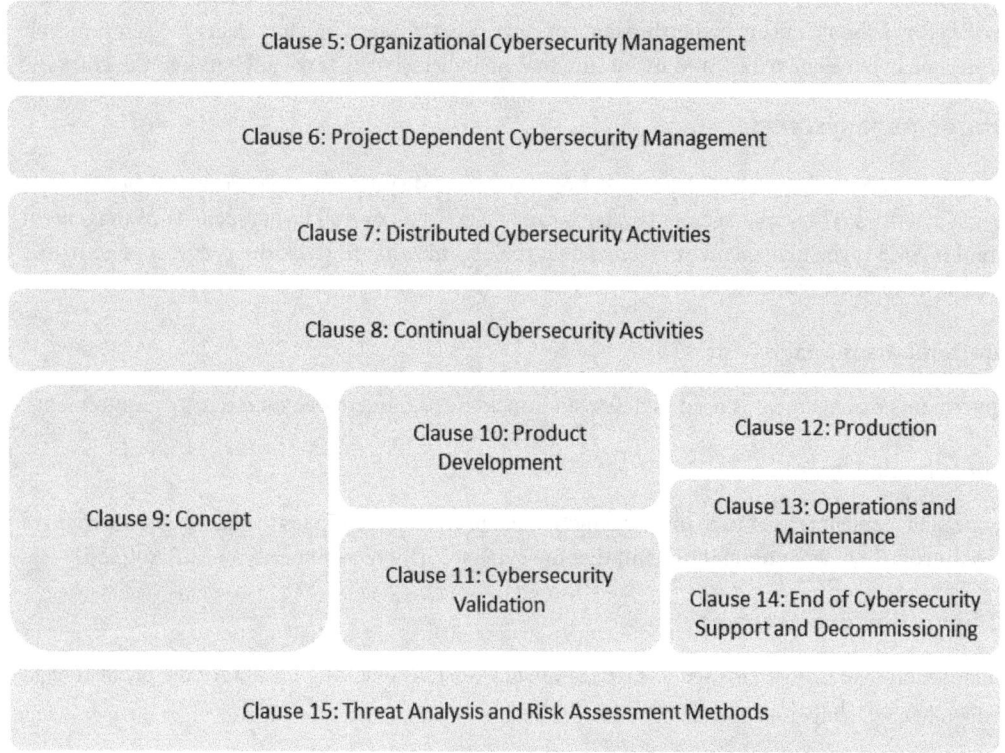

Figure 6.4 – Clauses from ISO/SAE 21434 (adapted from [1])

Rather than describing the contents for each clause, we instead briefly present relevant contents from the ISO/SAE 21434 standard grouped into organizational-level and project-level requirements.

ISO/SAE 21434 organizational-level requirements

There are a number of organizational-level requirements presented in ISO/SAE 21434. A few examples of activities that organizations need to consider are presented in the following subsections.

Policies and processes

An organization needs to define relevant **cybersecurity policies** and **processes** within the organization. These policies and processes should emphasize the importance and prioritization of cybersecurity within the organization, help establish a cybersecurity culture, and provide guidance to team members on relevant cybersecurity activities.

Roles and responsibilities

An organization needs to define specific cybersecurity **roles** and **responsibilities**. These roles and responsibilities should ensure ownership and coverage of activities so that it is clear who is responsible for a certain cybersecurity task and make sure that no cybersecurity tasks fall between the cracks.

Management systems

The organization needs to establish a number of **management systems** to help manage cybersecurity-relevant activities. These management systems can be part of an overall **Cybersecurity Management System** (**CSMS**). The management systems help the organization to track the progress of activities, as well as manage traceability.

Requirements management

Requirements management is used to define and track the fulfillment of cybersecurity requirements.

Change management

Change management is used to manage changes to the product or processes to ensure that the changes do not introduce new vulnerabilities and that the cybersecurity requirements are still fulfilled.

Configuration management

Configuration management is used to manage various versions or configurations of the product (e.g., software versions, hardware configurations, etc.).

Documentation management

Documentation management is used to manage and track all relevant documents that are created and used in the organization (for example, policy and process documents).

Competency management

Competency management is used to manage and track relevant cybersecurity skills and competencies in the organization and can be mapped to the roles and responsibilities to ensure that the appropriate competency and training are fulfilled for a specific role.

Tool management

Tool management is used to manage and track all the tools used in the organization that can have an impact on cybersecurity, including cybersecurity testing tools but also tools used for designing, coding, and building the software.

ISO/SAE 21434 project-level requirements

Besides the organizational-level requirements, there are a number of **project-level** specific requirements as discussed in the following sections.

Cybersecurity plan

The **cybersecurity plan** is extremely important as it serves as a guide to the project team on what cybersecurity activities should be performed and what work products should be produced during the life cycle of the product.

TARA

The project team or security team should perform a TARA to identify relevant threats against common security properties such as confidentiality, integrity, and availability of the defined assets.

Cybersecurity goals and concept

Cybersecurity goals to address the risks identified in the TARA and, consequently, a **cybersecurity concept** that fulfills the cybersecurity goals should be defined.

Cybersecurity specifications and requirements

Cybersecurity specifications and **requirements** specific to the project based on the cybersecurity concept should be defined.

Secure development

Secure development activities should be performed, including approaches for secure design and implementation.

Verification and validation

Verification and **validation** activities, including various testing approaches and reviews, should be performed to verify that the implementation and integration fulfill the cybersecurity specifications and goals.

Continual monitoring and updates

As part of **continual** cybersecurity activities and post-development activities, the organization should **monitor** for new threats and vulnerabilities specific to the project, and, where applicable, provide relevant project-specific **updates** to address cybersecurity issues.

ASPICE for Cybersecurity overview

This section focuses on the cybersecurity-specific activities in the ASPICE for Cybersecurity Standard [2], which was released in 2021. Please note that general activities for the ASPICE standard are presented in *Chapter 12, Processes and Practices*.

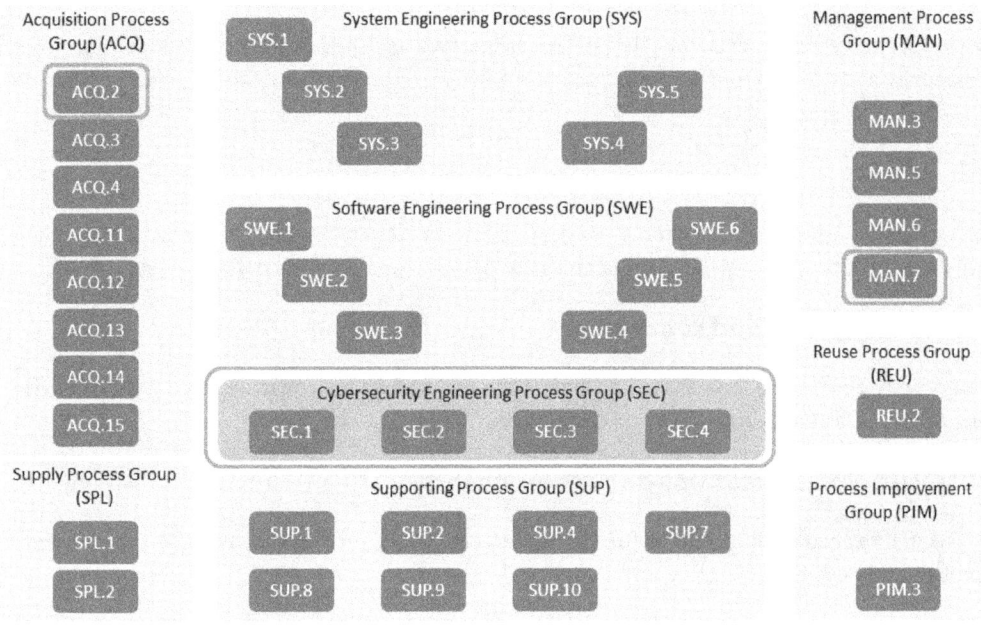

Figure 6.5 – Cybersecurity activities in ASPICE for Cybersecurity highlighted in red (adapted from [2])

As highlighted within the red boxes in *Figure 6.5*, there are several newly added process activities for ASPICE for Cybersecurity, which are explained in the following section.

ASPICE for Cybersecurity – security activities

These newly added security activities are described as follows.

MAN.7

MAN.7 is **Cybersecurity Risk Management**. Activities in this process include defining the scope of cybersecurity risk management, defining risk management processes, performing a TARA, and identifying risks. Moreover, the organization needs to determine risk treatment options, continuously monitor the risks, and take corrective actions if new risks emerge.

SEC.1

SEC.1 is **Cybersecurity Requirements Elicitation**. An organization performs this process early in the development phase. These activities include defining cybersecurity goals and requirements to address the risks identified in MAN.7. Additionally, the organization ensures consistency and traceability between cybersecurity goals and requirements.

SEC.2

SEC.2 is **Cybersecurity Implementation**. In this process, the organization takes the requirements from SEC.1 and creates architectural and detailed design documents. Appropriate security controls are selected and implemented. Additionally, the organization ensures consistency and traceability between the architectural and detailed design.

SEC.3

SEC.3 is **Risk Treatment Verification**. In this process, the organization verifies the effectiveness of the security measures implemented in SEC.2 based on a risk treatment specification. That is, the organization confirms that the implementation has been done according to the cybersecurity requirements, architectural design, and detailed design. Additionally, the organization ensures consistency and traceability between the cybersecurity requirements and the risk treatment verification specification.

SEC.4

SEC.4 is **Risk Treatment Validation**. An organization performs this process to confirm that the developed system fulfills the defined cybersecurity goals from SEC.1. Thus, this activity includes defining and executing a risk treatment validation strategy, such as performing testing according to a risk treatment validation specification to verify whether the cybersecurity goals are met. Additionally, the organization ensures consistency and traceability between the cybersecurity goals and the risk treatment validation specification.

ACQ.2

ACQ.2 is **Supplier Request and Selection**. This process supports an organization to perform supplier management with considerations for cybersecurity. The organization sets appropriate supplier evaluation criteria and assesses potential suppliers accordingly. This includes verifying the supplier's technical cybersecurity capabilities and evaluating their overall cybersecurity maturity. This process also supports managing the request for quotation and establishing the supplier contract, including defining a **cybersecurity interface agreement** and assigning responsibilities.

It is important to note that ISO/SAE 21434 and ASPICE for Cybersecurity provide guidance on cybersecurity engineering for **automotive systems**; however, the development of automotive IoT applications also requires organizations to consider security processes and best practices for **non-vehicle systems**, which will be presented in more detail in the following section.

NIST Cybersecurity Framework, ISO 27001, SOC 2, and OWASP

While ISO 21434 and ASPICE for Cybersecurity focus on the security processes for vehicle development, organizations must also consider security processes for the cloud and mobile app development related to automotive IoT applications.

Common security frameworks and approaches for such environments include the NIST Cybersecurity Framework, ISO 27001, SOC 2, and OWASP.

NIST Cybersecurity Framework

An overview of the NIST Cybersecurity Framework [3] is shown in *Figure 6.6*.

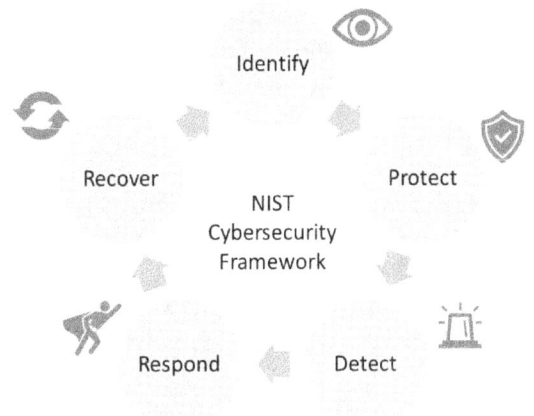

Figure 6.6 – Core functions in the NIST Cybersecurity Framework

There are five core functions organizations can follow to better manage their cybersecurity risks: **identify**, **protect**, **detect**, **respond**, and **recover**.

Identify

The identify function helps the organization to better understand the organization's cybersecurity landscape and potential threats. This includes mapping out critical assets, data, and functions that require cybersecurity protection. For example, automotive IoT backend solutions, mobile apps, and relevant assets to those automotive IoT use cases should be considered.

Protect

The protect function focuses on implementing safeguards to prevent attacks from occurring by reducing vulnerabilities in the ecosystem. It serves to protect critical assets identified in the identify function. It also serves to provide resilience against attacks to ensure that systems can continue to function even during an attack. For example, the organization should implement safeguards to protect the critical assets of automotive IoT applications.

Detect

The detect function assists an organization in identifying potential cybersecurity attacks in a timely manner. This includes performing continuous monitoring of relevant assets, detecting anomalies, and assessing the potential impact of cyberattacks. For example, monitoring of critical assets both in the backend solution and mobile devices should be considered.

Respond

The respond function focuses on taking action to address the detected cybersecurity events in the detect function. It includes incident response activities to help an organization contain the incident and limit the potential damage. It also includes communication processes to help navigate how to share information about an incident both to internal and external stakeholders. For example, this should include responding to cybersecurity attacks targeting automotive IoT use cases.

Recover

The recover function is the last function and serves to restore operations after a cybersecurity attack and to minimize the long-term impact on the use cases. Thus, this includes recovery plans and approaches to perform continuous improvement to processes and policies to avoid similar incidents from occurring again. For example, this function should consider restoring functionality to automotive IoT use cases after a successful cybersecurity attack on the backend solution or mobile app and making appropriate changes to processes, such as updating testing processes to prevent similar attacks from occurring.

ISO 27001

An overview of ISO 27001 [4] is presented as follows. ISO 27001 is an international standard for **information security management**. It provides a general framework and guidelines for establishing and managing a so-called **information security management system** (**ISMS**). There is actually a whole collection of ISO 27000 standards that provide guidance on information security management but for the sake of brevity, we focus on ISO 27001, which is the core standard for specifying the requirements for establishing, implementing, maintaining, and continually improving an ISMS.

As previously explained, a CSMS is used for managing cybersecurity-relevant activities for the development of automotive products. In contrast, an ISMS is used for managing general information security in the organization.

The main purpose is to protect an organization's confidential information, maintain its integrity, and ensure its availability. It follows a risk-based approach, which focuses on identifying and prioritizing information security risks and implementing appropriate security controls to mitigate them.

An overview of the main clauses in ISO 27001 is shown in *Figure 6.7*.

Figure 6.7 – Main clauses of ISO 27001

The main clauses of ISO 27001 are described as follows. **Context of the organization** defines the scope of the ISMS based on the organization's context and identifies the needs and expectations of stakeholders. **Leadership** covers top-down commitment from leadership to help with establishing policies, roles, and responsibilities. **Planning** involves determining actions to address risks as well as defining information security objectives and how to achieve them. **Support** handles resources, competence, awareness, and communication to maintain the ISMS. **Operation** focuses on operational planning and control, including defining risk treatment plans to implement security controls. **Performance evaluation** is done to monitor, measure, and evaluate the effectiveness of controls. It also covers internal audit and management reviews. Finally, **Improvement** involves continual improvement to detect non-conformity and apply corrective action.

ISO 27001 is a general security framework not specific to the development of automotive IoT applications; however, it is important for the organization to consider how to include the coverage of automotive IoT applications in their ISMS.

SOC 2

An overview of SOC 2 [5] is described as follows. SOC 2 is a cybersecurity compliance framework developed by the **American Institute of Certified Public Accountants** (**AICPA**). For completeness, there are three main types of SOC: SOC 1, SOC 2, and SOC 3. SOC 1 focuses on financial reporting controls and can be used by a service organization to demonstrate its commitment to financial reporting accuracy and completeness. SOC 2 evaluates the effectiveness of controls related to the five **Trust Services Criteria** (**TSC**) described later in this section. SOC 3 is a variant of SOC 2, where results from the SOC 2 report are summarized, and confidential and sensitive information is redacted in order to generate a more general public-facing report.

SOC 2 is highly relevant for organizations providing automotive IoT solutions, where security for cloud service providers and **software-as-a-service** (**SaaS**) providers need to be considered. Therefore, we will focus on SOC 2 in the rest of this chapter. Some SOC 2 requirements overlap with other frameworks, such as ISO 27001, and can be considered in conjunction when defining security processes in the organization.

The five principles for TSC are depicted in *Figure 6.8*.

Figure 6.8 – Five principles for SOC 2 compliance

Let us understand each of these principles in detail:

- **Security**: This principle refers to safeguarding data and systems from unauthorized access, use, disclosure, disruption, modification, or destruction. Access controls, firewalls, two-factor authentication, and intrusion detection are examples of solutions that can help achieve this principle.

- **Availability**: This principle aims to ensure that systems and data are accessible and operational when needed. The minimum acceptable performance level for system availability should be agreed upon in a **service-level agreement** (**SLA**). Using fault-tolerant systems, redundancy solutions, scalable solutions, and network monitoring systems can help achieve this principle.

- **Processing Integrity**: This principle refers to the system achieving its purpose by ensuring that the data processing is correct, complete, valid, timely, and authorized. Using methods for quality assurance and performance monitoring can help with fulfilling this principle.

- **Confidentiality**: This principle focuses on protecting sensitive information from unauthorized access or disclosure. Data is classified as confidential if only a specific group or set of users should have access to it. Examples of confidential data include source code, passwords and other credentials, and intellectual property (e.g., proprietary algorithms). Common approaches to achieving this principle include encryption of data in rest and during transit, and access control to only provide access to the data to authorized users.
- **Privacy**: This principle is about the collection, usage, disclosure, and retention of **personally identifiable information** (**PII**) in the system. This should conform with the organization's privacy policies as well as the criteria in the **Generally Accepted Privacy Principles** (**GAPP**) defined by the AICPA. PII refers to information that can be used to uniquely identify a specific individual, including name, date of birth, phone number, address, driver's license number, passport number, and social security number. Organizations that collect and store PII data have a responsibility to protect the data from unauthorized access by applying various security measures such as data encryption and access control.

Moreover, regarding SOC 2 compliance, there are two main types of compliance: **Type 1** and **Type 2**. Type 1 focuses on attesting an organization's use of compliant systems and processes at a specific point in time. Type 2 builds on Type 1 and further includes an attestation of compliance over a certain time period, for example, 12 months. This provides a stronger assurance since it assesses the effectiveness of controls operating over time.

When deploying automotive IoT applications, it becomes imperative for automotive organizations to consider cloud providers that meet SOC 2 Type 2 compliance.

OWASP

OWASP [6] provides guidance for the secure development of software with a focus on web apps and mobile apps.

It offers a collection of resources, including guidelines in the form of **cheat sheets** for cloud and mobile app development, lists of **top ten risks**, and **testing guides**.

Cheat sheets

OWASP provides cheat sheets [14], which are practical guides for specific application security topics. They provide practical steps and best practices that can be implemented into the development workflow to help developers and security professionals.

One practical advantage is that the cheat sheets focus on real-world applications rather than on theoretical topics, and therefore can have an immediate impact on improving security. Examples include cheat sheets for access control, authentication, authorization, cryptographic storage, key management, mobile application security, secure cloud architecture, and threat modeling, among others.

Top ten risks

OWASP provides several lists of the top ten risks that are regularly updated. These lists can be used, for example, as training material and to improve awareness about security. The **Top 10 Web Application Security Risks** list [15] contains the most common and critical **web application** risks. Examples of risks include broken access control, cryptographic failures, injection, insecure design, and security misconfiguration.

Furthermore, OWASP manages a list called **Cloud-Native Application Security Top 10** [16] that is currently under development. It covers risks to **cloud-native technologies** used in modern environments including considerations for public, private, and hybrid clouds, containers, and microservices. Examples of risks to consider are insecure cloud, container, or orchestration configuration, injection flaws (app layer, cloud events, cloud services), improper authentication and authorization, **continuous integration/continuous delivery** (**CI/CD**) pipeline and software supply chain flaws, and insecure secrets storage.

There is a separate list called **Mobile Top 10** [17] that contains the most common risks for **mobile apps**. Improper credential usage, inadequate supply chain security, insecure authentication/authorization, insufficient input/output validation, and insecure communication are some examples of risks.

Another relevant list is the **API Security Top 10** [18], which covers common risks for APIs. APIs are a critical part of modern web and mobile applications and could be targeted by attackers. Examples of risks are broken object-level authorization, broken authentication, broken object property-level authorization, unrestricted resource consumption, and broken function-level authorization.

Moreover, establishing CI/CD environments using various tools for automated testing is common practice for software organizations. OWASP also provides a **Top 10 CI/CD Security Risks** list [19] that helps organizations secure their **CI/CD environments** to reduce attack vectors and avoid potential security incidents. Examples of risks include insufficient flow control mechanisms, inadequate identity and access management, dependency chain abuse, poisoned pipeline execution, and insufficient pipeline-based access controls.

For automotive IoT applications, these top ten lists can help developers and security professionals focus on mitigating the most common risks to improve overall security.

Testing guides

There are two main testing guides provided by OWASP. The **Web Security Testing Guide** [20] provides a framework that helps developers and security professionals to test their web applications and web services. It covers steps to be performed before development, during definition and design, during development, during deployment, as well as during maintenance and operations. The testing guide covers various testing techniques, including authentication testing, authorization testing, input validation testing, and testing for weak cryptography.

Conversely, the **Mobile Application Security Testing Guide** [21] provides a comprehensive manual that covers processes, techniques, and tools used for the security testing of mobile apps. For example, it provides specific test cases for both Android and iOS testing, including testing of storage, cryptography, authentication, and network.

These testing guides can assist organizations in performing comprehensive testing of automotive IoT applications.

With this overview of security processes and practices for cloud and mobile app development related to automotive IoT applications, we will now move on to discussing development methodologies and approaches.

DevSecOps and agile development

It is important to note that automotive organizations need to change the traditional way of doing automotive development to support the development of new automotive IoT use cases. We will discuss the changes in automotive development required to enable the development of these new automotive IoT solutions.

Armed with this knowledge, engineers and managers can start preparing and establishing the required development environments, including defining and rolling out appropriate development processes, providing training to achieve necessary skills for developers, and incorporating suitable technologies to support the development of automotive IoT applications.

V-model

The V-model development approach has served the automotive industry well by ensuring that a set of rigorous steps of activities are performed in a specific order. The V-model is illustrated in *Figure 6.9*.

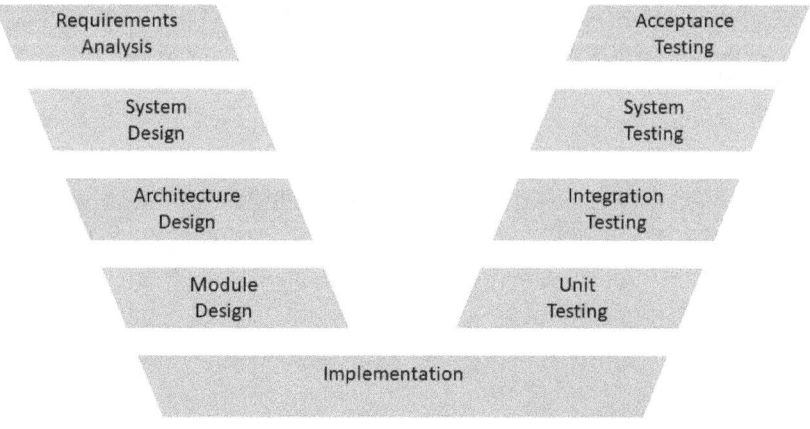

Figure 6.9 – Overview of V-model used traditionally in automotive development

A simplified description of the V-model is as follows. On the left side of the V are all the preparatory activities in terms of defining requirements and deciding on the design. These activities include requirements analysis, system design, architecture design, and module design. At the bottom of the V is the implementation activity based on the design documents.

On the right side of the V model are the verification and validation activities. These activities include unit testing, integration testing, system testing, and acceptance testing. These activities map back to the left side of the V and are performed in sequential order. This ensures that what has been implemented is first verified to have been implemented according to the design, and secondly validated to ensure that the implementation fulfills the initially defined requirements.

Agile

With a need to accommodate rapid changes to requirements during development in order to support new use cases and features for automotive IoT, some automotive organizations have started to embrace the agile methodology for software development.

It consists of a set of values and principles that emphasize flexibility, adaptability, and continuous improvement. Using the agile methodology, large complex projects can be broken down into smaller manageable tasks. These tasks are performed and completed during shorter cycles, so-called **sprints**, that typically last two to four weeks. The sprint **tasks**, which generally can be performed in hours, are mapped to **stories**, which can be completed in days, which in turn are mapped to **epics**, which are large bodies of work that can span weeks or months, which are then organized on a story map.

Performing tasks in sprints allows teams to produce deliverables iteratively and incrementally. Producing these deliverables iteratively as part of continuous delivery allows for earlier feedback and, if needed, allows for changes to scope or requirements. Thus, this approach allows us to easily adapt quickly based on the feedback.

While many automotive organizations are already following agile development approaches, it is imperative to ensure that cybersecurity activities are also performed in an agile and incremental fashion where possible.

> **Tip**
> The mantra, *"security must be built in, not bolted on,"* is worthwhile to remember.

While it is common to add cybersecurity activities to an epic, it is beneficial for organizations to consider how to assign tasks related to these activities to sprints early in development where possible, as well as how to perform the activities repeatedly or continuously during development and post-development. Performing cybersecurity activities in an agile manner allows for improved effectiveness, scope, and coverage to find and fix security issues earlier in development, leading to reduced costs. As a result, automotive organizations can incorporate cybersecurity activities into the development process and follow an agile development methodology to improve efficiency and reduce costs to better support the development of automotive IoT applications.

Scrum

While the agile methodology provides the backbone, there are several different frameworks that can be used with agile. One such noteworthy framework is **scrum**.

There are different roles defined in the scrum framework, such as product owner, scrum master, and development team. These roles are then assigned to team members in the organization. There are also several activities, so-called events, that are performed, such as daily scrums, sprint planning, and sprint reviews. These activities help to ensure that appropriate work is planned for the shorter iterations and is progressing according to the plan.

Scrum is a popular framework for software development because it is flexible, lightweight, and easy to implement.

DevSecOps

First, it is important to note that **DevSecOps** is an extension of **DevOps**, so, let us start by discussing what DevOps is.

The purpose of DevOps is to break down the barriers between **development** (**Dev**) teams and **operations** (**Ops**) teams. In many organizations, these are two separate teams with different objectives. This creates challenges when, for example, issues occur on the operations side that require the development side to be involved and create a fix. If these two teams work independently, there is often a misalignment of stakeholder objectives and a lack of prioritization. As a result, issues take longer to resolve, causing a potential negative impact and increased costs for the organization.

While DevOps aims to solve these issues by aligning development and operations, a new challenge that emerged in the past few years is misalignment with **security** (**Sec**) teams. As a result, the notion of DevSecOps was born. With DevSecOps, DevOps is expanded to break down barriers between development, security, and operations teams. In short, this means that there should be an alignment of stakeholder objectives and priorities.

In other words, DevSecOps is an organizational culture and mindset that integrates security into the **software development life cycle** (**SDLC**) as presented in *Figure 6.10*.

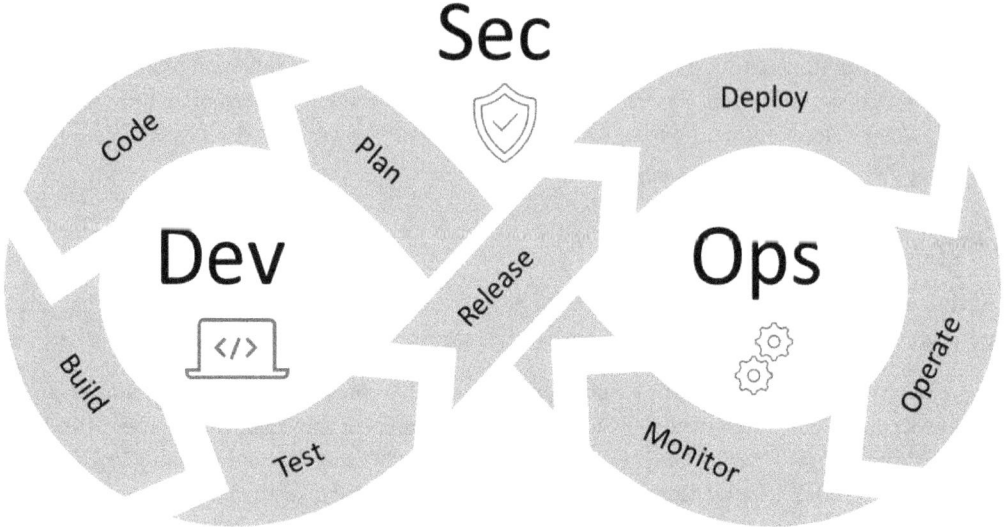

Figure 6.10 – DevSecOps applied to the SDLC

It aims at integrating security from the early phases of requirements and design, and continues with using automated application security tools during development and integration to find and fix vulnerabilities as early as possible.

DevSecOps follows a number of principles, for example, **shift-left** and **automation**. Shift-left allows us to find and fix vulnerabilities earlier in the SDLC, reducing costs. Automation allows a set of application security tools to run continuously throughout the SDLC, allowing us to catch more software defects and vulnerabilities over time, which improves overall software quality. By following these DevSecOps principles, automotive organizations can change their company culture to embrace a cybersecurity mindset.

> Note
> The life cycle of DevSecOps will be covered in *Chapter 11*, *Deploying and Maintaining an Automotive IoT Application*.

Summary

Automotive IoT applications are being deployed to provide various benefits and improved user experience; however, it is crucial to recognize that these automotive IoT applications also often introduce new cybersecurity attack vectors. As such, this chapter first explained the need for cybersecurity and secure software development practices by reviewing recent cyberattacks and guiding the reader through a simplified threat model of the automotive IoT ecosystem.

Next, we reviewed security-relevant development processes and frameworks. We divided the discussion into vehicle-related and non-vehicle-related development processes.

First, regarding the vehicle-side development, we covered ISO/SAE 21434 and ASPICE for Cybersecurity. We went into specifics on organizational-level requirements and project-level requirements, as well as reviewing specific security activities to help organizations understand what to focus on.

For the non-vehicle side development, targeting the backend solution and mobile device, we explained the relevant security processes and frameworks for web apps, cloud, and mobile app development. In particular, we covered the NIST Cybersecurity Framework, ISO 27001, SOC 2, and OWASP to provide guidance to organizations on how to manage and improve security for non-vehicle development.

Finally, we highlighted the importance of changes in the development methodology to support automotive IoT applications. We gave a brief background to the traditional V-model approach, and further discussed changes required to support agile, scrum, and DevSecOps.

Tooling and best practices commonly used in organizations that adopt DevSecOps include **static application security testing (SAST)**, **software composition analysis (SCA)**, **dynamic application security testing (DAST)**, fuzz testing, and establishing CI/CD pipelines [22]. How organizations can best utilize these tooling by establishing a secure development platform will be covered in more detail in *Chapter 7, Establishing a Secure Software Development Platform*.

References

- [1] ISO/SAE, *ISO/SAE 21434:2021 Road Vehicles Cybersecurity Engineering*, 2021
- [2] VDA, *Automotive SPICE for Cybersecurity Process Reference and Assessment Model*, 2021
- [3] NIST, *Framework for Improving Critical Infrastructure Cybersecurity*, version 1.1, 2018
- [4] ISO, *ISO/IEC 27001:2022 Information security, cybersecurity and privacy protection, Information security management systems requirements*, 2022
- [5] AICPA, `https://us.aicpa.org/interestareas/frc/assuranceadvisoryservices/users`
- [6] OWASP, `https://owasp.org/`

- [7] Upstream, https://upstream.auto/reports/global-automotive-cybersecurity-report/
- [8] VicOne, https://documents.vicone.com/reports/automotive-cyberthreat-landscape-report-2023.pdf
- [9] Synopsys, https://www.synopsys.com/content/dam/synopsys/sig-assets/reports/rep-ossra-2023.pdf
- [10] Sam Curry, https://samcurry.net/web-hackers-vs-the-auto-industry/
- [11] NIST, *NIST SP 800-30 Rev. 1, Guide for Conducting Risk Assessments*, 2012
- [12] Microsoft, https://www.microsoft.com/en-us/security/blog/2007/09/11/stride-chart/
- [13] ISO/IEC, *ISO/IEC 15408 Information security, cybersecurity and privacy protection – Evaluation criteria for IT security*, 2022
- [14] OWASP, https://cheatsheetseries.owasp.org/
- [15] OWASP, https://owasp.org/Top10/en/
- [16] OWASP, https://owasp.org/www-project-cloud-native-application-security-top-10/
- [17] OWASP, https://owasp.org/www-project-mobile-top-10/
- [18] OWASP, https://owasp.org/www-project-api-security/
- [19] OWASP, https://owasp.org/www-project-top-10-ci-cd-security-risks/
- [20] OWASP, https://owasp.org/www-project-web-security-testing-guide/
- [21] OWASP, https://owasp.org/www-project-mobile-app-security/
- [22] Dennis Kengo Oka, *Building Secure Cars: Assuring the Automotive Software Development Lifecycle*, Wiley, 2021.

Get This Book's PDF Version and Exclusive Extras

Scan the QR code (or go to packtpub.com/unlock). Search for this book by name, confirm the edition, and then follow the steps on the page.

Note: Keep your invoice handy. Purchases made directly from Packt don't require one.

7
Establishing a Secure Software Development Platform

After establishing secure development processes, as presented in the previous chapter, this chapter will focus on practical activities and tooling to help you establish a secure software development platform. Using this platform helps organizations to develop secure software for automotive **Internet of Things** (**IoT**) use cases systematically and efficiently.

This chapter starts by discussing relevant cybersecurity activities that automotive organizations should perform during the **secure software development life cycle** (**SSDLC**). Then, we'll review the importance of establishing a project inventory to achieve a better overall understanding of the security posture of all relevant automotive IoT projects. The project inventory helps manage activities during the SSDLC. Please note that the focus of this chapter is on the development phase. Thus, for the development phase, we'll provide step-by-step practical guidance on how to establish a secure development platform and explain the benefits of using this platform approach. This platform approach employs several different **application security** (**AppSec**) testing techniques and tooling based on the activities from the SSDLC, all of which will be described in detail in the rest of this chapter. Finally, we'll discuss how to handle vulnerability management and how to automate security testing using this platform.

In this chapter, we will cover the following topics:

- Activities in the SSDLC
- How to establish a project inventory
- Practical guidance on how to establish a secure software development platform
- Common AppSec tooling and test approaches

Activities in the SSDLC

To establish a secure software development platform, we have to understand what activities need to be performed in the SSDLC. With this understanding, we can then define which activities can be integrated and performed using the platform, especially when it comes to managing the activities and allowing for automation.

This section focuses on describing specific activities in the SSDLC. Various standards can be considered to help build a framework around SSDLC [1-3].

> **Note**
> The following is not an exhaustive list but provides an overview of common cybersecurity activities that can be used to establish a baseline of security. Please also note that some of the following activities require specified workflows with defined interactions between project teams and security teams to be established.

These cybersecurity activities have been mapped to a simplified V-model in *Figure 7.1*.

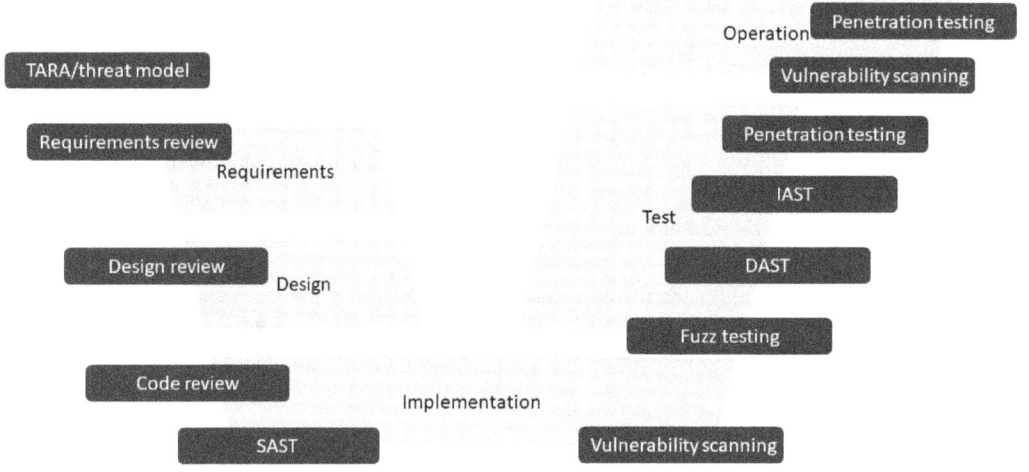

Figure 7.1 – Cybersecurity activities mapped to a simplified V-model

This figure shows an example of the phases in which the activities can be performed during the development life cycle.

TARA/threat model

Generally, one of the first activities for an organization to perform is a **Threat Analysis and Risk Assessment** (**TARA**) or a **threat model** for each project. Where applicable, the scope should encompass the entire ecosystem for the automotive IoT application – that is, the scope should cover both the vehicle side, backend solution, and mobile device side. Its purpose is to get a full understanding of the specific assets, threats, potential consequences, and corresponding risks for the automotive IoT use case. The project teams can then consider applying different mitigations to counter and manage the identified risks.

Requirements review

The countermeasures and security controls to mitigate the identified risks are defined as **security requirements**. In this activity, the security team reviews these requirements to check for appropriateness, consistency, and completeness.

Design review

Based on the security requirements, the project team defines **architectural and detailed designs** of the security solution. The project team can also consider secure coding guidelines for the design to help reduce potential rework by making sure the design allows for coding that's aligned with the guidelines. The security team reviews the design documents to identify any potential design flaws or weaknesses.

Code review

As the project teams start implementing features based on the design and start writing **code**, it is time to start performing code reviews. Code reviews can typically be performed using two different approaches. For example, it can be performed manually by a human, such as a security expert, who has experience in reviewing code. They will review the source code line by line to identify vulnerabilities. The second approach is using automated code reviews, which is explained next.

Static application security testing

Static application security testing (**SAST**) is a static analysis approach that can be used to perform **automated code reviews** early in the development life cycle. The SAST tool can be configured to scan source code and check for violations against certain coding rules such as **CERT C/C++** [4,5], **MISRA C/C++** [6], or **AUTOSAR C++** [7], as well as detect common weaknesses in the code. There is a catalog of common weaknesses that can lead to vulnerabilities defined as **Common Weakness Enumerations** (**CWEs**) [8]. SAST tools can help identify CWEs within the code base without running the application.

Vulnerability scanning

There are different types of vulnerability scanning tools. Their general purpose is to scan a target software or system to detect outdated versions, misconfigurations, or **known vulnerabilities** that are defined in a vulnerability database.

Fuzz testing

In contrast to vulnerability scanning, which may rely on a database of known vulnerabilities, fuzz testing is performed to test the target software or system to detect **unknown vulnerabilities**. Fuzz testing is a dynamic analysis approach that feeds the target software or system with malformed input to detect crashes or other unexpected behavior.

Dynamic application security testing

While static analysis tests software in source code without executing the code, **dynamic application security testing** (**DAST**) is a dynamic analysis approach where a **running system** is targeted for testing to find vulnerabilities. This approach is more of a form of **black-box testing** and simulates attacks from outside the application using pre-defined attack patterns and vulnerability databases.

Interactive application security testing

Interactive application security testing (**IAST**) is also a dynamic analysis approach that's used to test running applications to find vulnerabilities or unexpected behavior. In contrast to DAST, this approach is more of a **white-box approach** where the target application is monitored from inside the running application by using an agent, for example, to track data flows and memory usage to help identify issues.

Penetration testing

Penetration testing is performed by a human tester to **validate** that the cybersecurity goals are fulfilled. Various steps of penetration testing can be automated using different tools but generally, the test results are analyzed by a human tester to identify vulnerabilities.

Project inventory

An organization can create and maintain a project inventory as part of the development process to better understand its security posture and manage the overall security activities for the different projects. The project inventory helps the organization provide a clear and structured way of knowing what type of projects they are developing and maintaining, the respective risk levels, and relevant security activities to perform.

Project information and risk level

First, to get started with a project inventory, some general information about each project is collected. *Figure 7.5* shows an overview of an example project inventory.

Project name is the project name of the automotive IoT application. **Exposure** represents how accessible the automotive IoT application is. This can be divided into five categories and defined as shown in *Table 7.1*.

Level	Network Accessibility	User Accessibility
Very High	Public/internet-facing	By anyone
High	Public/internet-facing	A large number of authorized users
Medium	Public/internet-facing	A limited number of authorized users
Low	Private/restricted network	A large number of authorized users
Very Low	Private/restricted network	A limited number of authorized users

Table 7.1 – Example of Exposure categories

Impact denotes the potential damage that could occur if the automotive IoT application is exploited and compromised by a cyberattacker. It can be categorized into the following five types and defined as as shown in *Table 7.2*.

Level	Safety	Financial Damage	Operational	Privacy
Very High	Fatal injuries	> 30% of annual revenue	A large number of vehicles are inoperable	A complete loss of control over personal information that could lead to identity theft and financial loss
High	Severe injuries	20% > 30% of annual revenue	A limited number of vehicles are inoperable	Persistent surveillance and tracking of a large number of users
Medium	Moderate injuries	10% > 20% of annual revenue	Important function is unusable	Intermittent surveillance and tracking of a limited number of users
Low	Minor injuries	5% > 10% of annual revenue	Important function is negatively affected but operational	Disclosure of limited personal information
Very Low	No injuries	< 5% of annual revenue	None or an unimportant function is affected	No or insignificant personal data

Table 7.2 – Example of Impact categories

The different factors, such as safety, financial damage, operational, and privacy, can carry different weights to help determine the final impact scoring.

> **Note**
> These tables are merely examples and organizations need to create their own exposure and impact definitions that are suitable for their applications and requirements.

Based on the exposure and impact values, the organization can then calculate the **risk** value for each project in the project inventory list using the **risk matrix** defined in the organization. An example of a risk matrix is shown in *Figure 7.2*.

Impact	Exposure				
	Very Low	**Low**	**Medium**	**High**	**Very High**
Very High	Low	Moderate	High	Very High	Critical
High	Low	Moderate	High	High	Very High
Medium	Very Low	Low	Moderate	Moderate	Moderate
Low	Very Low	Low	Low	Low	Low
Very Low	Very Low	Very Low	Very Low	Low	Low

Figure 7.2 – Example of a risk matrix

For example, a project that is determined to have **High exposure** and **Medium impact** would be calculated to have a **Moderate risk** level.

Cybersecurity assurance level and activities

As a next step, the organization can then map the risk levels to **cybersecurity assurance levels** (**CALs**). Taking inspiration from the CAL definition from ISO/SAE 21434 [9], we can expand it so that it covers a broader scope of applications. Thus, we apply the concept of CALs to help us define relevant cybersecurity activities for automotive IoT applications, and consider the end-to-end use case.

First, we map risk levels to CALs, as shown in *Figure 7.3*. We simply map each risk level to a corresponding CAL level from one to six. Once again, this is merely an example and the organization developing automotive IoT applications would need to define the mapping.

Risk Level	Cybersecurity Assurance Level
Critical	6
Very High	5
High	4
Moderate	3
Low	2
Very Low	1

Figure 7.3 – An example of mapping a risk level to CAL

Second, once we've mapped the risk level to the corresponding CAL, we can define what cybersecurity activities and test approaches should be performed for each CAL. An example of this mapping is presented in *Figure 7.4*. Each higher level of CAL indicates a more rigorous and comprehensive approach to cybersecurity. That is, for each increasing CAL, we can achieve a higher level of assurance that weaknesses and vulnerabilities have been identified and mitigated. We can specify whether an activity is **Required**, **Recommended**, or **Optional**. The definitions of these terms are as follows:

- **Required** is defined as mandatory for the project
- **Recommended** is defined as suggested to be performed for the project, and if it is not performed, a justification is required
- **Optional** is defined as the project can decide whether to perform the activity or not, and if it is not performed no justification is required

Additionally, parameters for the cybersecurity activities, such as the test rigor, scope, depth, how often the activity is performed, the level of independence required for the activity, and others, should also be defined for different CALs. However, defining how to adjust these parameters for each cybersecurity activity would require an extensive discussion of the details and is therefore omitted. This is a complex topic that's being discussed in joint working groups currently developing a new standard called **ISO/SAE 8475** [*10*]. This upcoming standard will guide organizations on how to use CAL in more detail. Instead, in this chapter, to simplify this, we'll use **L1**, **L2**, and **L3**, as depicted in *Figure 7.4*, to denote different levels of test rigor, scope, depth, how often the activity is performed, the level of independence required for the activity, and so on. L3 is the highest, most rigorous, and most comprehensive level of the parameters. These are then gradually reduced for L2. Finally, L1 has the lowest, least rigorous, and least comprehensive level of parameters.

Cybersecurity activities	CAL					
	1	2	3	4	5	6
TARA/threat model	Recommended	Required L1	Required L1	Required L2	Required L2	Required L3
Requirements review	Optional	Recommended	Required L1	Required L1	Required L2	Required L3
Design review	Optional	Recommended	Required L1	Required L1	Required L2	Required L3
Code review	Optional	Recommended	Required L1	Required L1	Required L2	Required L3
SAST	Optional	Recommended	Required L1	Required L1	Required L2	Required L3
Vulnerability scanning	Optional	Recommended	Required L1	Required L1	Required L2	Required L3
Fuzz testing	Optional	Optional	Recommended	Required L1	Required L2	Required L3
DAST	Optional	Optional	Recommended	Required L1	Required L2	Required L3
IAST	Optional	Optional	Recommended	Required L1	Required L2	Required L3
Penetration testing	Optional	Recommended	Required L1	Required L2	Required L2	Required L3

Figure 7.4 – Example mapping of CAL to cybersecurity activities

> **Note**
> This figure is merely an example. Organizations will need to define the actual cybersecurity activities and parameters that apply to their development.

Example project inventory

Finally, all of these steps are put together to create the project inventory in an organization, as depicted in *Figure 7.5*. Here, we have two categories of information: **project-related information** and associated **cybersecurity activities**.

Moreover, the figure contains several rows that are used to manage this information. First, regarding project-related information, we have the following rows:

- Project name
- Exposure
- Impact
- Risk level
- CAL

Next, the associated cybersecurity activities include the following rows:

- TARA/threat model
- Requirements review
- Design review
- Code review
- SAST
- Vulnerability scanning
- Fuzz testing
- DAST
- IAST
- Penetration testing

An organization could then review all their automotive IoT projects and fill in the relevant information for the respective project, as shown here.

Project-related information					
Project name	<Project A>	<Project B>	<Project C>	<Project D>	...
Exposure	High	Medium	Very Low	Very High	
Impact	Medium	High	High	High	
Risk level	Moderate	High	Low	Very High	
CAL	3	4	2	5	

Cybersecurity activities					
TARA/threat model	Required L1	Required L2	Required L1	Required L2	
Requirements review	Required L1	Required L1	Recommended	Required L2	
Design review	Required L1	Required L1	Recommended	Required L2	
Code review	Required L1	Required L1	Recommended	Required L2	
SAST	Required L1	Required L1	Recommended	Required L2	
Vulnerability scanning	Required L1	Required L1	Recommended	Required L2	
Fuzz testing	Recommended	Required L1	Optional	Required L2	
DAST	Recommended	Required L1	Optional	Required L2	
IAST	Recommended	Required L1	Optional	Required L2	
Penetration testing	Required L1	Required L2	Recommended	Required L2	

Figure 7.5 – Example of a project inventory list

> **Note**
> The project inventory list is a living source of information that an organization can reference for risk management and use to track the progress of cybersecurity activities and verify their cybersecurity posture. It can also be used as part of the cybersecurity project plan to help plan when certain activities need to be performed. Please also note that the project inventory list needs to be periodically reviewed as **Exposure** or **Impact** and, as such, the corresponding risk level may change over time based on project scope changes or new types of threats and attacks.
>
> Depending on the project, some of the cybersecurity activities that are defined for a certain CAL may not be possible. For example, IAST may not be possible for certain embedded software as it may not be possible to attach agents to the application; however, IAST may be possible for the web application included in the project. For completeness, the reason why a certain activity is not possible or why the activity is skipped should be recorded so that there is proper justification to explain why the activity wasn't performed.

Some of these activities require personnel with appropriate cybersecurity, embedded, and cloud expertise and may be performed manually, such as the TARA/threat model, code review, design review, and penetration testing.

Other types of activities can generally be performed by using appropriate automated application security tooling, and therefore be executed periodically or continuously as a set of automated tools in a secure software development platform. The next section describes this platform in more detail.

Practical steps for establishing a secure software development platform

This section presents practical steps to establishing a **secure software development platform**.

First, we will introduce the purpose and need for a secure development platform. Then, we will provide an overview of the secure software development platform. In particular, we will dive into how to define requirements and policies for development projects and ensure compliance. Moreover, we'll briefly present how to handle vulnerability management during development. Finally, we'll introduce AppSec tooling that can be deployed in the secure development platform. More in-depth descriptions of individual AppSec tooling approaches will be provided in the following section.

Automotive organizations can establish this secure software development platform to improve the efficiency and effectiveness of testing by shifting left and performing automated testing using a set of consolidated AppSec tools.

Purpose and need

While different development teams can run individual tools independently, there may be challenges to achieving a structured approach to testing and achieving an appropriate level of cybersecurity assurance across the different development teams.

This means that each team might be running their own tools, and there may be different levels of maturity among the teams that affect the type of testing and tooling used. There may also be difficult development schedules and time pressures that dictate what type of testing may or may not be performed in a specific development project. There would typically also be overhead in managing multiple different tools in different teams.

As organizations become responsible for the development of more complex automotive IoT systems with end-to-end functionality spanning across vehicles, backend solutions, and mobile devices, it becomes equally imperative to **consolidate** cybersecurity activities and tooling to ensure that all relevant projects are following appropriate security processes and achieving appropriate levels of cybersecurity assurance. As such, to improve efficiency and security during the development life cycle, organizations can consolidate the large set of automated **AppSec** tools that they are currently using in their development environment by establishing a secure software development platform.

This platform allows software development to shift toward **DevSecOps**, fostering a cybersecurity culture within the organization to enable faster release cycles and more automated testing. Consolidating on tools and focusing on automation allow the organization to achieve a common baseline of cybersecurity. Moreover, the platform allows management to have a better overall view of all relevant projects to ensure the organization's security posture.

Overview of the secure software development platform

An overview of the secure software development platform is provided in *Figure 7.6*.

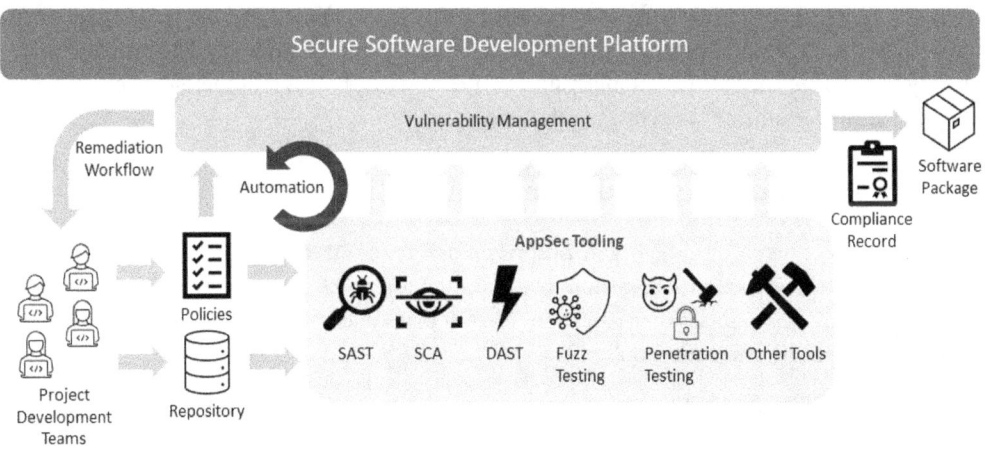

Figure 7.6 – Overview of the secure software development platform (adapted from [11])

Let's take a closer look at this workflow. First, on the left-hand side, we have multiple different **project development teams** that develop and manage their software in source code **repositories**.

These projects have different requirements for cybersecurity activities based on the security risks that are managed in the project inventory. These requirements can be used to define **policies**. How you can use these requirements and policies is described in more detail in the following subsection.

In the middle of the figure, there is a set of automated **AppSec tooling**. These tools will scan the source code or perform various security tests on the target software. AppSec tooling will be described in more detail in the following subsection.

Following the **automation** principle from DevSecOps, various tools for security test approaches, such as SAST, SCA, DAST, fuzz testing, penetration testing, and others, are automatically triggered in a **continuous integration** (**CI**) pipeline at different points during the development of the software projects to perform automated testing as part of this secure software development platform.

Once the various AppSec tools have been executed, the results are uploaded to a **vulnerability management** system. Vulnerability management will be described in more detail in the following subsection.

The organization then performs triaging by reviewing the top-priority findings and determines which issues to fix. This starts the **remediation workflow**, where developers are assigned tickets with issues to fix.

After the developers have fixed the issues, the workflow repeats once again, automatically executing various AppSec tools to scan and test the updated target software. This step includes verifying that the found issues have been fixed.

Once the software fulfills certain criteria defined by policies for the project, a **compliance record** is generated, and the corresponding **software package** is ready to be released and delivered. It will then be deployed to the target system, where applicable, through a secure software update process. There are already various approaches for establishing a secure update process and therefore the software update process itself is outside the scope of this book.

Requirements, policies, and compliance

Based on the project inventory, the organization can define requirements and policies on which automated application security tooling should be executed for which projects, how often, and with what configuration. These policies then determine the test plans and configuration settings to be used by the different application security tools.

The examples that were provided for the project inventory described earlier in this chapter used the notations L1, L2, and L3 for the test parameters. *Figure 7.7*, provides some examples of how to define L1, L2, and L3 for a few security test approaches.

Test Approach	L1	L2	L3
SAST	OWASP Top 10	OWASP Top 10 SANS Top 25 CWE	OWASP Top 10 SANS Top 25 CWE CISQ CWE
	Coding guidelines (all mandatory)	Coding guidelines (all mandatory and selected recommended)	Coding guidelines (all mandatory and all recommended)
SCA	Vulnerable software versions	Verify included vulnerabilities	Verify exploitability of vulnerabilities
Fuzz Testing	In-band instrumentation	External instrumentation	External instrumentation
	16 hours or 100,000 fuzz test cases	40 hours or 250,000 fuzz test cases	160 hours or 1,000,000 fuzz test cases
Penetration Testing	Identify exploitable vulnerabilities	Exploit critical vulnerabilities	Chain multiple vulnerabilities
	Enhanced-Basic attack potential	Moderate attack potential	High and Beyond High attack potential

Figure 7.7 – Example definitions of L1, L2, and L3 (inspired by [12, 13])

These policies can then also serve as the compliance criteria to determine whether software is ready to be released. This means that the organization can verify whether the criteria, such as different types of scanning and testing, along with the appropriate set of parameters, have been performed and that the test results fulfill the defined criteria. This information is stored as a **compliance record**. Performing these steps ensures that only software that's compliant and passed the criteria is released. As a result, this approach helps organizations establish a baseline of cybersecurity assurance.

Vulnerability management

As organizations run multiple AppSec tools and collect numerous findings from each one, a new challenge emerges in managing all of these findings. There is a need to understand and manage whether a specific finding is a false positive or needs to be addressed, what findings may be caused by the same root cause but detected by different tooling, what the priority should be to address certain findings, and more.

While it may be possible to manually review all the findings and manually perform the aforementioned activities, it is extremely tedious and error-prone. Instead, there is a specific type of tooling called **application security posture management** (**ASPM**) that can help with automated vulnerability management and risk visibility.

These ASPM tools can import test results from a large collection of scan and test tools. The findings in the test results that are collected from multiple tools are then analyzed and aggregated, deduplicated, correlated, and prioritized. This allows the organization to have a single consolidated overview of all the findings.

For example, there are open source tools such as **OWASP Defectdojo** [14] and commercial tools such as **Software Risk Manager** [15] that can be used. These tools often allow you to import test results from a diverse set of AppSec tools such as SAST, SCA, DAST, and IAST and common test result formats such as **Static Analysis Results Interchange Format (SARIF)**, **eXtensible Markup Language (XML)**, and **JavaScript Object Notation (JSON)**.

Besides vulnerability management during development, organizations also need to establish solutions for **security information and event management (SIEM)**. SIEM solutions are used to collect, analyze, and monitor security-related data from various sources, such as cloud, servers, network devices, and endpoints. These solutions help organizations analyze the collected logs from various systems within the automotive IoT ecosystem to identify potential security incidents, anomalies, and vulnerabilities. Moreover, these SIEM solutions can also be used for security auditing purposes and compliance reporting as they can provide a consolidated and comprehensive view of the organizations' vulnerability landscape. Examples of solutions include Security Onion, Wazuh, Splunk, and Graylog.

AppSec tooling

Some relevant AppSec tooling applies to the development of automotive IoT applications. A few common AppSec tools and test approaches will be described in more detail in the next section.

Common AppSec tooling and test approaches

There are several types of AppSec tooling and test approaches, as discussed earlier in this chapter. A few of the most common ones are SAST, SCA, DAST, fuzz testing, and penetration testing. This section will describe these types in more detail.

SAST

This section provides a brief overview of SAST.

What is it?

One of the foundational pillars of AppSec is SAST. From the early days of linters in the 1970s, which checked for coding style violations, static analysis has evolved into complex scanning techniques that are capable of identifying a wide array of issues in software.

These tools scan source code **statically** – that is, the target application isn't running but rather the source code is scanned. SAST tools generally serve two main purposes: identifying **weaknesses** and **vulnerabilities** and checking for **coding rule violations**.

SAST tools use various techniques, such as pattern matching, data flow analysis, and taint analysis, to find issues such as buffer overflows, resource leaks, hardcoded credentials, memory corruption, null pointer dereference, weak encryption algorithms, dead code, and more.

Common weaknesses and vulnerabilities are defined in the **OWASP Top 10** [16], **OWASP Mobile Top 10** [17], **SANS Top 25** [18], and **CISQ CWE** [19]. Moreover, common coding guidelines that are applied in the automotive industry include **CERT C/C++** [4,5], **MISRA C/C++** [6], and **AUTOSAR C++** [7].

How to do it...

SAST tools can be run during different phases of the development, as illustrated in *Figure 7.8*.

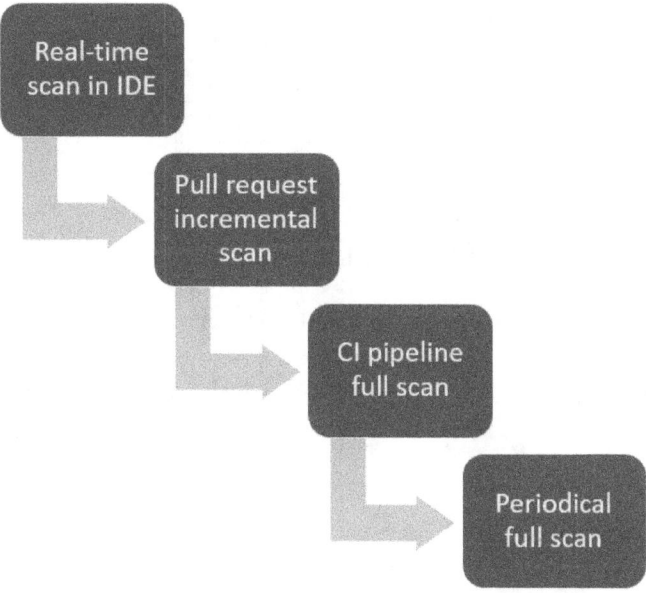

Figure 7.8 – SAST scan during different phases of development

For example, SAST tools can run as a plugin in the **integrated development environment** (IDE) and scan the code in **real time** as the developer is writing it to give immediate feedback. This helps the developer to immediately fix the issues and prevents such issues from being checked into the repository. These tools can also run during **pull requests** as part of an **incremental scan** so that the scan focuses on identifying only issues in the software changes made from the previous scan. Furthermore, as part of the **CI pipeline**, **full scans** of the application can be performed during the integration and build process. Finally, additional scans can be performed **periodically** as part of scheduled **full scans** of the complete software package to help maintain long-term quality and security assurance [20].

Table 7.3 provides some examples of common SAST tools and contains both commercial and open source tools.

Tool	Type
Bearer	Open source
CodeSonar	Commercial
Checkmarx	Commercial
Coverity	Commercial
Cppcheck	Open source
FindBugs	Open source
Fortify	Commercial
Horusec	Open source
Klocwork	Commercial
Semgrep	Open source
SonarQube	Commercial
Veracode	Commercial

Table 7.3 – Examples of SAST tooling

More information on SAST for automotive software development can be found in the *References* section [21].

SCA

This section provides a brief overview of SCA.

What is it?

As software systems become larger and more complex and comprise a large number of smaller software components, it becomes increasingly important to understand exactly what software components are included in the final software product. This can be achieved by preparing a so-called **software bill of materials** (**SBOM**), which is a list of included software components. An analogy is a list of contents for a processed food product at the supermarket, showing what the ingredients are.

SCA is a tooling approach that generally serves three main purposes. One, it scans software to **identify** the included **software components**, in particular **open source software** (**OSS**) components. Secondly, it maps **known vulnerabilities**, which are assigned a **common vulnerabilities and exposures** (**CVE**) number, from a vulnerability database associated with the identified software components and versions. Thirdly, it maps **license information** from a license database to the associated identified software components and versions. This information helps an organization with **license compliance**.

It is worth noting that some tooling focuses on achieving only two of these three purposes – that is, it either focuses on identifying security vulnerabilities or focuses on license compliance.

A simplified example of the results from an SCA tool is shown in *Figure 7.9*.

Software name and version	Known vulnerabilities	License
Linux kernel 4.14.48	CVE-2022-43945 CVE-2020-27068 CVE-2020-29569 CVE-2019-9500	GPL
zlib 1.2.8	CVE-2022-37434	zlib
shim 15.8	N/A	BSD
curl 7.58.0-r0	CVE-2019-3822 CVE-2018-14618 CVE-2018-16839	MIT
busybox 1.31.1	CVE-2022-48174 CVE-2022-28391	GPL

Figure 7.9 – Simplified example results from SCA tooling

The results show the identified software components' names and versions, any known associated vulnerabilities, and the associated license information.

How to do it...

Different SCA tools support different and sometimes multiple techniques for achieving the three purposes mentioned previously.

Common approaches include scanning the source code as-is, similar to SAST tools, and identifying various software components through **pattern matching** in filenames and folder names. Another approach is **identifying dependencies** by examining the package manager and its manifest (for example, JSON) to see what software components are being included.

Moreover, sometimes, developers may copy and paste certain parts of the code from OSS projects without including the entire software components. To detect such cases, **snippet analysis** can be performed, which can identify snippets of code that are associated with certain OSS components.

Finally, in some situations, developers do not have access to the source code, and the aforementioned approaches aren't applicable. Instead, if only the software binary is available, **binary scanning** can be applied. This method uses techniques to break apart the binary and identify the contents [22].

Some examples of SCA tools are presented in *Table 7.4*. Please note that this list contains examples of both free, open source tools as well as commercial tools.

Tool	Type
Black Duck	Commercial
Cybellum	Commercial
Foss-ID	Commercial
FOSSology	Open source
Snyk	Commercial
OWASP Dependency-Track	Open source

Table 7.4 – Examples of SCA tooling

More information on SCA for automotive software development can be found in the *References* section [21].

DAST

This section provides a brief overview of DAST.

What is it

In contrast to statically scanning source code, dynamic application security testing is an approach to perform testing **dynamically** – that is, on a running application.

Thus, DAST tooling typically requires a certain understanding of how to interact with the running application. For example, the tools need to know the communication protocols, interfaces, or **application programming interfaces** (**APIs**) to provide input to the application and would benefit from having a certain understanding of the behavior of the target application to test it effectively.

Typically, the target software or system is tested to detect **known vulnerabilities** or potentially detect **unknown vulnerabilities** based on **known attack patterns**. Thus, DAST tooling typically uses vulnerability databases and generates attack messages and sequences based on known approaches or common attack signatures. Common techniques include injection types of attacks, such as **Structured Query Language** (**SQL**) injection, **cross-site scripting** (**XSS**), and unauthorized access attempts. DAST tools can provide comprehensive coverage and test for a wide range of vulnerabilities, including findings issues that may not be detectable via static analysis.

How to do it...

Based on their capabilities, different DAST tools can be used to test automotive IoT applications. Some DAST tools can also be integrated into the development workflow so that they can be triggered automatically in a CI pipeline as part of the secure development platform. However, note that some DAST tools may need to be executed in specific test environments and require more human interaction to execute effectively. Thus, some of these tools are more suitable to be executed during penetration testing.

Most DAST tools are executed against a running application, software, or system while following a **black-box approach**, where the tools do not understand the internal application logic. Thus, there may be some challenges in detecting certain types of vulnerabilities.

Table 7.5 shows some common DAST tools, both commercial and open source, and their respective focus area for testing that can be performed. Please note that this list is non-exhaustive and only serves to give an example of the types of DAST tools that exist.

Tool	Type	Focus Area
Acunetix Web Vulnerability Scanner	Commercial	Web
Burp Suite	Commercial	Web
Invicti	Commercial	Web
OWASP ZAP	Open source	Web
CloudSploit	Open source	Cloud
Mobile Security Framework (MobSF)	Open source	Mobile

Table 7.5 – Examples of DAST tooling

Please note that at the time of writing, not many DAST tools exist that have been specifically designed for testing embedded applications on the vehicle side. Most of the tools are more generic IT security tools that can be used for testing the cloud, web apps, and mobile apps.

Fuzz testing

This section provides a brief overview of fuzz testing.

What is it?

Fuzz testing is a test technique that involves sending **malformed** input to a target system or software and observing the target to detect unexpected behavior. Fuzz testing is a **dynamic test** approach where the target application is executed and processes input that is invalid, unexpected, or semi-random – so-called **fuzzed data** – that is provided by a fuzz testing tool. In contrast to DAST tools, which focus mostly on detecting known vulnerabilities or unknown vulnerabilities using known attack patterns, fuzz testing focuses on finding **unknown vulnerabilities** using **unknown** or **new attack patterns**.

There are different approaches for performing fuzz testing and different types of tools that can be used depending on the approach. Fuzz testing can be performed while following a **black-box**, **gray-box**, or **white-box approach**.

For the **black-box approach**, the tool does not require access to the source code and does not need to understand the internals of how the target system works. Instead, the fuzzing tool provides fuzzed data for different protocols or file formats that are input to the target system. For example, for an automotive IoT application that uses **Message Queuing Telemetry Transport** (**MQTT**) for communication or processes JSON files, a black-box fuzzing tool can generate fuzzed MQTT frames or fuzzed JSON files. This fuzzed data is then provided to the target application without knowing the internals of the application.

For the **white-box approach**, the tool requires access to the source code of the target application. The tool can analyze the internal structure of the application and generate fuzz test cases for different functions to be tested. This approach requires the tool to understand the internals of the application to perform fuzz testing effectively. The tool can continuously track code that has been executed using intelligent instrumentation and therefore assist in maximizing code coverage.

The **gray-box approach** is somewhere between the black-box and white-box approaches. It utilizes partial knowledge about the target system to increase coverage and efficiency. The gray-box approach can be used to perform coverage-guided or feedback-based fuzzing – that is, the fuzz test cases are based on information from fuzzed data that was previously executed. Deep analysis of the source code is not required as in the white-box approach; instead, the fuzz testing tool can track which branches have been executed and how many times. This allows the tool to generate appropriate fuzzed data to explore branches that haven't been explored thoroughly yet [23, 24].

How to do it...

For embedded software development on the vehicle side, fuzz testing is typically performed on a hardware device, such as a target **electronic control unit** (ECU). A black-box approach is common when testing such systems as the software needs to be tested dynamically. Fuzz testing can be performed as a one-time activity at the end of development. However, to achieve better coverage, it is possible to perform continuous testing if the organization has established a test setup using, for example, a **Hardware-in-the-Loop** (HIL) test environment [25, 26]. It is also possible to shift left and perform continuous fuzzing on an embedded software target by using a virtual platform. This approach requires the organization to establish a **Software-in-the-Loop** (**SIL**) test environment [27]. These types of test environments can be integrated into the secure software development platform to allow for continuous testing.

To fuzz test software that's been developed for web and mobile apps, it is possible to follow a white-box or gray-box approach and perform testing automatically and continuously in the secure software development platform.

Table 7.6 provides some examples of fuzz testing tools, both commercial and open source, and indicates whether the tools are applicable for black-box, white-box, or gray-box testing:

Tool	Type	Test Approach
AFL++	Open source	Gray-box testing
Boofuzz	Open source	Gray-box testing
Code-Intelligence	Commercial	White-box testing
Defensics	Commercial	Black-box testing
Honggfuzz	Open source	Gray-box testing
OSS-Fuzz	Commercial	Black-box and gray-box testing
Radamsa	Open source	Black-box testing

Table 7.6 – Examples of fuzz testing tooling

More information on fuzz testing for automotive software development can be found in the *References* section [21].

Penetration testing

This section provides a brief overview of penetration testing.

What is it?

Penetration testing is performed to simulate real-world attackers and try to break the cybersecurity goals of the target system. This could include compromising the system and taking unauthorized control of functions, intentionally changing the behavior of the system, or extracting confidential information such as encryption keys or other credentials from the system.

For embedded targets on the vehicle side, it is generally more difficult to automate penetration testing because there are not many tools available. Also, penetration testing typically requires physical access to the target or specific automotive and embedded knowledge to analyze results from one activity before moving on to the next.

In contrast, for web apps and mobile apps, there are more tools available for automated testing. Therefore, this section focuses more on penetration testing the web apps and mobile apps in the automotive IoT ecosystem rather than penetration testing embedded systems on the vehicle side (since the focus is on automated software testing). Please note that it is important to consider how to perform continuous penetration testing during both the development and operation phases. While many tools can be run automated, they typically require a human tester to configure the tools and analyze the results to effectively perform penetration testing. This means that human interaction is required to validate the cybersecurity goals and make sure that the focus of testing is on the high-risk areas and to analyze and utilize results from one type of test to further determine attack vectors for subsequent testing.

How to do it...

Penetration testing generally follows some common steps, as illustrated in *Figure 7.10*:

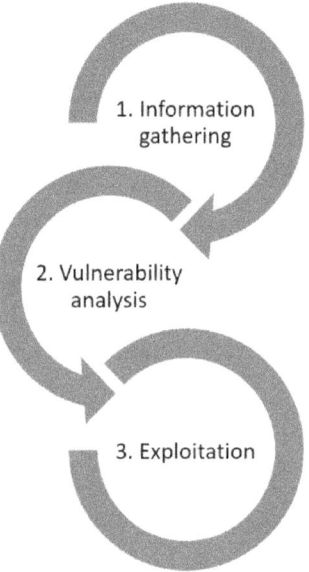

Figure 7.10 – Simplified flow for penetration testing

This is a very simplified flow; let's take a closer look. The first step is **information gathering**. Here, information about the target system, such as interfaces and communication protocols, applications, operating systems, open ports, and services, are identified. The second step is **vulnerability analysis**, where potential vulnerabilities are identified through various test approaches, such as SAST, SCA, DAST, and fuzz testing. It may also be possible to use specific vulnerability scanners or to reverse-engineer the software to analyze and find vulnerabilities. The third step is **exploitation**, where the identified vulnerabilities are exploited. This could include applying known exploit payloads or creating new exploits. It could also include bypassing some access control by brute-forcing a key or cracking a password.

A plethora of tools exist that can be used for penetration testing web apps and mobile apps. The intention is not to list all of them here as this list may easily become obsolete as new tools are released and other tools become ineffective. Instead, we will provide some examples of tools and their associated types and categories, as well as which penetration testing step such tools are generally used (that is, Step 1. Information gathering, Step 2. Vulnerability analysis, or Step 3. Exploitation). Some tools that can be used for penetration testing of embedded systems are also included.

Moreover, where applicable, the SAST, SCA, DAST, and fuzz testing tools mentioned in the previous subsections can also be used for penetration testing as part of the vulnerability analysis step. Since we covered these tools previously, we'll be omitting them from this list.

The list of available penetration testing tools is presented in *Table 7.7*. Please note that the list contains examples of both open source tools as well as commercial tools:

Step	Tool	Type	Category
1. Information gathering	Nmap	Open source	Network/port scanner
1. Information gathering	Wireshark	Open source	Packet analyzer
2. Vulnerability analysis	Nessus	Commercial	Vulnerability scanner
2. Vulnerability analysis	OpenVAS	Open source	Vulnerability scanner
2. Vulnerability analysis	Radare2	Open source	Reverse-engineering
2. Vulnerability analysis	IDA Pro	Commercial	Reverse-engineering
3. Exploitation	Metasploit	Open source	Penetration testing framework and exploit modules
3. Exploitation	John the Ripper	Open source	Password cracker
3. Exploitation	Hashcat	Open source	Password cracker

Table 7.7 – Examples of penetration testing tooling

It is suggested that the organization investigates what penetration testing tooling is appropriate for testing their automotive IoT applications and ensuring end-to-end coverage. The project inventory list described earlier in this chapter can be used as a reference to determine the type of test tools that can be applied to each target application or system for the automotive IoT use cases.

Summary

This chapter provided the foundation for establishing a **secure software development platform** in any organization.

First, we reviewed the various activities to perform in the SSDLC, such as the TARA/threat model, requirements review, design review, code review, SAST, vulnerability scanning, fuzz testing, DAST, IAST, and penetration testing.

Then, we discussed how to create a **project inventory** as this helps an organization better understand the security posture and plan, manage, and track cybersecurity activities for each project based on the risk level.

This chapter also provided a step-by-step practical guide on how to establish a **secure software development platform**. Specifically, we described the workflow where different project development teams can run a set of automated AppSec tooling in a CI pipeline. The different project development teams can define specific requirements and criteria for when which types of tools should be executed. Examples of AppSec tooling that can be applied to the secure development platform include SAST, SCA, DAST, fuzz testing, and penetration testing. Moreover, the findings from these AppSec tools can be managed in a centralized **vulnerability management** system.

As a natural next step, we will continue discussing how to secure the software supply chain in the next chapter – that is, *Chapter 8*, *Securing the Software Supply Chain*.

References

- *[1]* ISO/IEC/IEEE, *ISO/IEC/IEEE 12207: Systems and software engineering – Software life cycle processes*, 2017
- *[2]* ISO/IEC/IEEE, *ISO/IEC/IEEE 15288: Systems and software engineering – System life cycle processes*, 2023
- *[3]* ISO/IEC, *ISO/IEC 27002: Information technology – Security techniques – Security controls*, 2022
- *[4]* SEI, *SEI CERT C Coding Standard: Rules for Developing Safe, Reliable, and Secure Systems*, 2016
- *[5]* SEI, *SEI CERT C++ Coding Standard: Rules for Developing Safe, Reliable, and Secure Systems in C++*, 2016
- *[6]* MISRA: `https://misra.org.uk/`
- *[7]* AUTOSAR, *Guidelines for the use of the C++14 language in critical and safety-related systems*, 2019
- *[8]* MITRE: `https://cwe.mitre.org/`
- *[9]* ISO/SAE, *ISO/SAE 21434:2021 Road Vehicles Cybersecurity Engineering*, 2021
- *[10]* ISO/SAE, *ISO/SAE CD PAS 8475 Road Vehicles Cybersecurity Assurance Levels (CAL) and Targeted Attack Feasibility (TAF)*, `https://www.iso.org/standard/83187.html`
- *[11]* Dennis Kengo Oka, *AI and SDV are Transforming the Automotive Industry - But What are the Cybersecurity Considerations?*, 6th IoT Security Forum, Taipei, Taiwan, 2023
- *[12]* Dennis Kengo Oka, Tommi Makila, Rikke Kuipers, *Integrating Application Security Testing Tools into ALM Tools in the Automotive Industry*, A3S: IEEE International Workshop on Automobile Software Security and Safety, Sofia, Bulgaria, 2019

- [13] Madeline Cheah, Dennis Kengo Oka, *Cybersecurity Metrics for Automotive Systems*, SAE Technical Paper 2021-01-0138, 2021, doi:10.4271/2021-01-0138

- [14] OWASP, *Defectdojo*: https://owasp.org/www-project-defectdojo/

- [15] Synopsys, *Software Risk Manager*: https://www.synopsys.com/software-integrity/software-risk-manager.html

- [16] OWASP: https://owasp.org/Top10/en/

- [17] OWASP: https://owasp.org/www-project-mobile-top-10/

- [18] SANS: https://www.sans.org/top25-software-errors/

- [19] CISQ: https://www.it-cisq.org/coding-rules/

- [20] Synopsys, *Static Application Security Testing (SAST)*: https://www.synopsys.com/software-integrity/static-analysis-tools-sast.html

- [21] Dennis Kengo Oka, *Building Secure Cars: Assuring the Automotive Software Development Lifecycle*, Wiley, 2021, DOI:10.1002/9781119710783

- [22] Synopsys, *Black Duck Software Composition Analysis*: https://www.synopsys.com/software-integrity/software-composition-analysis-tools/black-duck-sca.html

- [23] Code Intelligence, *What Is Fuzz Testing?*: https://www.code-intelligence.com/what-is-fuzz-testing

- [24] Synopsys, *Defensics Fuzz Testing*, https://www.synopsys.com/software-integrity/security-testing/fuzz-testing.html

- [25] Dennis Kengo Oka, Aurore Yvard, Stephanie Bayer, Tobias Kreuzinger, *Enabling Cyber Security Testing of Automotive ECUs by Adding Monitoring Capabilities*, escar Europe, Munich, Germany, 2016

- [26] Dennis Kengo Oka, Toshiyuki Fujikura, Ryo Kurachi, *Shift Left: Fuzzing Earlier in the Automotive Software Development Lifecycle using HIL Systems*, escar Europe, Brussels, Belgium, 2018

- [27] Dennis Kengo Oka, *Fuzz Testing Virtual ECUs as Part of the Continuous Security Testing Process*, SAE International Journal of Transportation Cybersecurity and Privacy 2(2), 2020, DOI:10.4271/11-02-02-0014

Get This Book's PDF Version and Exclusive Extras

Scan the QR code (or go to `packtpub.com/unlock`). Search for this book by name, confirm the edition, and then follow the steps on the page.

Note: Keep your invoice handy. Purchases made directly from Packt don't require one.

8
Securing the Software Supply Chain

Following security processes and best practices as described in *Chapter 6, Exploring Secure Development Processes for Automotive IoT*, as well as performing automated application security testing by establishing a secure development platform as described in *Chapter 7, Establishing a Secure Software Development Platform*, will help organizations improve security with a focus on their in-house-developed software.

However, applications that support automotive **Internet of Things** (**IoT**) use cases typically encompass a wide variety of software, coming from numerous sources and provided through a so-called **software supply chain**. For example, a diagnostics application running on the vehicle may have parts of the software developed by the automaker and other parts of the software developed by a supplier. As part of the backend solution, **open source software** (**OSS**) frameworks and libraries may be used. These OSS components are generally developed and maintained by open-source communities. These communities are made up of volunteer software developers who collaborate with a shared purpose to develop, maintain, and extend a specific software project. Moreover, there are also commercial OSS options that are usually developed and maintained by a single vendor according to its commercial interests and often release software with a limited set of features for free with an option to pay for additional features or services.

As security is only as strong as its weakest link, it is imperative for the automotive industry to secure the software supply chain in order to achieve a certain level of cybersecurity assurance in the final product – that is, the end-to-end software for the automotive IoT use case. Using one vulnerable software component, or one supplier that does not meet the baseline security requirements, may invalidate the overall security assurance level and result in cyberattacks with disastrous consequences.

This chapter discusses the risks in the software supply chain and presents several practical suggestions on how to address these risks. For example, topics on how to manage OSS and **software bill of materials** (**SBOM**) will be covered.

In this chapter, you will learn about the following:

- Software supply chain and distributed development
- RASIC, vendor security assessments, and **cybersecurity interface agreements for development (CIADs)**
- Managing risks with OSS
- SBOM
- Secure software supply chain risk management

Software supply chain and distributed development

The automotive industry has a long history of using a supply chain, where suppliers provide components and hardware units that are finally assembled together to build a vehicle. With more software being used to support new advanced use cases, including automotive IoT use cases, the automotive supply chain also includes managing software in a so-called software supply chain.

While automotive organizations also need to consider securing the hardware supply chain, this topic is out of scope for this book. With the increased usage of software to enable automotive IoT use cases, we'll focus our attention on the software supply chain.

Overview of the software supply chain

While there are changes happening to the supply chain in the automotive industry as a response to accommodate current trends of **software-defined vehicles (SDVs)**, a generic view of a simplified supply chain is illustrated in *Figure 8.1*.

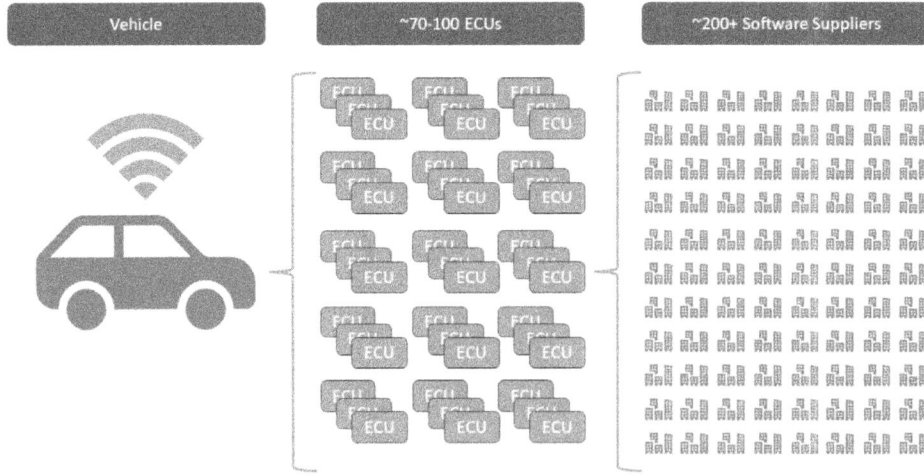

Figure 8.1 – Automotive supply chain example

A vehicle could contain, for example, 70-100 **electronic control units** (**ECUs**) coming from a large number of suppliers. Furthermore, the software running on those ECUs could come from 200+ software suppliers. This shows the complexity of the software going into a single vehicle.

The automotive supply chain typically also comprises multiple levels, called **tiers**, as shown in *Figure 8.2*. For example, using a **Tier 3 Supplier** as a starting point, it provides certain base software components to a **Tier 2 Supplier**, which adds some other software components integrated into a software package and passes it on to a **Tier 1 Supplier**, which in turn adds some additional software packages and assembles it into another software package integrated with hardware. This software and hardware package/system is then handed over to the automaker or the **original equipment manufacturer** (**OEM**), who assembles all these systems into the final product – that is, the vehicle. The OEM may also develop its own systems and/or add some OEM-specific software on top of supplier-provided systems.

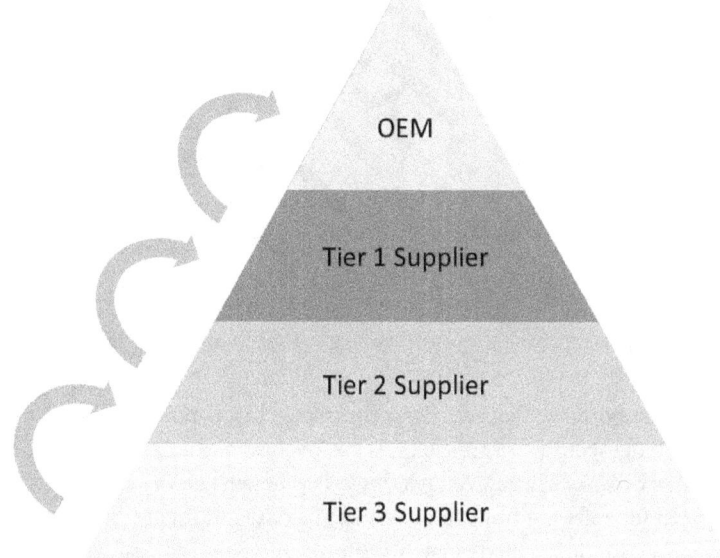

Figure 8.2 – Levels in the automotive supply chain

Thus, as part of development, it is common for different parts of software to be developed by different suppliers following a **distributed development approach**. This distributed development approach requires that the procurer and supplier have agreed expectations and understanding of the level of security. It is also important to note that certain software that is included could be OSS components. These components are developed by the community and may follow completely different development methodologies and approaches to security testing, depending on the maturity of the contributing development team. As a result, how to manage **cybersecurity assurance** for OSS components would also need to be considered.

As previously mentioned, because security is only as strong as its weakest link, as illustrated in *Figure 8.3*, one poorly designed or developed software component or one supplier with lower security maturity – for example, employing poor security processes and practices – could lead to the entire automotive IoT application or use case becoming compromised.

Figure 8.3 – Security is only as strong as its weakest link

Therefore, it is of utmost importance that security is considered throughout the entire software supply chain. This includes defining security responsibilities for involved parties and stakeholders, assessing vendors, and establishing cybersecurity agreements, as well as understanding and managing OSS risks, understanding the need for software transparency and SBOMs, and finally, putting all these pieces together to ensure an organization can adequately perform software risk management.

These different topics will be covered in the rest of this chapter.

RASIC, vendor security assessments, and CIADs

Different approaches can help with improving security in the software supply chain. First, it is crucial for the organization to define **roles and responsibilities** for the different cybersecurity activities during development and operations. One general approach that can be taken is establishing a **responsible, accountable, support, inform, and consult (RASIC)** chart.

Moreover, organizations in the automotive industry typically work with a multitude of suppliers. To improve security in the supply chain, it is imperative that organizations follow a systemic approach to evaluating the cybersecurity capabilities and cybersecurity posture of different suppliers. One such approach is to perform a **vendor security assessment** and use that information as one factor when selecting a vendor.

Furthermore, already common in the automotive industry is to define a **development interface agreement** (**DIA**) between the procurer and the supplier. The DIA is used to specify certain terms and activities for the development and ensure a common agreed understanding between the involved parties. Specifically for cybersecurity, there is what is known as a CIAD. This agreement typically focuses on achieving a common understanding of cybersecurity activities and responsibilities between the involved parties. This understanding could be based on a RASIC chart.

RASIC, vendor security assessments, and CIADs are described in more detail in the following subsections.

RASIC

A common approach to define respective responsibilities among several parties and roles is to use a RASIC table. RASIC is an acronym and is defined as follows:

- **Responsible** (**R**): Organization or role that performs the activity
- **Accountable** (**A**): Organization or role that is accountable for the completion of the activity and approves the results of the activity
- **Support** (**S**): Organization or role that provides support to the responsible party performing the activity, and may perform part of the activity
- **Inform** (**I**): Organization or role that is informed about the activity and its progress and results
- **Consult** (**C**): Organization or role that does not directly or actively work on the activity but provides input or guidance to the responsible party where needed

Some common rules or guidelines for using a RASIC chart include the following. Each row defining an activity should have at least one **R** role. This role is responsible for performing the activity. A row may have multiple **R** roles that are responsible for different parts of the activity; however, this may cause confusion and it may be better to further split the activity into smaller tasks so that each row only has one **R** role. A row should have only one **A** role assigned within an organization so that it is clear who is accountable. It is important to note that **A** and **R** cannot be assigned to the same role within an organization. It is optional to include one or more **S**, **I**, and **C** roles in each row.

RASIC charts can be used throughout the supply chain, as illustrated in *Figure 8.4*.

Activity	OEM	Tier 1 Supplier
Activity 1	A	R
Activity 2	I	
Activity 3	A	R
Activity 4	A	R
Activity 5	A	R
Activity 6	A	R
Activity 7	A	R
...		

Tier 1 Supplier	Tier 2 Supplier
A	R
A	R
A	R
A	R

Tier 2 Supplier	Tier 3 Supplier
A	R
A	R

Figure 8.4 – RASIC charts between the OEM and tier 1 supplier, tier 1 and tier 2 suppliers, and so on

In the RASIC table, the organization specifies which activities will be handled by the OEM and which ones by the Tier 1 Supplier. As we traverse through the supply chain, similar RASIC charts would be established between the Tier 1 and Tier 2 Suppliers, and similarly the Tier 2 and Tier 3 Suppliers, and so forth. This allows certain activities to be passed down the supply chain. Additionally, certain activities may be required to be performed at several levels in the supply chain (for example, targeting different levels or components of the software stack).

More specifically, *Figure 8.5* shows a simplified example of a RASIC chart for activities between an OEM and a supplier.

Activity	OEM		Supplier	
	Project Team	Security Team	Project Team	Security Team
TARA/Threat Model	A	R	A	R
Code Review	I		A	R
SAST	I		A	R
Vulnerability Scanning	I		A	R
Fuzz Testing	I		A	R
DAST	I		A	R
Penetration Testing	A	R	S	
...				

Figure 8.5 – Simplified example of a RASIC chart between an OEM and a supplier

The figure shows the roles and responsibilities of the **OEM** on the left-hand side and the **Supplier** on the right-hand side. The list of **cybersecurity activities** is provided in the far-left column. This list of activities could be aligned with the activities defined in the project inventory described in *Chapter 7, Establishing a Secure Software Development Platform*.

RASIC can also help with clarifying responsibilities within the organization. According to the figure, for the **Threat Analysis and Risk Assessment (TARA)** or **Threat Model** activity, the **Security Team** at the OEM is responsible for performing this activity and the **Project Team** at the OEM is accountable. This could be a vehicle-level TARA or an end-to-end automotive IoT use case threat model. Moreover, the supplier also performs this activity. The **Security Team** at the supplier is responsible for performing this activity and the **Project Team** at the supplier is accountable. The scope would be limited to what the supplier is responsible for developing – for example, it could be a system-level TARA or threat model focusing on a specific application for an automotive IoT use case. In this example, let's consider that the supplier is developing an automotive IoT application, and therefore the supplier is responsible for the **Code Review**, **static application security testing (SAST)**, and **Vulnerability Scanning** activities, and informs the OEM about the results of those activities. Moreover, the supplier is responsible for certain security testing activities including **Fuzz Testing** and **dynamic application security testing (DAST)**. Next, as the supplier-provided application is then executed in the OEM test environment, the OEM is responsible for **Penetration Testing**. Where required, the supplier provides support to ensure that the application is properly functioning to enable such testing. Depending on the capabilities of the OEM and supplier and the required access to allow certain types of testing, the responsibilities for different activities may be assigned differently between the OEM and the supplier in the RASIC chart.

Please note that this is merely an example, and the actual activities and RASIC chart contents would have to be defined for involved parties based on the actual development project.

Vendor security assessments

As part of distributed development, an organization typically works with multiple vendors. When selecting a vendor, it is imperative for the organization to gain assurance that the vendor can handle cybersecurity responsibilities and successfully perform the assigned cybersecurity activities. Depending on the vendors' capabilities, it may be necessary to update the RASIC table to ensure that it is aligned with realistic expectations.

The process and workflow for vendor selection can be established by incorporating requirements from *ACQ.2 Supplier Request and Selection* in *ASPICE for Cybersecurity* [1] or *Clause 7 Distributed Cybersecurity Activities* in *ISO/SAE 21434* [2].

In order to perform a **vendor security assessment**, the organization first needs to establish the evaluation criteria and the decision process. That is, an organization needs to determine which factors at a vendor should be considered and how they should be evaluated, as well as how to make a decision on whether to proceed with a certain vendor or not.

One approach is to prepare a set of questions that can be provided to the vendor to fill out. Some example questions for a vendor security assessment form are provided in *Table 8.1*.

Category	Questions
Development	Does the organization follow secure development practices and any specific coding guidelines/standards – for example, CERT C/C++, MISRA C/C++, AUTOSAR C++?
	Does the organization perform code reviews?
	Does the organization perform SAST?
	Does the organization perform vulnerability scanning?
	Does the organization perform fuzz testing?
	Does the organization perform DAST?
	Does the organization perform penetration testing?
	Can the organization generate an accurate SBOM and provide it as part of the delivery?
Organization	Does the organization have a competency management program?
	Does the organization have a vulnerability management process?
	Does the organization have a CSMS?
	Is the organization ISO 21434 certified?
	Is the organization TISAX certified?
	Is the organization SOC 2 Type 2 compliant?
Continuous activities/ Post-development	Does the organization perform continuous cybersecurity monitoring of its products?
	Does the organization provide alerts on newly found vulnerabilities and have an SLA to fix vulnerabilities within certain time frames? (e.g., critical within 3 days, high within 10 days, medium within 30 days, etc.)
	Does the organization have a software update process and can it provide software updates (within a certain agreed timeline)?
	Does the organization have an incident response process?

Table 8.1 – Example questions for a vendor security assessment form

Moreover, answers to these questions could be selectable from a predefined list of answers, as shown in *Table 8.2*. These answers could have different scoring weights to help determine the level of maturity and capabilities.

Selectable answers	Scoring weight
Established and rolled out to relevant projects	5
Established but not rolled out to relevant projects yet	4
Currently establishing	3
Not yet established but planned	2
Not established and not planned	1

Table 8.2 – Example answers to questions in a vendor security assessment form

This type of weighted evaluation allows an organization to gain a better perspective and more accurate view of a vendor's cybersecurity maturity and capabilities.

All the individual scores can then be added to calculate an overall score that's used to assess the vendor. Furthermore, these scores could be used to evaluate and compare multiple vendors.

CIADs

After establishing a RASIC, defining and assigning cybersecurity responsibilities, and performing a vendor assessment to have a better view of the security maturity and capabilities of vendors in order to select a vendor, the next logical step is to establish a CIAD between the procurer and the vendor.

To achieve a common and agreed understanding in the supply chain, CIADs can be established between an OEM and a Tier 1 supplier, between Tier 1 and Tier 2 Suppliers, between Tier 2 and Tier 3 suppliers, and so forth, as depicted in *Figure 8.6*.

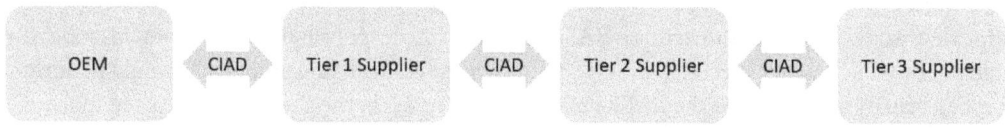

Figure 8.6 – CIAD between an OEM and a tier 1 supplier, between tier 1 and tier 2 suppliers, and so on

The CIAD serves to clarify and agree on the terms and expectations of cybersecurity between the procurer and the vendor. It should encompass and consider the outcome of the RASIC and vendor security assessment. For example, the CIAD may stipulate the following:

- Vendor to perform SAST to identify weaknesses and vulnerabilities
- Vendor to perform SAST to verify compliance with certain coding guidelines, such as CERT C/C++, MISRA C/C++, and AUTOSAR C++

In addition, the CIAD could stipulate something such as the following, where **X** could specify different criteria depending on the **cybersecurity assurance level** (**CAL**), described in more detail in *Chapter 7, Establishing a Secure Software Development Platform*, assigned to the specific project:

- Vendor to perform vulnerability scanning according to **X**
- Vendor to perform fuzz testing according to **X**
- Vendor to perform DAST according to **X**
- Vendor to perform penetration testing according to **X**

For example, **X** could specify how often the vendor should perform a certain activity and with what rigor, and to what extent and depth a certain testing activity should be performed. Specifying and defining different criteria and parameters for **X** is important, and organizations should use the CALs, the project inventory, and TARA to help determine appropriate definitions for **X**. The upcoming ISO/SAE 8475 standard should also be considered to help organizations when adopting CALs and appropriate test criteria.

Coming to a common and agreed understanding between the supplier and the procurer to ensure that the supplier performs the required cybersecurity activities is crucial. Using a systematic approach such as a CIAD allows the involved parties to have a clear view of the agreed activities.

Managing risks with OSS

For third-party-developed software, since an organization is dealing with a specific vendor, following an approach using the previously discussed RASIC, vendor security assessment, and CIAD is possible.

However, for OSS, an organization needs to handle this differently, since the organization is not typically dealing with a specific vendor that is developing the OSS component. Instead, the OSS component is developed and maintained by the community.

Although it would be possible to define a RASIC to clarify the responsibilities of the organization, it would not be beneficial to expect that the OSS project or the contributing individuals would perform the cybersecurity activities that the RASIC may indicate. As such, the RASIC may be used more to indicate any gaps in cybersecurity activities or to indicate what the organization will do.

Moreover, it would not be possible to establish a CIAD with the individual contributors to the OSS project, since most of them are contributing to the project on a volunteer basis or as a non-profit activity and, therefore, they would not be able to agree to take on that level of responsibility.

It would also not be possible to conduct a typical vendor security assessment either since there is no official vendor per se but rather, typically, many different code contributors. Depending on how the vendor security assessment is construed, there may, however, be certain questions or evaluation criteria that could be considered if the relevant information could be gathered. For example, the project may indicate how its contributors are doing certain security testing and how quickly they are able to issue patches and updates. This type of information could be considered when choosing and comparing different OSS components to incorporate in the project.

There are generally three main concerns to consider when using OSS: **security vulnerabilities**, **license compliance**, and **operational risk**.

Security vulnerabilities

As with any software development, there is a risk that software vulnerabilities are introduced in the code during development. OSS is no exception and over the years, several critical vulnerabilities have been detected in various commonly used OSS components. Some examples of OSS vulnerabilities include the following:

- The **Heartbleed** vulnerability in OpenSSL (CVE-2014-0160) [3,4], discovered in 2014, that affected more than half a million servers
- The **Blueborne** vulnerability [5] found in the Bluetooth code of Linux, Android, iOS, and Windows in 2017, that affected more than 5 billion devices, including in-vehicle infotainment systems
- The **Log4j** (CVE-2021-44228) [6] vulnerability detected in 2021 in the logging tool Log4j, used on millions of systems

A recent survey of OSS risks, reviewing over a thousand code bases across 17 industries, including automotive, found that 84% of the code bases contained at least one open source vulnerability [7].

These types of vulnerabilities can negatively affect systems involved in automotive IoT use cases. For example, attackers may be able to remotely take control of features and functions on the vehicle, or be able to extract confidential or sensitive information from the backend solution or mobile device.

As such, organizations need to have clearly defined workflows for how to handle vulnerabilities in OSS. An example workflow on how to **handle detection of new vulnerabilities in OSS** is shown in *Figure 8.7*.

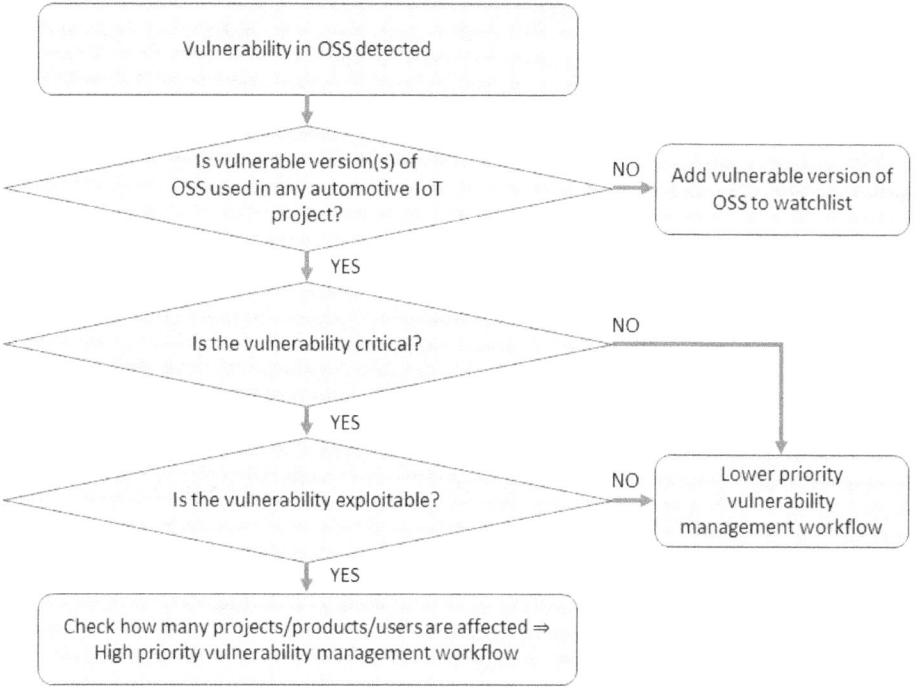

Figure 8.7 – Example workflow on how to handle detection of OSS vulnerabilities

As shown in the figure, when the organization first is notified that a new vulnerability in an OSS component has been detected, it investigates whether the vulnerable version(s) of the OSS component is used in any existing automotive IoT projects. If it is not, the vulnerable version is added to a **watchlist** to prevent developers from using this vulnerable version in future projects. It should be noted that this step should be aligned with the established policies in the organizations specifying whether vulnerable OSS components are allowed to be used in projects.

If the vulnerable version is found to be included in existing automotive IoT projects, the organization analyzes the **criticality** level of the vulnerability. If it is not critical, then the organization proceeds with the **lower priority vulnerability management workflow**. On the other hand, if the vulnerability is critical, then the organization evaluates whether the vulnerability is **exploitable**. If it is not exploitable, the lower priority vulnerability management workflow is initiated.

If the vulnerability is found to be exploitable, the organization continues with verifying how many of the projects, products, or users are affected by this critical, exploitable vulnerability and initiates the **high priority vulnerability management workflow**.

An example of how an **OSS vulnerability management workflow** could look is presented in *Figure 8.8*.

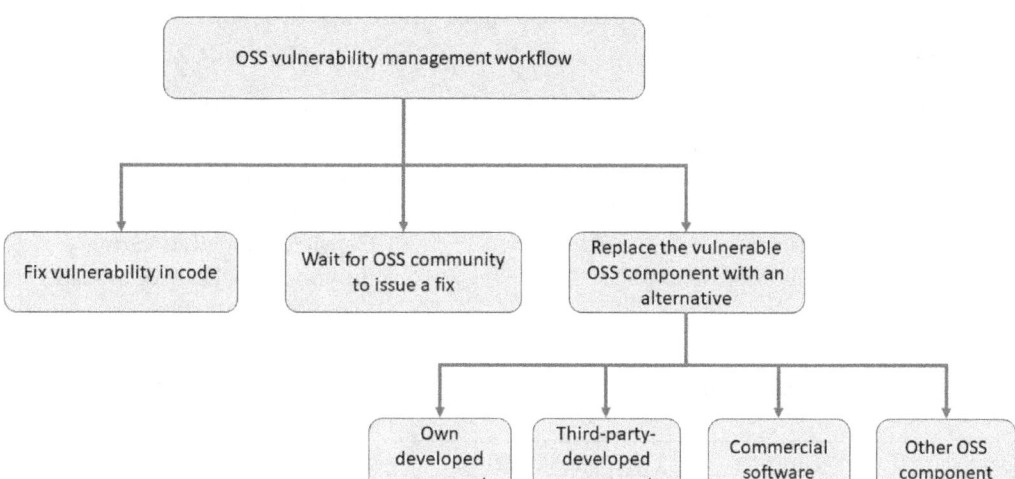

Figure 8.8 – Example workflow of OSS vulnerability management

As shown in the figure, the organization needs to establish a workflow for **OSS vulnerability management**. There are several approaches that can be taken to handle OSS vulnerabilities. For example, the organization can **fix the vulnerability themselves** in the code and use the patched OSS component. The organization could also **wait for the OSS community** to issue an updated, fixed version of the OSS component. Another approach is to **replace the vulnerable OSS component with an alternative**. There are multiple alternatives that can be considered. For example, depending on the type of features and functionality the OSS component is providing, the organization may decide to **develop its own component**. The organization may also contract a **third party to develop the component**. There may also exist **commercially available components** that the organization could choose from. Finally, there may exist **alternative OSS components** providing the same features and functionality that the organization could use instead.

License compliance

While not specifically related to security, one crucial aspect of incorporating OSS components in your products is to consider **license compliance**.

There exist some 3,000+ licenses with different terms and conditions on how the OSS is allowed to be used, what attribution is required, and in what format the software must be accessible to other users.

Some licenses such as **Apache** and **MIT** are known as **permissive**. Software with permissive licenses can be used freely. On the other hand, some licenses such as **GPLv2** are what is known as **copyleft**. This means that the terms and conditions applying to the OSS component should be preserved in any modified or derivative works. Thus, if a copyleft OSS component is modified and distributed as part of an automotive IoT system, those modifications would also need to be made available under the same copyleft license. That is, the organizations must ensure that the modifications are made available as source code.

Let's look at a few examples of license types for some common OSS components to get a better understanding. An overview is provided in *Table 8.3*.

OSS component	Description	License type	Implications
BusyBox	Toolset of common functions	GPLv2	Copyleft –modifications to BusyBox must be made available as source code
Linux kernel	Core software of operating system	GPLv2	Copyleft –modifications to Linux must be made available as source code
OpenSSL	Secure communication and encryption library	OpenSSL v3.0 and later: Apache License 2.0

OpenSSL before v3.0: Dual license (OpenSSL license and SSLeay license) | Apache License 2.0: Permissive, can be included in commercial projects, requires attribution

Dual license: Need to comply with both licenses, requires attribution, the OpenSSL license may have compatibility issues with some copyleft licenses |
| curl | Tool for transferring data | MIT | Permissive, can be used in commercial projects, does not require attribution |
| libpng | Library to read/write PNG image files | zlib/libpng | Permissive, can be used in commercial projects, does not require attribution but origin of software must not be misrepresented |

Table 8.3 – License types of common OSS components

Some OSS components have different licenses for different versions, as indicated by the OpenSSL example in *Table 8.3*. As such, it is important to verify and understand what terms and conditions are applicable to the newer versions of the OSS component before applying the updated versions of the software. For example, while the updated version may fix certain security issues, the associated terms and conditions for the new license type may be incompatible with the organization's policies.

One example where license types changed between versions is Redis. Redis is a popular in-memory storage solution used as a distributed, key-value database and cache and message broker, primarily functioning as an application cache or quick-response database. It was previously released under the **BSD 3-Clause license**, which is a permissive license. Starting with Redis 7.4, all future versions will be dual-licensed under the **Redis Source Available License (RSALv2)** and the **Server Side Public License (SSPLv1)**. These new licenses, in particular the **SSPLv1**, are more restrictive than the previous BSD 3-Clause license. For example, the SSPL requires organizations offering Redis as a service to release their entire source code, including potentially sensitive or proprietary code. This license change has been met with controversy by the open source community and led to debates about the sustainability of OSS.

Although these types of license changes between software versions are uncommon, they highlight the complexity around licensing considerations when using OSS components. Therefore, it is important that development teams are aware of potential changes in license types and the potential implications when using OSS components in their projects.

Operational risk

A third concern when using OSS is regarding maintainability and **operational risk**. In particular, considering the long life cycle of vehicles of 10-15 years, an organization needs to consider how to manage software including OSS for such a long period of time. Similar considerations need to be taken into account when developing automotive IoT applications.

First, the organization needs to define workflows to help guide developers in selecting appropriate OSS components to be included in their software. An example workflow is depicted in *Figure 8.9*.

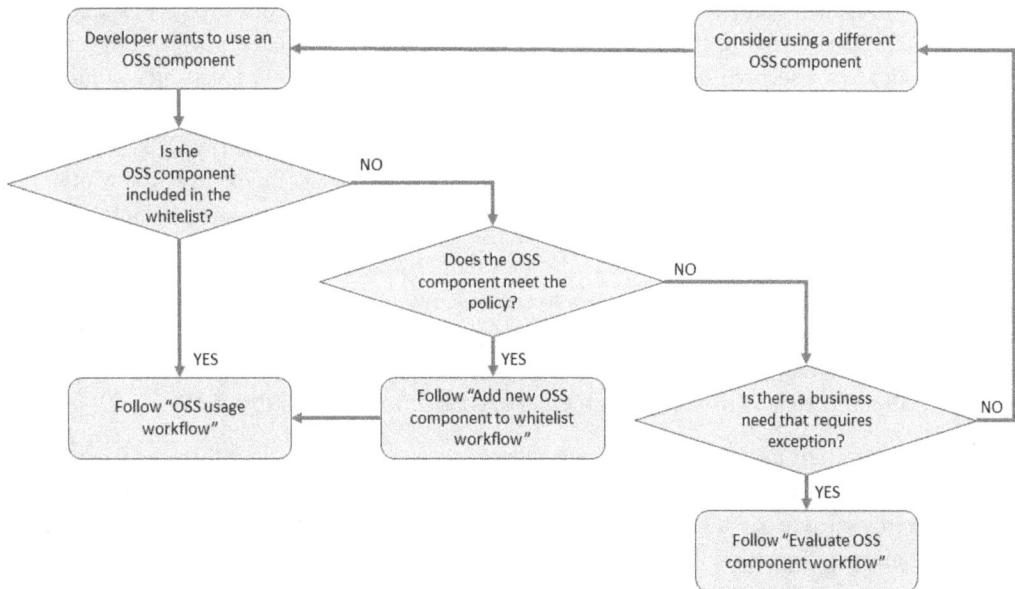

Figure 8.9 – Example workflow on selecting OSS components

For example, the organization may manage a whitelist of OSS components that developers can freely choose from. If a developer would like to use an **OSS component that is included in the whitelist**, the developer can follow a separate defined **workflow for OSS usage**. This is a separate workflow not described further in this book, but a workflow that the organization would need to define and would include activities the organization has to perform when including the OSS component (e.g., handling potential license attribution, cybersecurity monitoring of the OSS component, etc.).

On the other hand, if the OSS component is not included in the whitelist, the next step is to check whether the **OSS component meets the organization's policy**. This policy could specify, for example, acceptable licenses for OSS components, criteria for operational risk, and so on. If the component meets the policy, it triggers a separate workflow to **add a new OSS component to the whitelist**. After the OSS component has been added to the whitelist, the developer follows the OSS usage workflow.

However, if the OSS component does not meet the organization's policy, there is a final step that checks whether there exists a certain **business need that requires the organization to make an exception** to the policy. If there are compelling business needs for the OSS component, then it triggers a separate **workflow to evaluate the OSS component**. This workflow is not elaborated on further in this book, but the organization would need to define the activities and criteria for evaluating OSS components that do not meet the organization's policy.

If there is no business need that requires the organization to make an exception, the workflow continues to inform the developer to **consider using a different OSS component** and the workflow starts from the beginning again.

Managing the whitelist is extremely important and could potentially have disastrous consequences if not handled properly. Adding an OSS component to the whitelist requires approvals from various stakeholders such as the **security team** (responsible for security evaluation), the **legal and compliance team** (responsible for license compliance evaluation), and the **development team** (responsible for function/feature usability and maintainability evaluation). Please note that adding an OSS component to the whitelist means adding a specific version or versions of the OSS component to the whitelist, as different versions may have different license terms and associated known security vulnerabilities.

Equally important to the initial evaluation to decide whether to include an OSS component on the whitelist is to **periodically review** the whitelist to remove any software components or versions that were whitelisted in the past, but no longer fulfill the organization's criteria (e.g., new vulnerabilities may have been discovered in previously whitelisted OSS components or, functionality-wise, there may exist other OSS components that are more modular and easier to update and maintain).

Moreover, organizations need to define specific workflows on how to handle situations where, for example, **new versions of OSS components are released**, **new vulnerabilities in OSS components are discovered**, or there are needs for **new features in the currently used OSS components** (e.g., support for new cryptographic algorithms in an encryption library).

Updating software with new versions of OSS components also requires defined workflows. For example, the organization should track when new versions of the included OSS components are released. If there are no major updates or fixes, the existing versions could probably be used as is. If the new release includes important security fixes, then it should be investigated whether there are any new dependency requirements or potential compatibility issues when incorporating the updated OSS component with the existing software.

While incorporating OSS components in automotive products is becoming more common, it is imperative for organizations to consider and manage the potential risks.

SBOM

One common approach to help manage security in the software supply chain is the use of an SBOM. An SBOM is simply described as an inventory list of software components in a software package. An analogy for an SBOM is a list of ingredients for a processed food package that describes the contents. SBOMs are valuable for managing the software supply chain and are described in further detail as follows.

SBOM formats

There are various formats that can be used to express SBOMs. Some common formats include **Software Package Data Exchange (SPDX)** [8] and **CycloneDX** [9]. In its simplest form, an SBOM is nothing more than a **JavaScript Object Notation (JSON)** file. However, what types of fields and contents that are supposed to be included are defined by the SBOM format.

Please also note that there are different versions of the SBOM formats of SPDX and CycloneDX. Version 2.3 of SPDX is also defined as an ISO standard, namely ISO/IEC 5962:2021 [10]. The latest version at the time of this writing is SPDX version 3.0-rc1, released in May 2023. An alternative to SPDX is CycloneDX. The latest version of CycloneDX at the time of this writing is version 1.5, released in June 2023.

Using a defined SBOM format helps different teams and organizations to achieve a common and agreed understanding of the software components contained in a software package. The software package itself may have been produced by one team or organization and handed over to a different team or organization, and therefore, using an SBOM to explain the contents of the software package becomes crucial.

Common fields in an SBOM are as follows:

- **Component name**: What is the name of the component?
- **Version**: Which version of the component is used?
- **Supplier**: Who created the component?
- **License**: What is the associated license information of the component?

A comparison of SPDX and CycloneDX is given in *Table 8.4* to help you better understand the differences between the two formats.

Feature	SPDX	CycloneDX
First released	2011	2017
SBOM metadata	Author, creation date, creation tooling	Author, creation date, creation tooling
Format	JSON, SPDX, YAML, RDF/XML, tag/value	JSON, XML
Size	Heavyweight	Lightweight
Certification	ISO/IEC 5962:2021	None
License	Creative Commons	Apache 2.0
Component metadata	Name, version, type, package level, and so on	Name, version, type, file level, and so on
VEX support	Use external reference	Native support

Table 8.4 – Comparison between SPDX and CycloneDX

Considering security in the software supply chain, it is important to note that CycloneDX supports **Vulnerability Exploitability eXchange** (**VEX**). VEX support allows the exchange of component information and vulnerability information in a standardized manner. The VEX information includes, for example, **Common Vulnerabilities and Exposures** (**CVE**) IDs and **Common Weakness Enumerations** (**CWEs**) for vulnerability assessment. The integration of VEX into the CycloneDX SBOM format improves its ability to communicate the exploitability of components with known vulnerabilities and allows organizations to use this information for cybersecurity risk management.

On the other hand, SPDX does not natively support VEX. However, it is possible to include VEX information in SPDX documents externally by using references. This similarly achieves the same goal and allows for enhanced vulnerability analysis and reporting.

Executive Order 14028

Although it is not automotive-specific, it may be useful to have a brief overview understanding of **Executive Order 14028** (**EO14028**). This is a directive issued by the US government on May 12, 2021, in an effort to improve cybersecurity. Among other things, it mandates that all software supplied to the US government is provided with respective SBOMs. The specific formats that can be used include SPDX, CycloneDX, and **Software Identification** (**SWID**) tags.

Although this executive order does not apply to the automotive industry, there are organizations that look at EO14028 for guidance on how to manage SBOMs with their suppliers. As such, it is sometimes referred to in supplier discussions and thus worth being aware of.

NTIA

The **National Telecommunications and Information Administration (NTIA)** is an agency within the US Department of Commerce and has been active in establishing a common baseline for **software transparency** in the past few years. Similar to the EO14028, it is not specific to the automotive industry; however, it provides valuable information that sometimes is referred to by automotive organizations and is therefore worth being aware of. Software transparency covers how organizations can be transparent about the software that is used in their products and supplied through the software supply chain. There are mainly three areas that the NTIA focuses on.

- First, it provides guidance on *processes and activities* in the supply chain to help with software transparency.
- Second, it provides information on relevant *formats and information* that can be included in the SBOM and shared along the supply chain.
- Third, it provides guidance on the *tooling* that an organization can utilize in order to build up an environment to support software transparency.

OpenChain

Another community activity focusing on the software supply chain that has garnered traction lately, and is supported by major automotive companies such as Toyota and Bosch, is **OpenChain**. OpenChain also provides guidance on how to manage software in the supply chain. It focuses on the *processes and activities* an organization needs to establish in order to manage software in the supply chain. It has published a specification for managing OSS, focusing on *license compliance*. This specification was also issued as the **ISO/IEC 5230:2020** [11] international standard. Moreover, OpenChain also considers *cybersecurity* and how to manage vulnerabilities in OSS, and as such, published the **ISO/IEC 18974:2023** [12] industry standard for *open source security assurance programs*.

Secure software supply chain risk management

With this understanding of RASIC, vendor security assessment, CIAD, the risks of OSS, and SBOMs, let's put it all together to get a comprehensive view of secure **software supply chain risk management (SSCRM)**.

Generally speaking, SSCRM focuses on **identifying**, **assessing**, and **mitigating** the risks associated with the **software development life cycle (SDLC)**. These steps are briefly explained as follows.

Identifying the risks

The first step in SSCRM is **identifying the risks** in the software supply chain. An organization can start by creating a comprehensive *inventory* of all supply-chain-related software components used in their projects. Thus, this inventory includes all libraries, frameworks, OSS components, third-party-developed components, commercial components, external cloud services, and so on.

The organization then needs to analyze all these components in its inventory for potential weaknesses and vulnerabilities. This includes, for example, identifying known vulnerabilities in software components, poor security practices established at vendors, low security maturity or limited security capabilities at suppliers, license compliance and operational risks for OSS components, and so on.

Supply chain attacks are becoming more common, and it is imperative for organizations to be aware of the relevant risks. For example, malicious parties may be able to inject malware or create backdoors in software that is included somewhere along the supply chain. There may be unintentional vulnerabilities left in commercial or third-party-developed components or included in OSS components.

Assessing the risks

The next step in the SSCRM involves **assessing the risks**. Assessing the risks allows the organization to evaluate the criticality and help with prioritization for mitigating the risks. For example, the organization can perform a risk assessment considering the risks identified in the first step.

The risk assessment can consider various factors, such as whether the vulnerable version of the software is used in any product or system, whether the vulnerability is exploitable, the difficulty of exploiting the vulnerability, the likelihood of the vulnerability being exploited, the number of systems/products/users being affected by a potential attack, the potential impact (e.g., to safety, financial, operational, and privacy) of a successful attack, and so on.

Mitigating the risks

After identifying and assessing the risks, the last step in the SSCRM focuses on **mitigating the risks**. The different risks can be addressed or mitigated by implementing various activities and practices throughout the SDLC, as depicted in *Figure 8.10*.

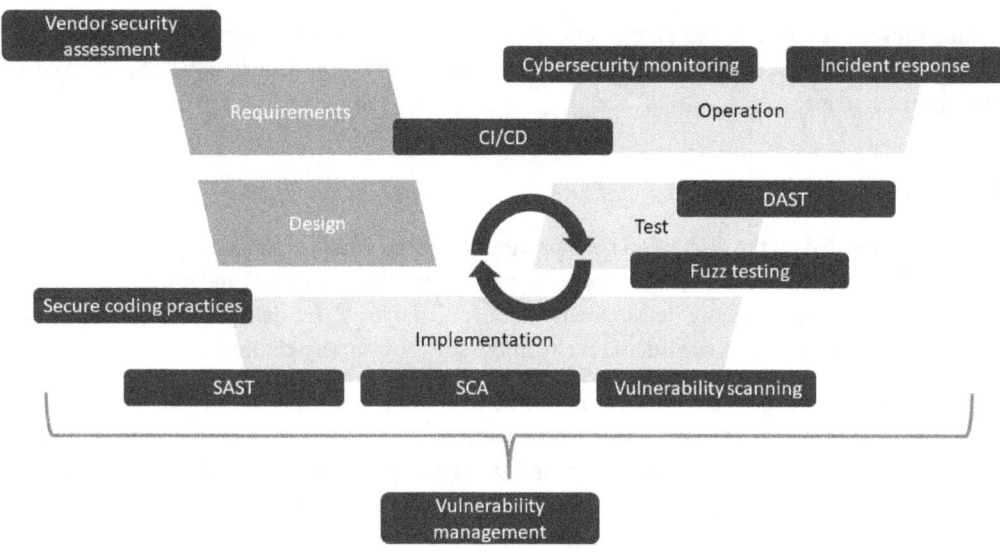

Figure 8.10 – Activities and practices throughout the SDLC to mitigate risks

Please note that the focus is on software development and testing, and therefore, activities related to requirements and design are excluded. Some of the activities and practices were discussed in *Chapters 6 and 7*. Examples include the following:

- **Vendor security assessment**: Evaluate the security practices, capabilities, and maturity of external software vendors
- **Secure coding practices**: Train developers on secure coding principles and best practices
- **SAST**: Analyze the code statically for weaknesses and vulnerabilities
- **Software composition analysis (SCA)**: Identify known vulnerabilities in OSS and third-party components
- **Vulnerability scanning**: Scan the software for known vulnerabilities or known attack patterns
- **Fuzz testing**: Perform fuzz testing to find unknown vulnerabilities
- **DAST**: Test the software dynamically for vulnerabilities
- **Continuous integration/continuous delivery (CI/CD)**: Integrate security scanning and testing approaches into the development pipeline to help identify and fix vulnerabilities early
- **Vulnerability management**: Manage, prioritize, and fix vulnerabilities during development as well as operation
- **Cybersecurity monitoring**: Continuously monitor for new security vulnerabilities and threats
- **Incident response**: Establish processes to respond to security incidents during operation

Adopting these types of activities and practices for SSCRM improves the organization's ability to mitigate software supply chain risks and helps organizations build more secure software and protect their data and systems.

Summary

This chapter expanded on the discussion from previous chapters, which explained high-level security processes (*Chapter 6, Exploring Secure Development Processes for Automotive IoT*) and provided practical guidance on establishing a secure development platform (*Chapter 7, Establishing a Secure Software Development Platform*), to also consider security in the **software supply chain**.

In particular, this chapter gave an overview of the software supply chain and distributed development in the automotive industry. Moreover, we reviewed specific approaches to defining responsibilities for cybersecurity activities between procurers and suppliers using a **RASIC** table. We also covered how organizations can evaluate their suppliers using a **vendor security assessment** approach and then, after selecting a vendor, establish a common understanding and agreement with suppliers on expectations for cybersecurity using a **CIAD**.

This chapter then presented how to manage **risks** when using **OSS** components, and explained the need for creating **SBOMs** in order to achieve software transparency. Finally, all of these topics were put together to discuss addressing risks for **SSCRM**.

Now you have a solid understanding of the cybersecurity topics presented in this part of the book, the next part of the book will provide a comprehensive view of the end-to-end design, development, and deployment of automotive IoT applications, starting with *Chapter 9, System Design of an Automotive IoT Application*.

References

- [1] VDA, "*Automotive SPICE for Cybersecurity Process Reference and Assessment Model*", 2021
- [2] ISO/SAE, "*ISO/SAE 21434:2021 Road Vehicles Cybersecurity Engineering*", 2021
- [3] NIST, "*CVE-2014-0160 Detail*", https://nvd.nist.gov/vuln/detail/CVE-2014-0160
- [4] Synopsys, "*The Heartbleed Bug*", https://heartbleed.com/
- [5] Armis, "*BlueBorne*", https://www.armis.com/research/blueborne/
- [6] NIST, "*CVE-2021-44228 Detail*", https://nvd.nist.gov/vuln/detail/CVE-2021-44228
- [7] Synopsys, "*2024 Open Source Security and Risk Analysis Report*", https://www.synopsys.com/software-integrity/resources/analyst-reports/open-source-security-risk-analysis.html

- [8] The Linux Foundation, "*System Package Data Exchange (SPDX®)*", https://spdx.dev/
- [9] OWASP, "*CycloneDX*", https://cyclonedx.org/
- [10] ISO/IEC, "*ISO/IEC 5962:2021 Information technology SPDX® Specification V2.2.1*", 2021
- [11] ISO/IEC, "*ISO/IEC 5230:2020 Information technology OpenChain Specification*", 2020
- [12] ISO/IEC, "*ISO/IEC 18974:2023 Information technology OpenChain security assurance specification*", 2023

Get This Book's PDF Version and Exclusive Extras

Scan the QR code (or go to packtpub.com/unlock). Search for this book by name, confirm the edition, and then follow the steps on the page.

Note: Keep your invoice handy. Purchases made directly from Packt don't require one.

Part 4: Automotive IoT Application Life Cycle

This part details the end-to-end design, development, and deployment of automotive IoT applications, encompassing the entire lifecycle from initial concept and requirement gathering through to the final deployment and ongoing maintenance of these connected vehicle solutions. This process includes software development, cloud integration, and rigorous testing, ultimately aiming to enhance vehicle functionality, safety, and efficiency.

This part of the book comprises the following chapters:

- *Chapter 9, System Design of an Automotive IoT Application*
- *Chapter 10, Developing an Automotive IoT Application*
- *Chapter 11, Deploying and Maintaining an Automotive IoT Application*

System Design of an Automotive IoT Application

System design refers to the process of defining and creating the architecture, components, modules, and interfaces for a system to fulfill specific requirements. It encompasses a structured and methodical approach to design and develop a system that effectively meets the desired objectives. It is an iterative process where the system engineering team collaborates with – not limited to – product management, **user experience** (**UX**), software, hardware engineering, validation, and manufacturing teams to come up with a system design that is feasible to implement, viable to the business and delivers values to the end customer.

In this chapter, we will focus on how a system engineering team comes up with a system design for an Automotive IoT application such as remote diagnostics.

In this chapter, we're going to cover the following main topics:

- System design process overview
- **UX-driven design (UXDD)**
- Use case – remote diagnostics
- System components
- Gateway design considerations
- Cloud design considerations
- Remote diagnostics applications
- Regulatory compliance
- Build versus buy

System design process overview

During system design, ensuring that a new product or feature is not only innovative but also practical and sustainable is crucial for success. This is where the **Desirability**, **Feasibility**, and **Viability** (**DFV**) model comes into play. Serving as a foundational framework, the DFV model guides teams in evaluating the potential of use cases, products, or features through three critical lenses: desirability, feasibility, and viability. By assessing these, system designers can make informed decisions that balance user needs with technical and financial realities:

- **Desirability**: From a user's perspective, it asks whether the product or feature fulfills a genuine need or desire. It emphasizes understanding and meeting the expectations of the target audience to ensure the solution is embraced by users when deployed in the field.

- **Feasibility**: From an engineering perspective, is it technically possible to create the product or feature? It examines whether the current technological capabilities and resources are sufficient to turn a concept into a reality, ensuring the product/feature can be completed as envisioned.

- **Viability**: From a business perspective, it scrutinizes the economic sustainability of the product/feature. It involves evaluating whether the idea can generate sufficient financial returns or cost savings to justify the investment, ensuring long-term success and sustainability.

The DFV Venn diagram shown in *Figure 9.1* shows the intersection of all three areas is the sweet spot suggesting that the idea has strong potential for success.

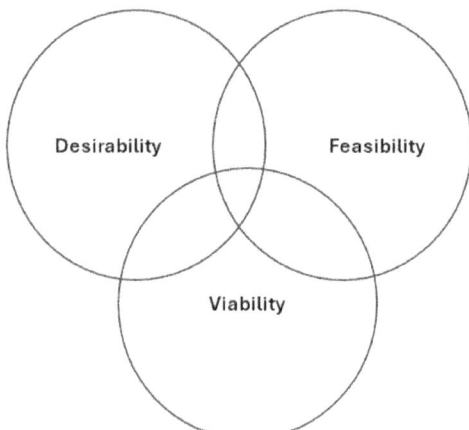

Figure 9.1 – The DFV model

Typically, when a product management team approaches the systems team with a product or a feature request, they have already established the desirability of a product/feature. What they are looking for is the feasibility and cost estimates of what it takes to implement, deploy, and operationalize so that they can determine the viability of the product/feature. *Figure 9.2* illustrates the different teams that interact with the system engineering team.

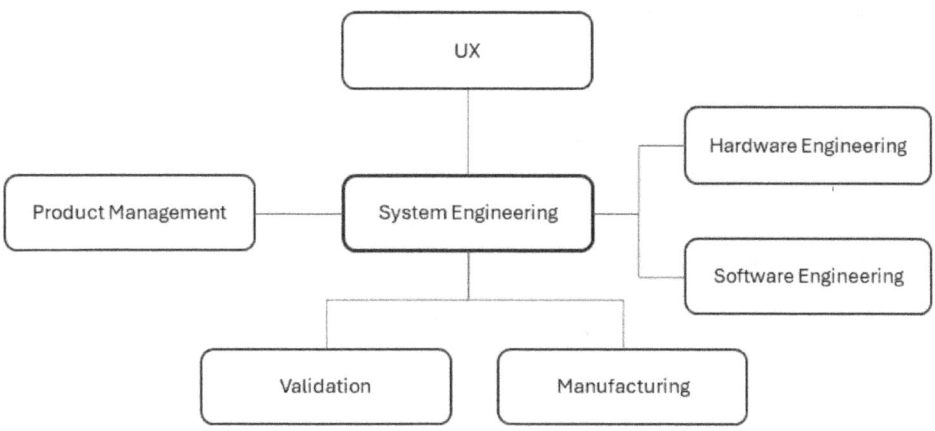

Figure 9.2 – System engineering interfaces

System engineers/architects serve as the crucial link between requirements, limitations, and available technologies, offering perspectives from both technical and business standpoints. They define system boundaries, managing diverse and often conflicting requirements while navigating complex design landscapes. Understanding, describing, and communicating relationships within and outside the system is key, considering customer environments and risk factors. They conduct life-cycle analyses, providing both short- and long-term visions while guiding the application of architecture principles to stakeholders. They gather and organize information by asking six critical questions: *who*, *what*, *when*, *where*, *why*, and *how* to bridge the gap between intricate system requirements and user-centric solutions. The UXDD approach emerges as a pivotal strategy.

UXDD

UXDD is an approach to design and development that prioritizes the UX throughout the entire process. It's not just about designing interfaces or interactions; it's about understanding the user's needs, motivations, and pain points and then using that understanding to inform every aspect of the design and development process.

Some key characteristics are the following:

- **User-centricity** places the user at the center of the design process. This means that every decision, from the initial concept to the final product launch, is made with the user in mind.
- **The holistic approach** considers the entire user journey, from the moment they first encounter the product to the ongoing interactions they have with it.
- **Data-driven** and is informed by data and research. This data can come from user interviews, surveys, usability testing, and analytics.
- **An iterative process** means that the design is constantly being tested and refined based on user feedback.
- **Collaboration** between different stakeholders, including designers, developers, product managers, and marketing professionals.

Figure 9.3 shows the user flow design, which encompass the user goal, task flow, wireflow, and user flow.

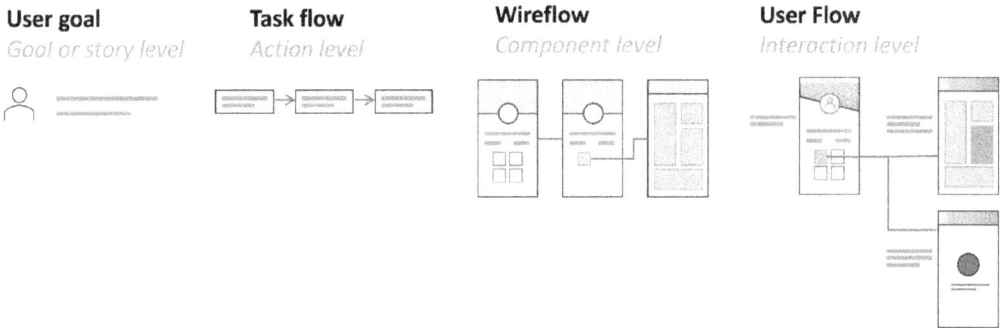

Figure 9.3 – User flow design

Some of the benefits of UXDD are the following:

- **Improved user satisfaction and engagement** by designing products that are easy to use and meet the needs of users
- **Increased adoption of products**, as those that are designed with UX in mind are more likely to be adopted by users and used regularly
- **Enhanced brand reputation** by delivering a positive UX that can help to enhance a company's brand reputation and customer loyalty

- **Reduced development costs** by identifying and addressing usability issues early in the design process
- **Improved return on investment (ROI)** by focusing on the UX; this can help to increase ROI for product development

Overall, UXDD is a powerful approach that can help in creating products and features that are both successful and user-friendly.

Use case – remote diagnostics

Let's explore how connectivity and remote diagnostics are reshaping the traditional vehicle service model. We will compare the conventional scenario with a new, more efficient approach that leverages connected vehicle technology. This comparison will highlight the benefits and efficiencies gained by both vehicle owners and service centers through innovative solutions:

- **Current scenario**: Traditionally when a vehicle malfunctions, the owner of the vehicle schedules a visit and takes it to a service center so that the service technician can diagnose the problem and repair the vehicle. In case the diagnosis needs parts replacement and if the parts are not available, the owner of the vehicle must schedule further visits to get the repair done. This current conventional scenario is illustrated in *Figure 9.4*.

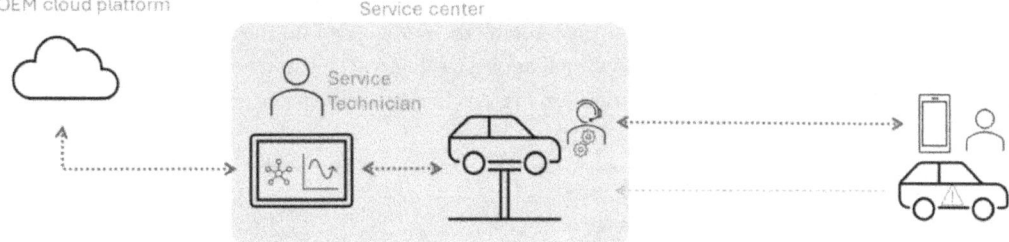

Figure 9.4 – Scenario where the user brings a vehicle to the service center for diagnostics and repair

The challenge with the current scenario is that the owner of the vehicle must visit the service center multiple times, and the service center should schedule multiple visits to resolve the same issue. Now, let us explore a more convenient and efficient approach; let us call it the new scenario.

- **New scenario**: With a connected vehicle, the owner of the vehicle for the same issue can request the service center to diagnose the problem and grant access to the service center to remotely connect to the vehicle. Once the issue is remotely diagnosed by the service technician, the parts can be ordered, and repair can be scheduled. *Figure 9.5* illustrates the vehicle being diagnosed remotely.

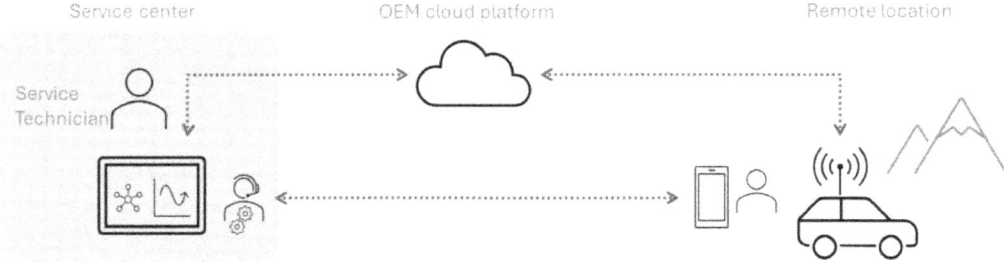

Figure 9.5 – Vehicle is diagnosed remotely

This makes it convenient for the vehicle owner by reducing the trips to the service center. The service center can be more efficient in resolving issues for more customers in the same amount of time.

Secondly, if the vehicle owner gives the OEM permission to collect diagnostics information, then the OEM can aggregate the diagnostic information across multiple vehicles/vehicle lines, analyze the data to identify issues, prioritize and root cause them efficiently based on the impact and spread of the issue across the vehicle lines. The following diagram depicts a scenario where aggregated vehicle diagnostic data in the cloud is analyzed to develop insights into the vehicle performance and warranty:

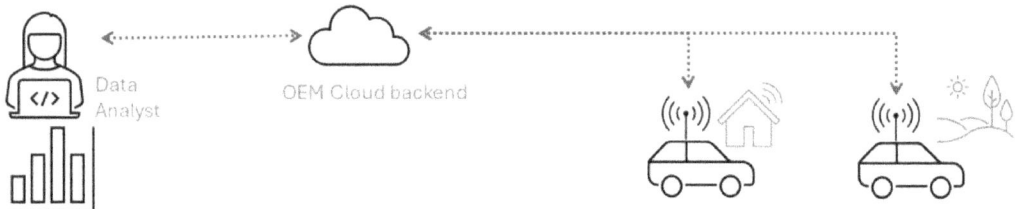

Figure 9.6 – Vehicle diagnostic information is aggregated in the OEM cloud platform for analysis

In the next section, let us explore the various system components required to realize the remote diagnostics use case.

System components

The system components of a typical automotive IoT system include a cloud platform, a vehicle gateway that connects the vehicle subsystems to the cloud platform via a Wi-Fi network or a cellular network, and the user's mobile device. *Figure 9.7* illustrates the various high-level system components of a typical automotive IoT application.

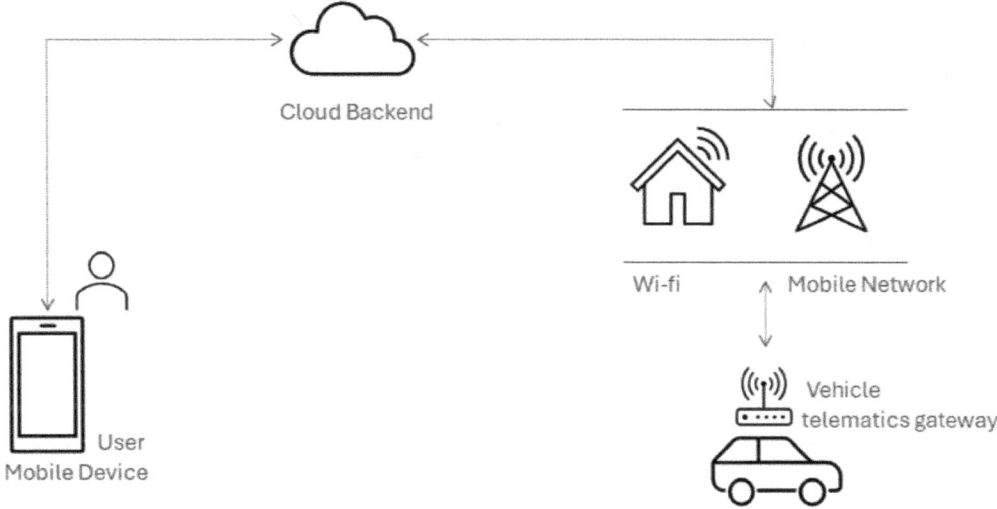

Figure 9.7 – High-level system components of a simple automotive IoT application

Now, let us explore each of the high-level system components and understand their unique roles and contributions to the overall functionality.

Vehicle telematics gateway

A **vehicle telematics gateway** is a computing platform that allows data exchanges between the cloud platform and multiple vehicle subsystems. It's the central hub for collecting, processing, and transmitting vehicle data to the cloud platform.

Some key functions of a telematics gateway are as follows:

- **Advanced data collection** from various sensors and systems within the vehicle, such as engine performance, fuel consumption, battery charge level, battery health, location, speed, driver behavior, and diagnostic information. This provides valuable insights into vehicle health and operation, enabling proactive maintenance and optimization.

- **Two-way communication**, allowing for remote commands to be sent to the vehicle. This enables features such as remote diagnostics, engine immobilization, and geofencing, allowing for better control and security.

- **Remote data transmission** of the collected data to a cloud platform using cellular, satellite, or Wi-Fi connectivity allows fleet managers and other stakeholders to access and analyze vehicle data in real time from any location, regardless of the vehicle's physical location.

- **Data processing and analysis** of the collected data before transmission could involve filtering irrelevant data, calculating key metrics, and generating reports. This provides preprocessed data that is more readily usable and insightful for users.

So, the vehicle telematics gateway plays a pivotal role in seamlessly bridging the gap between in-vehicle systems and the vehicle cloud platform. It acts as the nerve center, facilitating the flow of information to and from the vehicle.

Vehicle cloud platform

A **vehicle cloud platform**, also known as a connected vehicle platform, serves as a central hub, collecting data from various sources within the vehicle, processing it, and providing valuable insights and services.

Some key functions of a vehicle cloud platform are as follows:

- **Data ingestion** from various sources within the vehicle, including sensors, **electronic control units** (**ECUs**), and telematics gateways. This data can include engine performance, fuel consumption, location, speed, driver behavior, diagnostic information, and more.

- **Data processing and analysis** of the collected data using AI and **machine learning** (**ML**) algorithms. This helps uncover valuable insights into vehicle health, performance, driver behavior, and potential issues.

- **Data storage and management** of the collected data efficiently and securely, ensuring data accessibility and integrity. This often involves cloud storage solutions for scalability and accessibility.

- **Service orchestration** for developing and deploying various services related to the vehicle and its data. This includes services such as the following:
  - Remote diagnostics and maintenance
  - Predictive maintenance
  - Usage-based insurance
  - Fleet management
  - Driver behavior monitoring
  - In-vehicle entertainment and services

The vehicle cloud platform serves as the backbone for data management and service delivery, so we now shift our focus to the endpoint of this system; that is, the end-user mobile device.

End-user mobile device

The **end-user mobile device** plays a crucial role in providing a seamless and personalized experience.

Some key functions of an end-user mobile device are as follows:

- The primary **interface and control** for interacting with the connected vehicle platform. Users can access various features and services, such as remote diagnostics, vehicle tracking, route planning, and in-vehicle controls through the mobile app, providing users with convenient access to vehicle information and control even when they are away from the vehicle.
- **Security and notification** features, such as remote lock and unlock, geofencing alerts, and vehicle location tracking, which enhance the security of the vehicle and provide users with peace of mind.
- **Personalization and preferences** allow users to personalize their connected vehicle experience. They can set preferences for various features, such as climate control, audio settings, and driver profiles, which ensures that the vehicle is tailored to the individual's needs and preferences.

The role of the end-user mobile device in a connected vehicle platform is continually evolving as technology advances. However, its core function remains the same: to provide a convenient, personalized, and secure interface for interacting with the connected vehicle and its features. This ultimately contributes to an improved driving experience and a safer and more efficient transportation ecosystem.

Now that we understand the functionality of the high-level system components, let us dive deeper into each of the components and understand the various technologies and design considerations required to realize them.

Gateway design considerations

When designing a vehicle telematics gateway, it's crucial to integrate a variety of technologies to ensure comprehensive functionality and connectivity. Incorporating **Global Navigation Satellite System** (**GNSS**) technology enables precise location tracking and navigation, enhancing services such as fleet management and emergency response. The **Controller Area Network** (**CAN**) and various sensors facilitate real-time monitoring and reporting of vehicle performance metrics, including engine health, fuel consumption, and driver behavior. To ensure seamless connectivity, both wireless and wired communication protocols are employed, allowing the gateway to transmit and receive data over cellular networks, Wi-Fi, and dedicated physical connections. The inclusion of an **embedded Subscriber Identity Module** (**eSIM**) further revolutionizes this connectivity, offering flexibility in network provider selection and enabling remote subscription management. *Figure 9.8* shows symbols of different technologies used in the design of vehice telematics gateway.

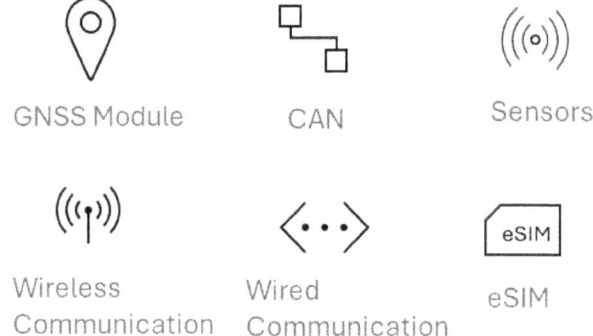

Figure 9.8 – Technologies used in the design of a vehicle telematics gateway

The gateway design is largely influenced by the available vehicle interfaces and the UX that needs to be delivered to the end user. The following sub-sections cover some of the technologies a telematics gateway uses.

GNSS receivers

GNSS is a generic term for satellite navigation systems that provide global coverage for **positioning, navigation, and timing** (**PNT**). GNSS includes systems such as **Global Positioning System** (**GPS**), **Globalnaya Navigatsionnaya Sputnikovaya Sistema** (**GLONASS**), Galileo, and BeiDou.

GPS is the most well-known and widely used GNSS system. It is operated by the US government and has been operational since 1995. GPS consists of a constellation of 24 satellites that orbit the Earth in **medium Earth orbit** (**MEO**).

GLONASS is the second-largest GNSS system operated by Russia. It began operation in 1982 and currently has a constellation of 24 satellites.

Galileo is a European GNSS system that began operation in 2016. It is composed of a constellation of 30 satellites that orbit the Earth in MEO. All GPS, GLONASS, and Galileo receivers can use signals from satellites to determine their position, velocity, and time.

BeiDou is a Chinese GNSS system that began operation in 2000. It is composed of a constellation of 35 satellites that orbit the Earth in MEO. BeiDou receivers can use signals from up to 24 satellites to determine their position, velocity, and time.

Table 9.1 shows the differences between various GNSS systems.

Feature	GPS	GLONASS	Galileo	BeiDou
Satellites	24	24	30	35
Orbit	MEO	MEO	MEO	MEO
Accuracy	2.5-8 meters	4 meters	1.5 meters	3 meters
Availability	Global	Global	Global	Global
Cost	High	Medium	High	High

Table 9.1 – Differences between various GNSS systems

Other regional GNSS systems include the following:

- **Quasi-Zenith Satellite System (QZSS)**: Japan (7 satellites)
- **Navigation with Indian Constellation (NavIC)**: India (7 satellites)

Just as GNSS systems have revolutionized how we locate and navigate vehicles globally, wireless communication technologies have equally transformed the exchange of information, which we will explore in the next section.

Wireless communication

Wireless communication technologies have become indispensable, not only enhancing UX but also improving vehicle functionality and safety. These technologies can be broadly categorized into short-range and long-range communications, each serving distinct purposes within automotive applications.

Short range

Short-range wireless technologies are Bluetooth, **Bluetooth Low Energy** (BLE), Wi-Fi, **ultra-wide band** (UWB), and **near-field communication** (NFC).

Bluetooth is widely used in vehicles for hands-free calling and streaming audio from smartphones to car audio systems. Its limited range of up to 33 feet is suitable for in-cabin communication, ensuring drivers can remain connected without distractions.

BLE is a low-power version of Bluetooth that is designed for applications that require long battery life and data exchange. It is used in keyless entry systems and for connecting the vehicle to the driver's mobile devices to exchange information such as vehicle status or for personalization settings. BLE operates at a lower frequency than Bluetooth Classic, which means it can penetrate walls and other obstacles more easily.

Wi-Fi in vehicles is used to provide internet access to passengers, enabling the download of entertainment content or software updates for the vehicle itself. Its longer range and higher speed compared to Bluetooth make it ideal for streaming and data-intensive applications.

UWB is particularly valued in the automotive sector for its precision in location and proximity detection. This technology is used for advanced keyless entry systems, where it enables passive entry and start functions by accurately determining the distance between the key fob and the vehicle. It offers high-speed data transmission and exceptional accuracy, enhancing security and convenience for users. NFC allows devices to communicate with each other by touching them together or bringing them within a few centimeters of each other. NFC is commonly used for contactless payment and for sharing data between devices. NFC is also used in some smartphones to unlock doors and start cars.

Table 9.2 shows a comparison of various short-range wireless technologies.

Feature	Bluetooth	BLE	Wi-Fi	UWB	NFC
Range	33 feet	100 feet	300 feet	10 feet	Upto 4 inches
Power consumption	High	Low	Medium	Medium	Very low
Speed	1 Mbps	10 Mbps	100 Mbps	400 Mbps	424 Kbps
Accuracy	Moderate	High	Low	High	Very high

Table 9.2 – Short-range wireless technologies comparison

Long range

Long-range wireless technologies include 4G **Long Term Evolution** (**LTE**), 5G, **Long Term Evolution for Machines** (**LTE-M**), and **Narrowband IoT** (**NB-IoT**).

4G LTE is the fourth generation of cellular technology and is the successor to 3G. It is the most widely deployed cellular technology in the world and is used to connect smartphones, tablets, and other devices to the internet. LTE is known for its high speeds and low latency, which is ideal for streaming video, downloading large files, and playing online games.

5G is the fifth generation of cellular technology and is the successor to 4G LTE. It is the next generation of mobile broadband technology that is designed to provide significantly faster speeds, lower latency, and higher capacity than 4G LTE. 5G is expected to be used for a wide range of applications, including **augmented reality** (**AR**), **virtual reality** (**VR**), and self-driving cars.

LTE-M is a **low-power, wide-area** (**LPWA**) cellular technology that is designed for **machine-to-machine** (**M2M**) communications. It is a more power-efficient and cost-effective alternative to 4G LTE for applications that require long battery life, such as **industrial IoT** (**IIoT**) devices, smart meters, and wearables.

NB-IoT is another LPWA cellular technology that is designed for **ultra-low-power, wide-area** (**ULPWA**) communications. It is the most power-efficient cellular technology available and is ideal for applications that require very long battery life, such as asset tracking and environmental monitoring. *Table 9.3* shows a comparison of various long-range wireless technologies.

Feature	4G LTE	5G	LTE-M	NB-IoT
Speed	Up to 100 Mbps	Up to 20 Gbps	Up to 1 Mbps	Up to 200 Kbps
Latency	Up to 50 ms	Up to 1 ms	Up to 20 ms	Up to 100 ms
Range	Up to 30 miles	Up to 6 miles	Up to 10 miles	Up to 15 miles
Power consumption	High	Medium	Low	Very low

Table 9.3 – Long-range wireless technologies comparison

After exploring various short-range and long-range wireless technologies, let us now transition to the realm of wired communication.

Wired communication

Wired communication technologies are integral to vehicle operation, diagnostics, and connectivity, ensuring the efficient and reliable exchange of data within a vehicle's complex network. We will explore the following wired communications: **Recommended Standard 232** (**RS232**), RS485, and Automotive Ethernet.

RS232 was introduced in 1960. RS232 is a standard for serial communication transmission of data. It connects computers and devices for data exchange, often used for short-distance, low-speed requirements between a computer and peripheral devices. RS232 is still widely used due to its simplicity and wide availability; it finds its application in automotive diagnostics and testing equipment.

RS485 is an upgrade to RS232. RS485 supports longer cable lengths and higher speeds, and it allows for multi-point systems where multiple devices can be connected to the same communication line. Its robustness against interference makes it suitable for automotive applications requiring reliable communication in electrically noisy environments. It is often used in larger vehicles, such as buses or commercial trucks, for networking sensors and control units across the vehicle due to its superior noise immunity and multi-point capability.

Automotive Ethernet is designed to meet the demands of modern vehicles' networking needs, supporting high data rates for applications such as infotainment systems, **advanced driver assistance systems** (**ADAS**), and **vehicle-to-everything** (**V2X**) communications. It offers a high-bandwidth, scalable network that can handle the vast data throughput required by current and future automotive applications. The difference between Ethernet and Automotive Ethernet is in the physical layer. The physical layer has been optimized around automotive use cases; the physical transceivers and cables are different. *Table 9.4* shows the differences between various wired communications.

Feature	RS232	RS485	Automotive Ethernet
Communication mode	Point-to-point (P2P)	Multi-point	P2P, multi-point
Max. number of devices	2	32	Many
Max distance	Up to 50 feet	Up to 4,000 feet	Up to 100 meters
Data rate	Up to 20 kbps	Up to 10 Mbps	Up to 10 Gbps
Noise immunity	Low	High	High
Cost	Low	Low	High
Applications	Legacy devices, short-range communication	Industrial automation, sensor networks	In-vehicle communication, ADAS

Table 9.4 – Comparison of various vehicle wired networks

As we advance from wired communication protocols such as RS232 and RS485 to Automotive Ethernet, it's essential to also consider CAN, which is specifically tailored to the automotive industry's demands for a resilient and efficient vehicle bus system.

CAN

CAN is a robust vehicle bus standard designed to allow microcontrollers and devices to communicate with each other. It is a message-based protocol, designed for automotive applications. It's known for its high reliability and simplicity in networking sensors, actuaries, and control units.

CAN Flexible Data-Rate (**CAN FD**) is an extension of the CAN bus standard that allows for higher data rates and longer message lengths. It was first standardized in 2011 by *ISO 11898-1:2015*, and it is widely used in automotive applications. CAN FD can support data rates of up to 8 Mbps, which is significantly faster than the standard CAN bus. It can also support message lengths of up to 64 bytes, which is 8 times the length of standard CAN messages.

High-Speed CAN (HS-CAN) is a variant of the CAN bus standard that specifically targets high-speed applications. It was first standardized in 2002 by *ISO 11898-2* and it is a subset of the CAN FD protocol. HS-CAN can support data rates of up to 1 Mbps, which is twice the speed of standard CAN. It is typically used in applications that require high-speed data transfer, such as automotive powertrain control and industrial automation.

Low-Speed CAN (LS-CAN) is another variant of the CAN bus standard that is specifically designed for low-speed applications. It was first standardized in 1993 by *ISO 11898-1*, and it is the most common variant of the CAN bus. LS-CAN can support data rates of up to 125 Kbps, and it is typically used in applications that do not require high-speed data transfer, such as automotive body control and electronic fuel injection. *Table 9.5* provides a comparison of the various CAN networks.

Feature	CAN	CAN FD	HS-CAN	LS-CAN
Data rate	Up to 250 Kbps	Up to 8 Mbps	Up to 1 Mbps	Up to 125 Kbps
Message length	Up to 8 bytes	Up to 64 bytes	Up to 8 bytes	Up to 8 bytes
Network latency	10 µs-20 µs	10 µs	2 µs-10 µs	10 µs-20 µs

Table 9.5 – Comparison of different CAN networks

The various CAN protocols ensure reliable and efficient data exchange between electronic components; **microelectromechanical systems (MEMS)** [1] sensors provide the critical data required for these communications to be meaningful and actionable.

Sensors

MEMS sensors are tiny devices that can measure a variety of physical phenomena, such as acceleration, gyration, and magnetism. These microscale sensors combine mechanical components with electronic circuitry, allowing them to measure physical phenomena with great precision. They are used in various automotive applications such as airbag deployment systems, electronic stability control to detect skidding and loss of traction, navigation systems to provide accurate heading information, automatic adjustments in suspension systems, and enhancing ride comfort by adapting to road conditions.

SIM/eSIM

A **SIM** is a small, removable chip that contains a unique identification number, account information, and carrier settings. An **eSIM** is a digital SIM card that is built into a device. It's a microchip that replaces the physical SIM card. eSIMs are programmable remotely and can download carrier data without the need for a physical chip.

Gateway hardware

Gateway hardware is an embedded system that aggregates data from various sensors and systems within the vehicle, processes the data, and then communicates it to external networks for a variety of applications.

The key functions of telematics gateway hardware are data aggregation, data processing, remote access control, power management, and security. *Figure 9.9* shows the typical architecture of telematics gateway hardware, highlighting its key components and their interconnections.

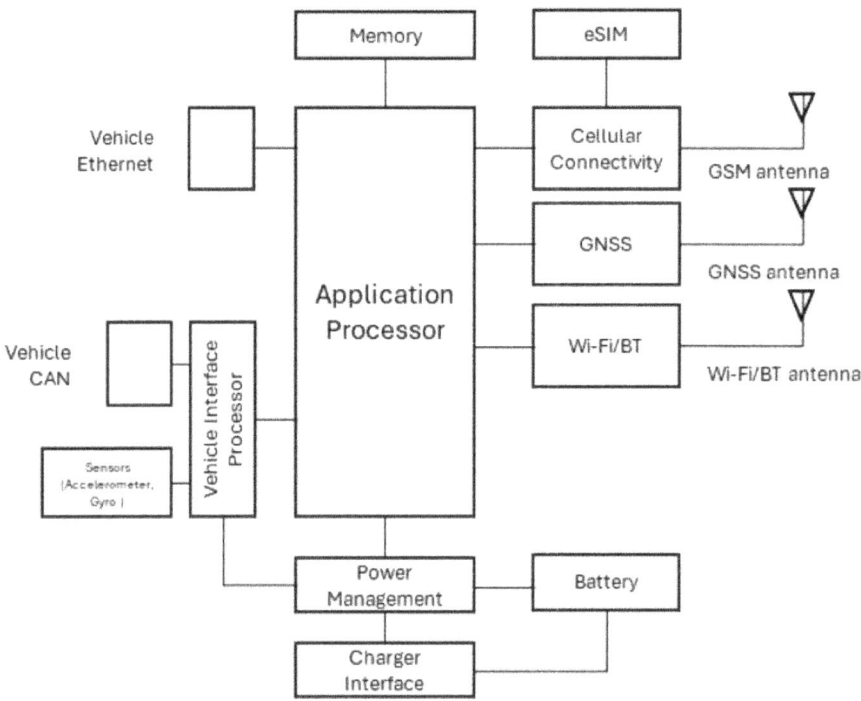

Figure 9.9 – Block diagram of a typical telematics gateway hardware

The preceding block diagram is that of a generic vehicle telematics gateway. Depending on the use case and the target vehicle line architecture, the system architect needs to determine the required sensors, wireless technologies, vehicle interfaces, and form factors based on performance, space, power consumption, and cost.

In the remote diagnostics use case, for the service technician to perform remote diagnostics, we need at least 10 Mbps dedicated cellular connectivity with the cloud. This eliminates long-range technologies such as LTE-M and NB-IoT and narrows down to 4G-LTE or 5G. If the OEM is focused only on aggregating diagnostic information for analysis, then LTE-M might suffice. If the target vehicle line is a four-wheeler, then the space and environmental constraints are less stringent compared to a two-wheeler vehicle line.

As we delve into the technologies and protocols that underpin a vehicle's telematics gateway hardware, it's equally important to consider how this gateway interacts with larger network infrastructures, specifically cloud-based platforms. Cloud design considerations are paramount when integrating telematics gateways with cloud services to enable advanced features such as real-time data analytics, remote vehicle diagnostics, **over-the-air** (**OTA**) updates, and comprehensive fleet management solutions. In the next section, we will explore the critical factors that influence cloud architecture and service design.

Cloud design considerations

The cloud solution requires several core components to be integrated to support remote diagnostics effectively. The **vehicle management** module contains essential metadata related to a vehicle, including details such as vehicle model, network structure, and ECUs. **Device management** functionality supports the setup and ongoing management of vehicle gateways that connect to the cloud platform, ensuring their provisioning and life cycle are maintained. A **connectivity management** solution is crucial for establishing and sustaining secure and dependable communication between devices and networks. **Vehicle telemetry** technology involves the remote collection and analysis of data from vehicles, enabling remote monitoring and analysis. The **vehicle OTA updates** process involves wirelessly delivering software updates and fixes to vehicles, ensuring they are updated without requiring physical intervention. **User management** functionality handles the authentication and authorization of users accessing different subsystems within the cloud environment. **Subscription management** encompasses various customer-related activities, from onboarding to billing, cancellations, and ongoing engagement. It enables the management of customer access to additional services and applications built upon these foundational modules. *Figure 9.10* shows the various functional blocks required in the cloud.

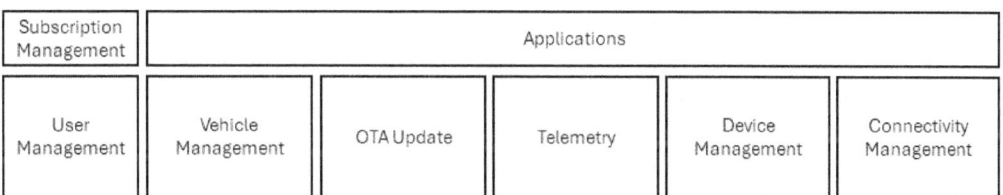

Figure 9.10 – The basic functional blocks required in the cloud

From a broad overview of the cloud solution's core components, we will now transition to a focused examination of device management and connectivity management in the next two sections.

Device management

Device management serves as a vital component that ensures the secure and efficient functioning of vehicle gateways connected to the cloud. This is accomplished through robust provisioning, configuration, life-cycle management, and security protocols that are designed to manage a vast array of devices seamlessly. Some key functions of a device management solution are as follows:

- **Device provisioning** is the process of onboarding and provisioning gateway devices securely onto the IoT platform.

- **Configuration and management** is a centralized management of device configurations, settings, and firmware updates across a large fleet of IoT devices. Administrators can remotely configure device properties, update firmware, and manage device policies.

- **Monitoring and diagnostics**: Providing real-time monitoring, telemetry data collection, and device health monitoring. This allows for proactive detection of issues, performance analysis, and troubleshooting of devices to ensure optimal operation.

- **Security management**: Implementing robust security measures for IoT devices, including device authentication, access control, encryption, and threat detection. It helps protect devices from potential cyber threats and ensures data integrity and confidentiality.

- **OTA updates**: Facilitating remote and secure firmware updates for IoT devices without physical intervention. This feature allows for the seamless deployment of software patches, bug fixes, and feature enhancements across distributed devices.

- **Life-cycle management**: Managing the entire life cycle of IoT devices, from initial provisioning to decommissioning. It includes monitoring device health, tracking device status, and retiring or replacing devices as needed.

Connectivity management

A connectivity management solution/platform is required to maintain reliable and secure communication between devices and networks, where a vast number of telematics gateway devices need to be connected and communicate seamlessly. Some key functions of a connectivity management platform are as follows:

- **Monitoring and provisioning**: Managing SIM cards and subscriptions for different devices and network operators. Monitoring device connectivity status, including uptime, signal strength, and data usage. Provisioning new devices by adding them to the network, configuring settings, and activating services.

- **Network selection and switching**: Choosing the best network operator for each device based on coverage, speed, and cost. Automatically switching between network operators to ensure seamless connectivity. Managing roaming charges and ensuring compliance with local regulations.

- **Cost optimization**: Choosing the most appropriate network operator and data plan based on device usage and location. Negotiating bulk discounts and managing service contracts. Optimizing data usage through techniques such as data compression and throttling.
- **Analytics and reporting**: Gathering data on device connectivity performance and usage patterns. Analyzing data to identify trends and potential issues. Generating reports to provide insights into the overall health of the network.

A **Mobile Network Operator** (**MNO**) owns and operates its own mobile network infrastructure (towers, antennas, and so on), manages its own spectrum licenses, and provides mobile services directly to subscribers under its own brand. Some examples of MNOs in the North America region are AT&T, Verizon, and T-Mobile.

A **Mobile Virtual Network Operator** (**MVNO**) does not own or operate its own network infrastructure. It leases network capacity from an MNO and resells it to subscribers under its own brand. It offers competitive pricing and customized services to differentiate itself from an MNO. Some examples of MVNOs in the North America region are Cricket Wireless, Mint Mobile, and Google Fi.

Table 9.6 compares MNOs' and MVNOs' key features and differences.

Feature	MNO	MVNO
Network ownership	Own	Lease
Spectrum ownership	Own	Lease
Service provider (SP)	Direct	Reseller
Brand	Own	Own
Services offered	Network, voice, data	Voice, data
Focus	Network infrastructure	Reselling mobile services
Examples	AT&T, Verizon, T-Mobile	Cricket Wireless, Mint Mobile, Google Fi

Table 9.6 – Comparing MNOs and MVNOs

Most MNOs operate in a few countries in the best-case scenario. If the solution needs to be deployed in a specific region or a few countries only and there is no roadmap to expand to other regions, depending on the scale of deployment, it will make economic sense to build integrations with specific MNOs, but if the solution needs to be deployed globally and the vehicle population is spread across multiple regions, then to reduce the complexity of the integration, it's advisable to work with an MVNO.

Connectivity management helps organizations ensure that their gateway devices are connected securely, reliably, and cost-effectively, which is essential for businesses that rely on real-time data and remote operations.

Now that we have established the building blocks to realize remote access to the vehicle, we will now explore the components required for a diagnostics application.

Remote diagnostics applications

There are multiple diagnostics application use cases – namely proximity, remote, and in-vehicle. In the **proximity use case**, the service technician uses their diagnostic tool and connects via a wireless or wired interface to the vehicle, in the **remote use case**, the technician connects via the cloud platform to the vehicle remotely, and in the **in-vehicle use case**, the vehicle dash or central-console interface is used for diagnostics. *Figure 9.11* illustrates the proximity use case and remote diagnostics use case.

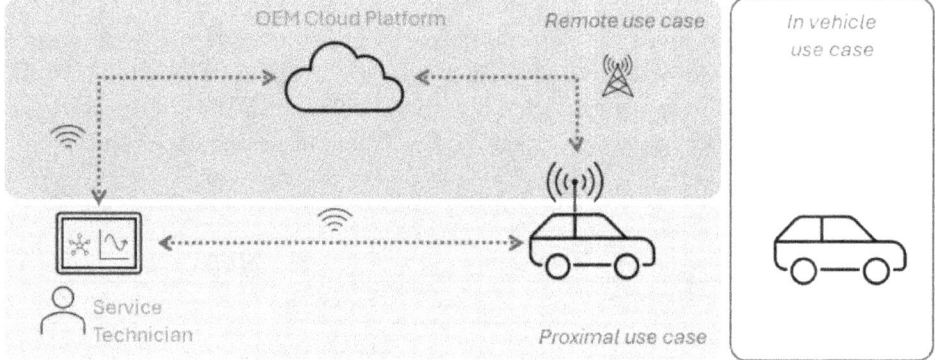

Figure 9.11 – Various diagnostic use cases

Vehicle diagnostics [2] can be classified into two types:

- Classic vehicle diagnostics based on **Unified Diagnostic Services (UDS)**
- Service-oriented vehicle diagnostics

While classic vehicle diagnostics rely on well-established protocols such as UDS, service-oriented vehicle diagnostics represent a more flexible and modern approach.

Classic vehicle ECU diagnostics

Classic vehicle ECU diagnostics are based on UDS and require a matching static offline description for communication, typically in **Open Dialogue data eXchange (ODX)** format.

UDS, as explained in *Chapter 3, Vehicle Architecture and Framework*, Shifting from Classic to Adaptive AUTOSAR Architectures is a protocol that is used to diagnose errors and reprogram ECUs.

Open Test Sequence eXchange (OTX) is a standardized format for describing automated diagnostic sequences, is standardized in *ISO 13209*, and supplements the ODX format. It allows users to write diagnostic sequences in XML.

ODX is an XML-based standard for describing ECU data, standardized in *ISO 22901-1*. ODX outlines methods for communication with an ECU, while OTX provides sequences of instructions that can be carried out based on the corresponding ODX information. Figure 9.12 shows the various components of the OTX and ODX based diagnostic stack.

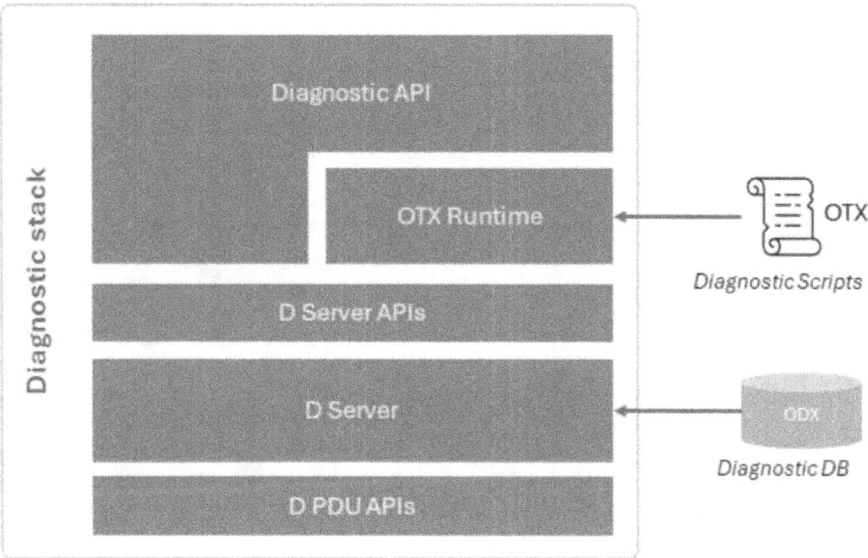

Figure 9.12 – Classic OTX- and ODX-based diagnostic stack

As you can see in *Figure 9.13*, there are a couple of ways the diagnostic stack can be deployed. Option A is on the edge, which in this case is on the telematics gateway, provided there are sufficient memory and computational resources available. If the telematics gateway has limited resources, then option B is to deploy part of the diagnostics stack on the cloud, and the **Diagnostic Protocol Data Unit (D-PDU)** is deployed on the gateway, which communicates over a reliable channel (Wi-Fi, 4G/5G).

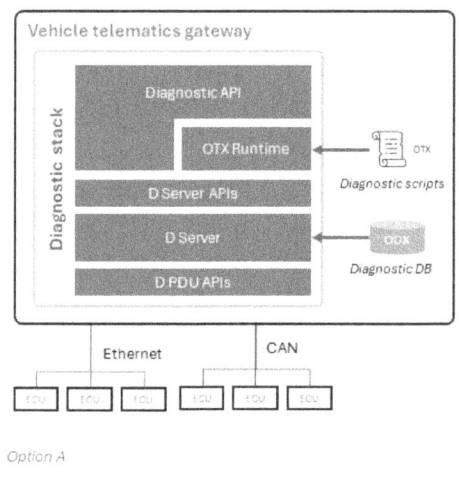
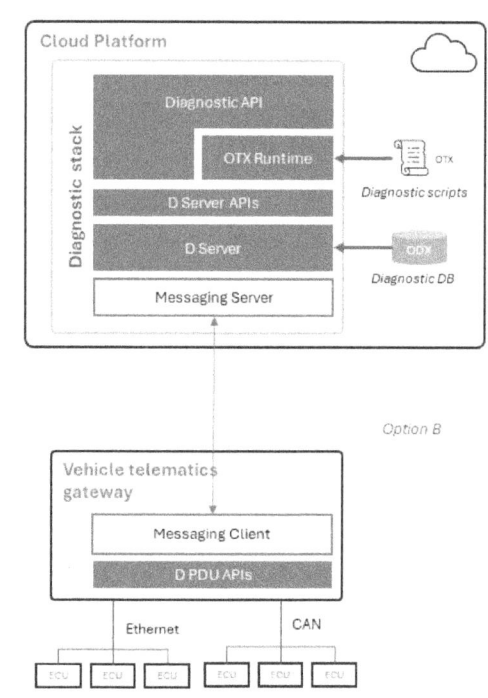

Figure 9.13 – Diagnostic stack deployment options

Option C [3] can be considered where there is a custom implementation on the vehicle telematics gateway that is optimal for specific diagnostic scenarios such as aggregating ECU fault codes and shipping that information to the cloud for further analysis.

Service-oriented vehicle diagnostics

The traditional vehicle ECU diagnostic methods, which rely on UDS, are not sufficient to meet the demands of the new vehicle architectures based on **high-performance computing** (**HPC**), multiple operating systems (OSs), different software applications, and their dependencies. To analyze software issues, we need different types of data such as logs, traces, process information, stack traces, and so on. To address the changing landscape, the automotive industry is working toward a new standard called **Service-Oriented Vehicle Diagnostics** (**SOVD**). *Figure 9.14* shows the SOVD reference architecture:

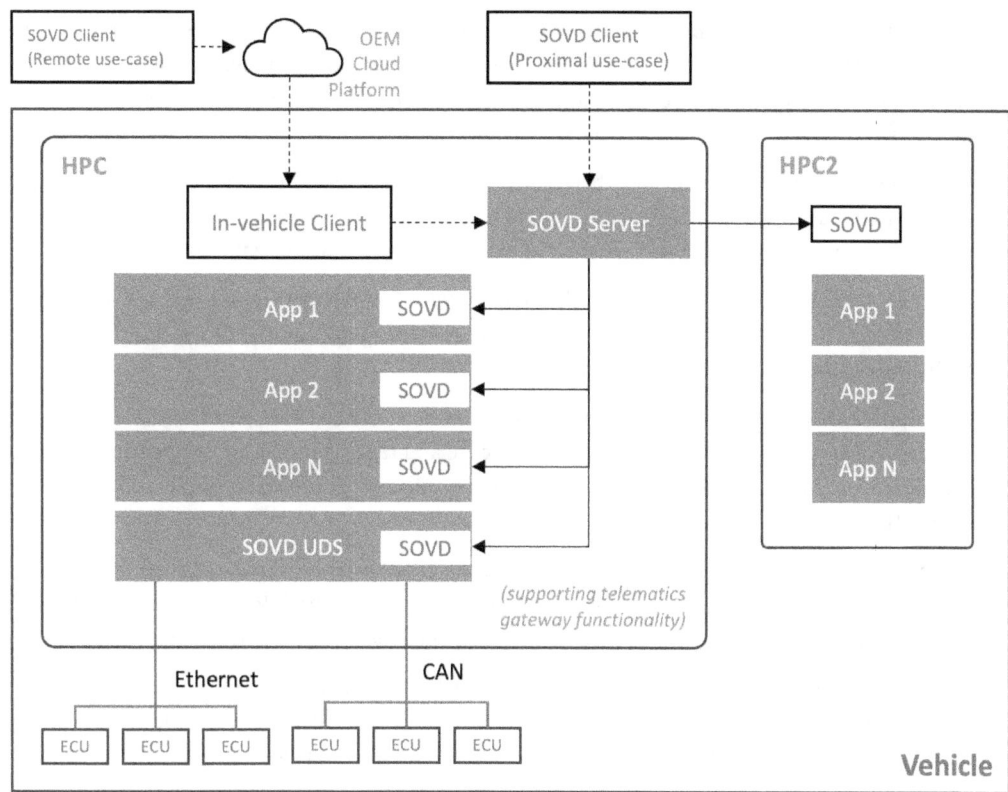

Figure 9.14 – SOVD reference architecture

SOVD standardization [4][5][6] aims to create a modern, simple diagnostic interface that equally enables access to classic ECUs and emerging software-based systems.

Regulatory compliance

Automotive IoT, connected vehicles, and smart vehicle regulations are evolving and primarily focus on ensuring safety, security, data privacy, and standardization in connected vehicles. These regulations vary significantly across different regions and countries.

Here are some common areas covered by these regulations:

- **Vehicle safety standards** for connected vehicles are established to ensure that IoT technology integration doesn't compromise vehicle safety. This includes regulations related to functional safety, crashworthiness, cybersecurity, and secure communication protocols.

- **Data privacy and protection regulations** govern the collection, processing, and sharing of data generated by connected vehicles. Manufacturers and **Service Providers** (**SPs**) may need to adhere to strict guidelines for data handling, user consent, and data anonymization to protect user privacy.

- **Cybersecurity regulations** mandate cybersecurity measures to safeguard connected vehicles against cyber threats and hacking attempts. Manufacturers are required to implement security features, regular software updates, and secure communication protocols to mitigate cybersecurity risks.

- **Telematics and remote access** regulate telematics systems and remote access to vehicle data. This involves securing remote access to vehicle systems, ensuring data authenticity, and protecting against unauthorized access or control of vehicles.

- **OTA updates and software management regulations** cover OTA software updates for connected vehicles, ensuring that manufacturers provide secure and reliable methods for updating vehicle software. Compliance might include guidelines for authentication, encryption, and integrity verification of software updates.

- **Standardization and interoperability regulations** encourage or require industry-standard protocols to promote interoperability among different automotive IoT systems. This helps ensure compatibility and seamless communication between vehicles and infrastructure.

- **Liability and responsibility** regulations address liability and responsibility issues related to accidents or malfunctions caused by IoT technology in vehicles. This might include determining the roles and responsibilities of manufacturers, SPs, and users in case of incidents involving connected vehicles.

- **Data localization and storage regulations** govern where and how data collected by IoT gateways should be stored and processed. Data localization requirements may dictate that certain data must remain within specific geographic boundaries.

Some common regulations are the following:

- The **General Data Protection Regulation** (**GDPR**) applies to the processing of personal data of individuals in the **European Union** (**EU**)

- The **California Consumer Privacy Act** (**CCPA**) gives California residents the right to access, delete, and opt out of the sale of their personal data

- The **Radio Equipment Directive** (**RED**) sets requirements for the radio spectrum used by wireless devices in the EU

- The **Federal Communications Commission** (**FCC**) regulates the use of the radio spectrum in the US

- The **International Organization for Standardization** (**ISO**) develops standards for a wide range of products and services

The regulatory landscape is complex and constantly evolving. Different regulations apply in different countries; some regulations are vague and lack specific requirements, and implementing measures to comply with regulations can be expensive. System engineers should stay up to date on the latest regulations, consult with regulatory experts, and conduct risk assessments. They should implement appropriate security measures and plan to test for compliance. Some of the regulations not only have an impact on the design and development aspects but also add operational overhead.

Now that we developed an understanding of the various system components that are required to develop an automotive IoT application such as remote diagnostics, we approach a critical juncture in the design and development process: whether we should build these various components or buy them.

Build versus buy

A **build-versus-buy** decision is a strategic choice when deciding whether to produce a product or service internally or purchase it from an external supplier. It's a **cost-benefit analysis** (**CBA**) to determine the most effective way to acquire a needed component or service.

Table 9.7 details the pros and cons of the build-versus-buy approach.

	Build	**Buy**
Pros	• Greater control over quality and production processes • Potential for cost savings if production is efficient • Protection of **intellectual property** (**IP**) • Increased flexibility to adapt to changing needs	• Lower upfront costs • Access to specialized expertise and economies of scale • Frees up internal resources for other activities • Reduces risks associated with production and inventory management
Cons	• Higher upfront investment in equipment and personnel • Requires ongoing management and expertise • May be less efficient than specialized suppliers	• Less control over quality and production processes • Dependence on the supplier's performance and pricing • Potential loss of IP

Table 9.7 – Pros and cons of build versus buy

The overall decision will depend on a variety of factors, including the following:

- The complexity of the product or service
- The **time to market** (**TTM**) of the product or service
- The cost of production versus purchase
- The company's existing capabilities and resources
- The importance of quality and control
- The supplier's track record and reputation

Here are a few examples to help illustrate the build-versus-buy decision process. In the case of a telematics gateway, one can start with **commercially available off-the-shelf** (**COTS**) generic hardware, which reduces the risks and helps with the TTM. As the volume of vehicles increases, COTS hardware might be cost-prohibitive, so investing in custom hardware might be an option. In the case of a cloud platform, the decision might not be as simple, as starting with a COTS-based platform and then migrating to a proprietary designed platform or to an application-ready platform might make the cost of migration prohibitively expensive, so an upfront strategic decision needs to be made.

Summary

System design is a complex and challenging task, but it is essential for creating systems that are reliable, scalable, and meet the needs of the users. In this chapter, we introduced the system design process of an automotive IoT application and explained the UXDD approach. We then dived into the remote diagnostics use case and understood the current and new UX that product planning teams want to deliver, then we identified the high-level system components required to realize the remote diagnostics use case. We then did a further deep dive into vehicle telematics gateways and cloud platforms. We explored various vehicle telematics gateway technologies and the design trade-offs and decomposed the cloud platform to understand the foundational modules/subsystems required to realize remote application use cases. Once we had established the building blocks, we explained the classic vehicle diagnostics approach and upcoming SOVD. Finally, we looked at other key aspects such as regulatory compliance and the process of a build-versus-buy decision.

In the next chapter, we will deep dive into the development of an automotive IoT application and understand the design, development, and testing process.

References

- [1] https://en.wikipedia.org/wiki/MEMS
- [2] https://www.vector.com/in/en/products/solutions/diagnosticstandards/
- [3] https://www.kpit.com/insights/future-of-diagnostics-here-and-now/
- [4] https://automotive.softing.com/fileadmin/sof-files/pdf/ae/interviews/2023_Softing_Automotive_Standardization_in_Vehicle_Diagnostics_Interview_EN_WEB.pdf
- [5] https://cdn.vector.com/cms/content/application-areas/diagnostics/2022-05-31_SOVD_1.0_The_standard_explained.pdf
- IoT Security Foundation: https://iotsecurityfoundation.org/
- **Global System for Mobile Communications Association (GSMA)**: https://www.gsma.com/
- The **European Telecommunications Standard Institute (ETSI)**: https://www.etsi.org/
- https://en.wikipedia.org/wiki/MEMS

Get This Book's PDF Version and Exclusive Extras

Scan the QR code (or go to packtpub.com/unlock). Search for this book by name, confirm the edition, and then follow the steps on the page.

Note: Keep your invoice handy. Purchases made directly from Packt don't require one.

10
Developing an Automotive IoT Application

In the previous chapter, we explored the system design of an automotive **Internet of Things** (**IoT**) application; we will now go through the software design and development process of the application. As part of the software design, we will identify various high-level software components required for both the cloud backend and vehicle telematics gateway.

In this chapter, we will focus on how a software engineering team comes up with a software design for an automotive IoT application such as remote diagnostics.

In this chapter, we're going to cover the following main topics:

- Cloud backend deployment and service models
- IoT application architecture
- Vehicle telematics gateway
- Remote diagnostics application
- Predictive maintenance
- Development process

Cloud backend deployment and service models

The cloud backend is the *cloud* section of a system that contains all cloud computing resources, services, storage, and applications offered by a **cloud service provider** (**CSP**). Cloud computing is categorized based on **deployment models** and **service models**.

Deployment models

A deployment model determines where and how cloud resources are hosted, managed, and accessed. Selecting the right model depends on the specific organization's policy and needs for control, flexibility, security, and cost. Popular deployment models are **public cloud**, **private cloud**, and **hybrid cloud**. Table 10.1 compares these three cloud deployment models.

Deployment Model	Public Cloud	Private Cloud	Hybrid Cloud
Characteristics	Shared infrastructure, highly scalable, readily available, pay-as-you-go	Dedicated infrastructure, less scalable, highly secure, and customizable	A combination of public and private clouds with the benefits of both models, but it's complex to manage
Pros	Low cost, easy to set up, highly scalable, elastic resources	Enhanced security, greater control, and customization options	Flexibility, scalability, and security control for sensitive data
Cons	Security concerns, less control over infrastructure, vendor lock-in	Higher cost, complex to manage, limited scalability	Increased complexity, management challenges, vendor lock-in
Use cases	Start-ups, non-critical workloads, e-commerce platforms, and web applications	Sensitive data workloads, financial institutions, and government agencies	Enterprises with diverse needs, healthcare organizations, and manufacturing companies
Examples	Amazon Web Service (AWS), Microsoft Azure, Google Cloud Platform (GCP)	On-premises data center with cloud-based management tools	Private cloud for core applications and public cloud for non-critical workloads

Table 10.1 – Comparing cloud deployment models: public, private, and hybrid clouds

The cost of building a private cloud might be prohibitively expensive as we need to consider the significant **capital expenditure** (**CapEx**) required in buying/renting data centers to host servers, switches, routers, storage, and making sure there is uninterrupted power supply, physical security, connectivity, and so on. We will also need maintenance personnel and cloud computing software to manage computing, storage, networking, connectivity, and other cloud services. The private cloud will have scalability challenges compared to hyper-scale public cloud providers such as Google, Amazon, and Microsoft.

The quickest way to operationalize any application development and deployment is to start with a public cloud; when it comes to the final production deployment, this should align with the organization's cloud strategy and policies. With the public cloud, the organization uses **operational expenditure (OpEx)**. Some popular public cloud providers are AWS, Microsoft Azure, and GCP.

A hybrid cloud is a combination of both public and private clouds where critical workloads and core applications will run on the private cloud and all non-critical workloads and applications will run on the public cloud. This might be a solution in case an organization already has an existing private cloud, or it reduces the risk of completely depending on a public cloud. The challenges with this approach are that it adds an operational overhead of maintaining different clouds and it's costly. Overall, this might be the right solution to meet certain infrastructure customization needs or meet regulatory and legal requirements.

Service models

A service model defines the level of responsibility required for managing the infrastructure that runs the application. They range from having complete control of the underlying **Infrastructure as a Service (IaaS)** to simply using a pre-built application **Software as a Service (SaaS)** and the **Platform as a Service (PaaS)**, which falls somewhere in between. *Figure 10.1* explains the different cloud service models.

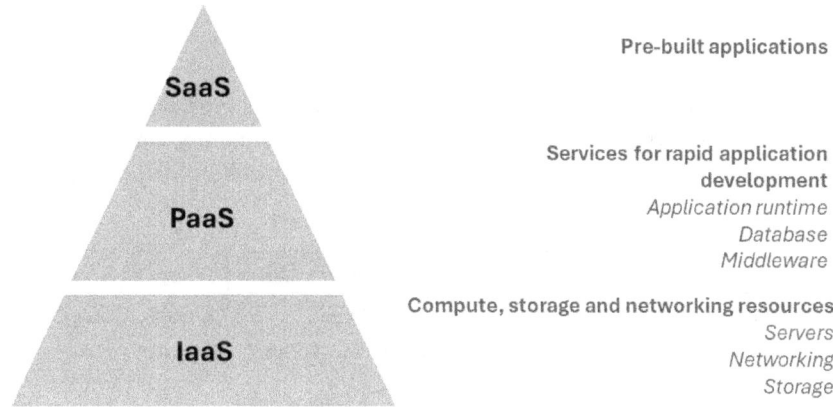

Figure 10.1 – Cloud service models: IaaS, PaaS, and SaaS explained

In the case of the IaaS model, we are renting raw computing resources such as virtual servers, storage, and networking from a cloud provider. While we have great control over the environment, the responsibility of managing, from operating system to application, is on us. In the case of the public cloud IaaS offering, the public cloud provider follows a **shared responsibility model**, where they take care of physical security and cloud infrastructure security, while consumers of the public cloud are responsible for all other aspects, such as operating systems of **virtual machines** (**VMs**), data security and privacy, applications, **identity and access management** (**IAM**), and so on.

PaaS provides us with a complete development and deployment environment. This includes things such as operating systems, databases, and development tools. Our focus will be on building and deploying applications while the cloud provider takes care of the underlying infrastructure.

The SaaS model is the simplest of the cloud service models; the cloud provider manages everything, including the infrastructure, the application, and the data. Table 10.2 compares the three cloud service models.

Feature	IaaS	PaaS	SaaS
Control	High	Moderate	Low
Responsibility	Manage everything	Manage applications	Use application only
Flexibility	High	Moderate	Low
Cost	Variable	Usage-based or monthly fixed	Per user/month Per asset/month
Examples	AWS, Microsoft Azure, GCP	Heroku, Google App Engine (GAE), AWS Elastic Beanstalk	Microsoft Office 365, Dropbox

Table 10.2 – Comparing cloud service models: IaaS, PaaS, and SaaS

Here is a comparison of the various IaaS offerings from various cloud providers. AWS provides the broadest range of instance types, mature features, and a strong community, while Azure shines in hybrid cloud integration and is good for Windows workloads. Google offers competitive pricing with a strong focus on containerization and **machine learning** (ML).

Feature	AWS	Azure	GCP
Compute services	Amazon **Elastic Compute Cloud** (EC2) (VMs), Lambda (serverless)	Azure VMs, Functions (serverless)	**Google Compute Engine** (GCE) (VMs), Cloud Functions (serverless)
Instance types	Diverse configurations: general purpose, compute-intensive, memory-optimized, storage-optimized	Similar to AWS with additional VM series such as burstable	Similar to AWS and Azure with a focus on sustained use and ML
Scaling options	Auto Scaling, **Elastic Container Service** (ECS)	Autoscaling, App Service plans (scaled app service instances)	Autoscaler, **Google Kubernetes Engine** (GKE)

Feature	AWS	Azure	GCP
Storage options	**Simple Storage Service (S3)** (object storage), **Elastic Block Store (EBS)** (block storage), **Elastic File System (EFS)** (shared file storage)	Blob Storage, Azure Files (shared files), Disks (managed disks)	Cloud Storage (object storage), Persistent Disk (block storage), Filestore (shared file storage)
Networking options	**Virtual Private Cloud (VPC)**, Direct Connect (dedicated connection)	**Virtual Network (VNet)**, ExpressRoute (dedicated connection)	VPC, Cloud Interconnect (dedicated connection)
Pricing models	On-demand, reserved instances, spot instances	On-demand, reserved VMs, pay-per-use for functions	On-demand, sustained use discounts, committed use discounts

Table 10.3 – Comparison of IaaS offerings from various cloud providers

As we navigate the landscape of cloud computing, it's evident that the shift from traditional infrastructure management to more abstracted services has paved the way for innovations in server-based and serverless computing. This directly correlates with the service models we discussed, where the responsibility and management overhead transitions from the user to the provider. Now, let us explore how server-based and serverless computing frameworks are built upon these models.

Server-based and serverless computing

Server-based computing is a traditional approach to building applications or services in the cloud. It's the developer's responsibility to install and configure operating systems, software libraries, and application code on the virtual servers in the cloud (IaaS). A server-based approach provides fine-grained control and flexibility to customize and can be cost-effective for long-running applications with consistent resource demands.

Serverless computing is a modern approach to building an application or a service. Developers focus on writing code and deploying without worrying about the underlying infrastructure. Developers package their code in containers or functions, and the cloud provider takes care of provisioning, scaling, and managing the servers. It is faster to develop and deploy, thereby reducing operational overhead for managing servers. It can be cost-effective for applications that have variable workloads. Table 10.4 compares the various features of server-based and serverless applications.

Feature	Server-based Applications	Serverless Applications
Developer focus	Infrastructure management and code deployment	Code development and deployment
Development time	Longer due to infrastructure setup and maintenance	Faster due to focus on code and simplified deployment

Feature	Server-based Applications	Serverless Applications
Control and flexibility	High control over servers, software, and environment	Limited control over underlying infrastructure
Scalability	Manual scaling or limited auto-scaling	Automatic scaling based on demand
Security	Requires separate security management for servers and applications	Cloud provider handles some security aspects, but shared responsibility
Infrastructure responsibility	Developer/IT team	Cloud provider
Operational overhead	High for server management, patching, security	Low to minimal, less operational burden
Troubleshooting	More complex due to direct server interaction	Can be challenging due to abstraction from infrastructure
Suitable for	Long-running applications, consistent resource needs, high control requirements	Event-driven applications, microservices, variable workloads, cost efficiency
Cost	Can be more expensive for idle or low-traffic applications	Pay only for executed code and consumed resources
Examples	Web servers, databases, VMs, containerized applications	API gateways, event-driven workflows, data processing pipelines, serverless functions

Table 10.4 – Comparison of server-based and serverless computing

There are many application architectures out there, but we will focus on modern, cloud-native approaches such as **event-driven**, **microservices**, and **serverless** architecture. From a deployment perspective, **containers** are well suited to deploying microservices-based applications. Let us understand the terms *event-driven*, *microservices*, and *containerization* in the context of the cloud backend.

In microservices, instead of building a monolithic application where all components are tightly integrated, they advocate breaking down the application into loosely coupled and independently deployable services that can be owned by small cross-functional teams. While the obvious benefits are increased agility, resilience, and improved scalability, they also come with some challenges, such as increased complexity, distributed data management, and testing challenges. *Figure 10.2* shows the difference between monolithic and microservices architectures.

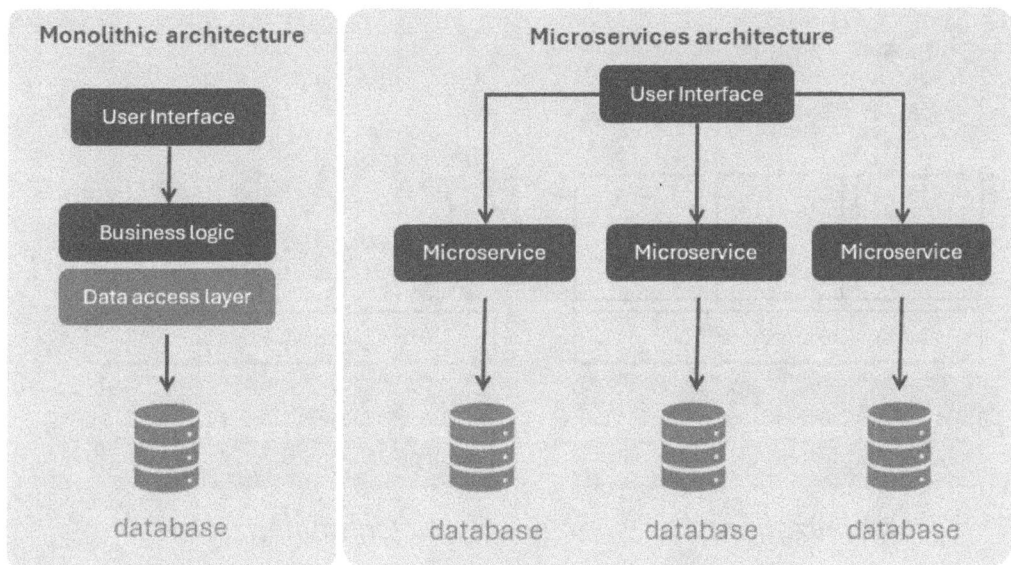

Figure 10.2 – Monolithic versus microservices architecture

Event-driven architecture (**EDA**) is a software design paradigm that revolves around reacting to events as they occur in the system. In a traditional request-response model, one service directly requests something from another service and waits for a response. In EDA, services publish events when something interesting happens, and other interested services subscribe to those events and act accordingly.

Containerization is a lightweight and portable technology compared to VMs that allows us to package and isolate applications and their dependencies, along with the necessary runtime, libraries, and system settings, into a single container. Containers provide a consistent and reproducible environment, ensuring that an application runs consistently across various environments, such as development, testing, and production. Docker is a popular platform used to build, run, and manage containers. Cloud providers provide a range of solutions that offer a scalable and secure platform for running containerized applications, such as GKE, Amazon ECS, and Microsoft **Azure Kubernetes Service** (**AKS**), and if we want to take a platform-agnostic approach, Rancher delivers a unified platform for managing Kubernetes clusters across multiple cloud environments. *Figure 10.3* illustrates the differences between virtualization and containerization.

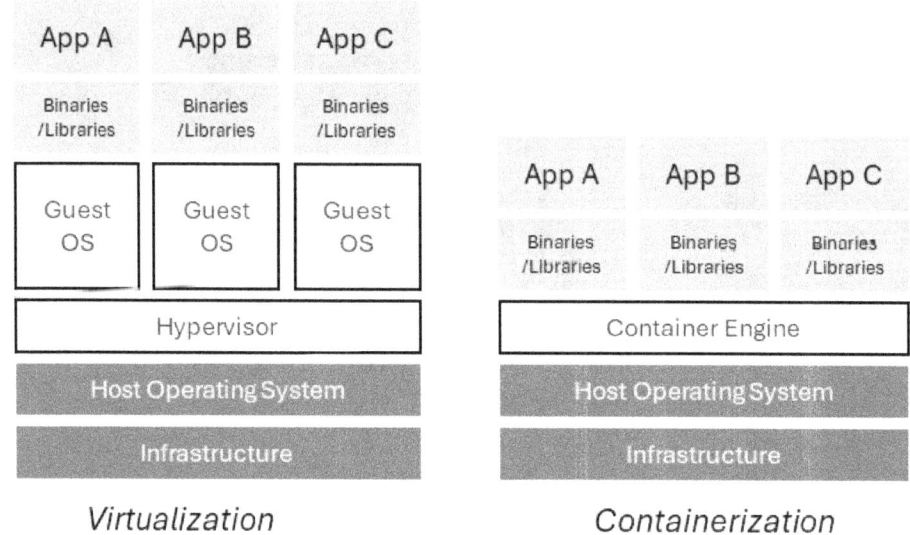

Figure 10.3 – Difference between virtualization and containerization

In the following section, we will delve into the IoT application architecture, building upon the cloud concepts outlined and the comparative analysis of server-based versus serverless computing presented earlier.

IoT application architecture

To realize the functional blocks in the cloud that we identified in the previous chapter, we will start with an IoT reference architecture. *Figure 10.4* provides a visual guide to the typical structure and flow within an IoT ecosystem.

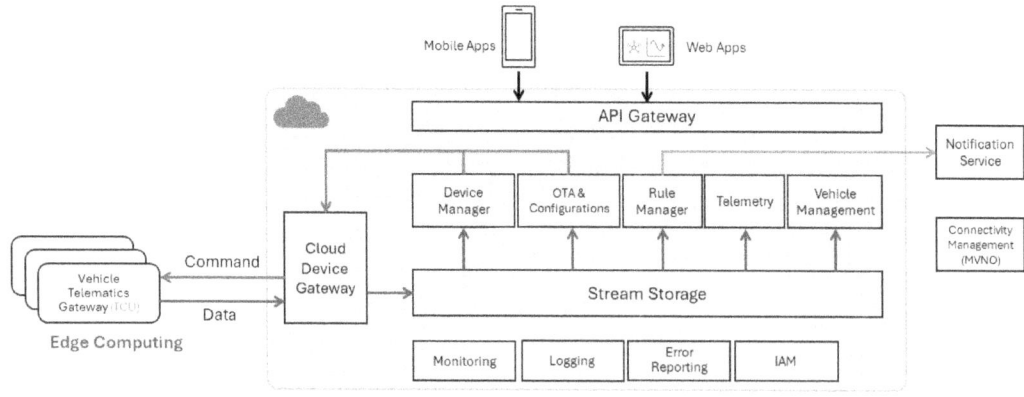

Figure 10.4 – IoT reference architecture

We will identify the technology behind each of the components where applicable and explore the offerings from the cloud providers.

Cloud device gateway

The vehicle telematics gateway interacts with the cloud via a cloud device gateway. The cloud device gateway is a message broker that acts as a mediator between the cloud and the device by sending and receiving messages and facilitating secure and reliable communication.

One of the most popular message brokers is **Message Queuing Telemetry Transport (MQTT)** [1], which is a **publish/subscribe (pub/sub)** messaging protocol that is good for IoT scenarios where lightweight communication, minimal overhead, and low power consumption are important. MQTT supports persistent sessions, which reduces the time to reconnect the client with the broker. MQTT is an **Organization for the Advancement of Structured Information Standards (OASIS)** standard. The specification is managed by the OASIS MQTT Technical Committee. *Figure 10.5* depicts the MQTT pub/sub model.

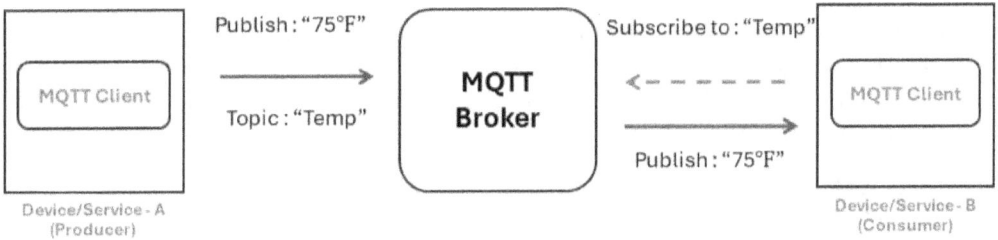

Figure 10.5 – MQTT client and broker pub/sub model

Let us understand how the MQTT broker works based on the preceding diagram:

1. Devices/services (*A* and *B*) connect to an MQTT broker, and the broker acts as a central hub that routes messages between devices or services.
2. Devices/services (*A*) publish messages to topics that are like labeled mailboxes.
3. Devices/services (*B*) subscribe to topics of interest in receiving messages.
4. The broker routes messages; when a device publishes a message to a topic, the broker sends the message to all devices subscribed to that topic.

MQTT brokers can handle thousands of messages at a time due to their simplified architecture. One common challenge in the case of automotive IoT is the network can get disconnected from time to time. The broker, in this case, can persist the messages and deliver them once the client reconnects. The broker is also responsible for handling authentication and authorization.

There are various open source MQTT deployments widely used in building scalable and efficient IoT applications, such as Mosquitto, HiveMQ, RabbitMQ, and so on. Before choosing a specific MQTT deployment, it's essential to consider factors such as scalability, ease of integration, and enterprise support.

AWS provides **AWS IoT Core**, a managed service for connecting and managing billions of devices. It handles device authentication, secure communication, and message routing.

Azure provides **Azure IoT Hub**, a similar service for connecting and managing devices. It also offers advanced features such as device twin management and built-in analytics.

Edge computing

The vehicle telematics gateway is the bridge between all vehicle **electronic control units** (**ECUs**) and the cloud device gateway (MQTT broker). The vehicle telematics gateway is responsible for data aggregation, data processing (cleaning, filtering, transforming data), local analysis (applying ML models), connectivity management (managing connections for efficient data transmission), device management (device configuration, updates, and monitoring), to name a few.

If our strategy is to build from the ground up, then all the aforementioned capabilities must be developed using an MQTT client from scratch or by utilizing open-source stacks such as **OMA Lightweight M2M** (**OMA-LwM2M**) [2] for device management, which will be explored further in the *Device management* section.

AWS provides **AWS IoT Greengrass** for running applications and processing data on edge devices. It allows for offline operation and reduces latency.

Azure provides **Azure IoT Edge**, with similar edge computing capabilities. It integrates with Azure services such as **Azure Machine Learning** (**AML**) and Digital Twins.

Stream processing

Stream processing is the practice of acting on a series of data at the time the data is created. Actions that stream processing takes on data include aggregations, predictive analytics, transformations, enrichment, and ingestion. There are various stream processing architectures, such as Lambda, Kappa, Micro-stream, and Stream-ETL.

Let us dive into Lambda architecture, which is designed to handle massive quantities of data by combining batch and stream processing methods.

IoT application architecture   201

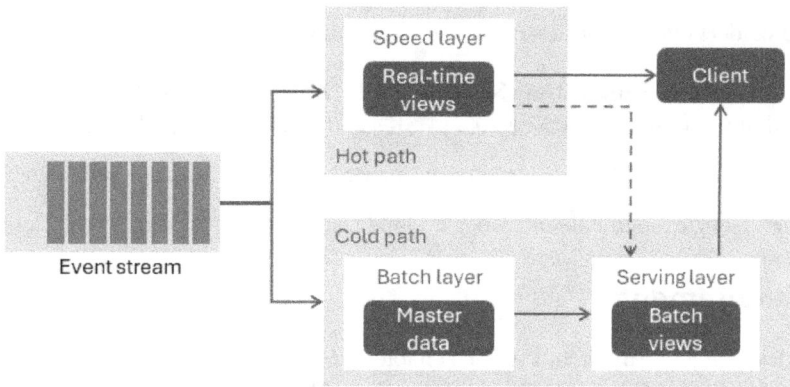

Figure 10.6 – Lambda data processing architecture

As you can see in *Figure 10.6*, the **batch layer** processes all incoming data in a scheduled manner, creating a comprehensive, accurate, and immutable master dataset. The **speed layer** processes data in real time as it arrives and handles time-sensitive analysis and actions, providing low-latency views. The **serving layer** indexes and stores the outputs from both batch and speed layers and responds to user queries efficiently, providing a unified view of data.

Apache Kafka [3] is a widely used open source platform for processing and preserving event-based messages in a sequential format. Kafka scales horizontally across multiple servers for high-velocity, high-volume data. The messaging system is known as a pub/sub system. It consists of publishers, which are data sources, and subscribers, which are data consumers. *Figure 10.7* illustrates the architecture of Apache Kafka.

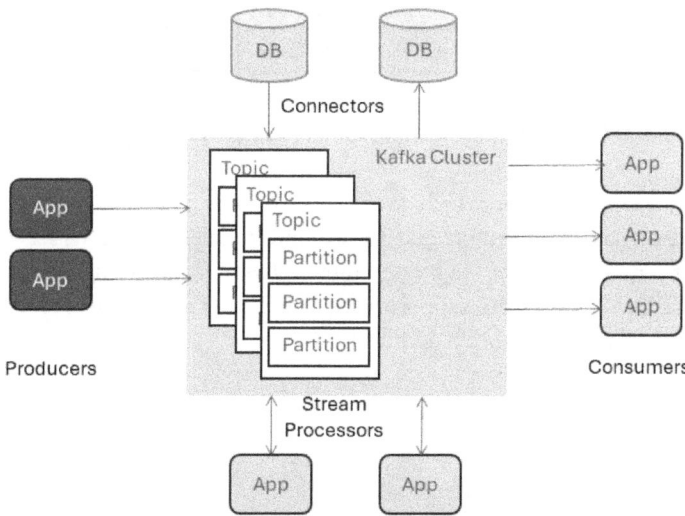

Figure 10.7 – Apache Kafka architecture

All the cloud providers offer various managed data streaming platforms:

- AWS offers Amazon **Kinesis Data Streams** (**KDS**), Amazon Kinesis Data Firehose, Amazon **Managed Streaming for Apache Kafka** (**MSK**), and Amazon Kinesis Data Analytics
- Azure offers Azure Event Hubs, Azure Stream Analytics, Azure Data Factory, and Azure Databricks
- GCP offers Google Cloud Pub/Sub, Google Cloud Dataflow, Cloud Spanner, and Cloud Bigtable

Device management

LwM2M is a device management and service enablement protocol from **Open Mobile Alliance** (**OMA**) for **machine-to-machine** (**M2M**) or IoT. The LwM2M standard defines the application layer communication protocol between the LwM2M server and the LwM2M client. LwM2M was originally built on **Constrained Application Protocol** (**CoAP**), but later versions support new transports for messaging, which allows it to be conveyed over MQTT and **Hypertext Transfer Protocol** (**HTTP**). *Figure 10.8* shows the LwM2M protocol stack and various transport binding options.

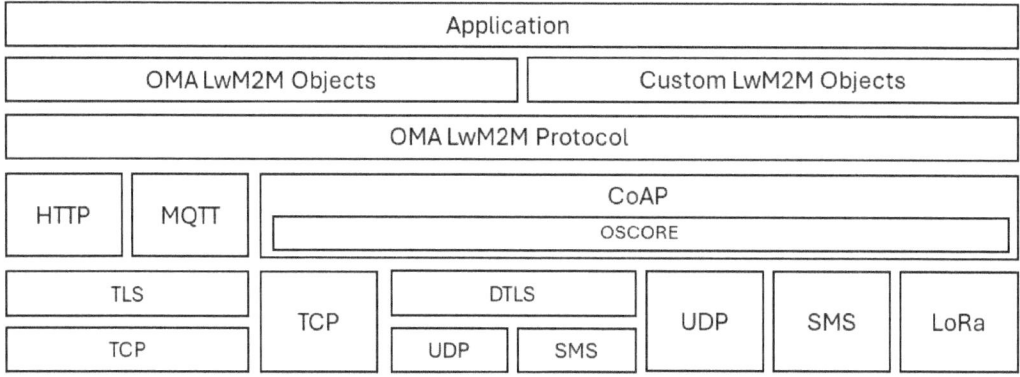

Figure 10.8 – LwM2M protocol and various transport binding options

LwM2M's device management capabilities include remote provisioning of security credentials, firmware updates, connectivity management, remote diagnostics, and troubleshooting. Its service enablement capabilities allow for sensor and meter readings, remote actuation, and configuration. The LwM2M data model, known as **LwM2M Objects**, supports various use cases and can be extended for different industries. *Figure 10.9* depicts the LwM2M architecture.

IoT application architecture 203

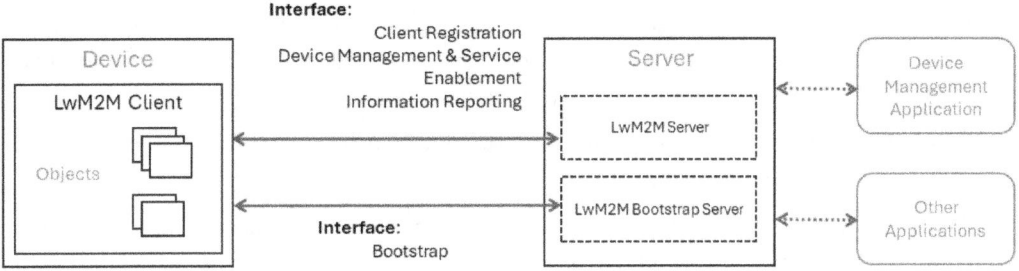

Figure 10.9 – LwM2M architecture

In combination with the LwM2M protocol, the LwM2M Objects data model supports various LwM2M use cases. The data model can be extended and support applications for various kinds of industries.

OTA solutions

Over-the-air (OTA) solutions are essential in maintaining the security, performance, and functionality of devices/ECUs in a vehicle. They allow **Original Equipment Manufacturers** (**OEMs**) to deliver improvements, bug fixes, and security patches to their vehicles remotely, efficiently, and securely.

An OTA updater is a software or firmware component in a gateway device or in another vehicle ECU that allows for the **remote updating** of the device's software or firmware, without requiring physical access to the device. *Figure 10.10* illustrates various components of an OTA solution.

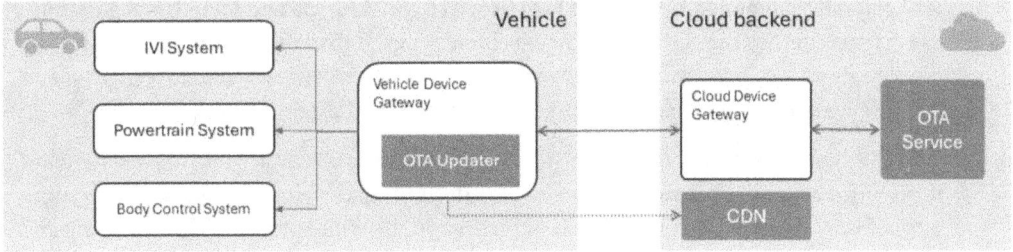

Figure 10.10 – Various components of an OTA solution

Here are some key aspects of OTA updaters:

- **Rollback capability**: In case an update fails or causes issues, OTA updaters should support rollback mechanisms to revert the device to a previous working version.
- **Efficiency**: They often use differential or incremental updates, sending only changes or additions to the new version of the software. This reduces the amount of data that needs to be transferred and saves bandwidth.

- **Security**: OTA updaters need to implement robust security measures to ensure that updates are delivered securely and that the authenticity and integrity of updates are verified. Security is crucial to prevent unauthorized updates or tampering.

- **Scheduling and control**: Administrators can schedule updates at convenient times and have control over when and how updates are applied.

- **Automatic updates**: In some cases, OTA updaters can be set to automatically download and install updates without user intervention.

- **Redundancy**: To ensure reliability, OTA updaters may use redundancy and failover mechanisms to handle interruptions or failures during the update process.

- **Notification**: OTA updaters often include mechanisms for notifying users about available updates and their importance.

The automotive industry has implemented several frameworks, regulations, and standards for software updates in vehicles:

- **Uptane** [4] is an open source security framework specifically designed for secure software updates in automobiles. It builds upon the foundation of **The Update Framework (TUF)** [5] but adds several key features to address the unique challenges of car software updates, such as resilience to compromise, multi-party trust, and flexible implementation. While not itself a standard, Uptane plays a significant role in achieving compliance with regulations such as *UN R155* and *R156*.

- **United Nations Economic Commission for Europe (UNECE) R155 and R156 UN regulations**: These are mandatory regulations for new vehicle types globally (from July 2024) that set requirements for cybersecurity and software update management systems, covering aspects such as secure delivery mechanisms, update verification, and user notifications.

- **ISO 24089**: *Vehicle Software Update Engineering Standard* is a recently introduced standard that provides a comprehensive framework for designing, developing, and implementing secure software updates for road vehicles. It focuses on the entire process, from planning and implementation to testing and monitoring.

- **ISO/SAE 21434**: The *Road vehicles — Cybersecurity engineering* standard outlines requirements for managing cybersecurity risks across the life cycle of automotive electronic systems. It provides a comprehensive guide for cybersecurity risk management in the design, development, production, operation, and decommissioning of vehicles' electrical and electronic systems.

Some popular commercial vehicle OTA solutions are Airbiquity's OTAmatic Software Management Platform and HARMAN's OTA solutions (Red Bend).

All cloud providers offer OTA updates to the device gateway; AWS provides them through AWS IoT Device Management, Azure provides them via Azure IoT Hub Device Management, and GCP provides them via Cloud IoT Core.

Telemetry datastore

Optimizing the real-time processing of vehicle telemetry within a Lambda architecture for low latency and high throughput requires careful selection of the underlying database technology. Several types of databases are commonly used in this context:

- **Time series databases**: **InfluxDB** is optimized for timestamped data and is often used in real-time analytics and monitoring applications. Cloud providers offer fully managed time series databases, such as Amazon Timestream by AWS, Azure Time Series Insights by Azure, and Firebase Realtime Database by GCP.

- **NoSQL databases**: **Apache Cassandra** is known for its **high availability** (**HA**) and scalability; it is often used in real-time data processing scenarios where high write and read throughput is required. **MongoDB** offers real-time processing capabilities and is particularly useful when dealing with **JavaScript Object Notation** (**JSON**)-like, document-oriented data structures. **Redis** is an in-memory data structure store, often used as a database, cache, or message broker. Redis is known for its low-latency data access, making it suitable for real-time processing tasks. Cloud providers offer various fully managed NoSQL databases, as detailed in Table 10.5.

Feature/Service	AWS	Azure	GCP
Key-value store	Amazon DynamoDB	Azure Table Storage	Google Cloud Bigtable
Document store	Amazon DocumentDB	Azure Cosmos DB (document API)	Firestore
Column-family store	Amazon Keyspaces (for Apache Cassandra)	Azure Cosmos DB (Cassandra API)	Google Cloud Bigtable
Graph store	Amazon Neptune	Azure Cosmos DB (Gremlin API)	Not offered directly
Managed Apache Cassandra	Amazon Keyspaces	Azure Cosmos DB (Cassandra API)	Not offered directly
In-memory data store	Amazon ElastiCache (Redis, Memcached)	Azure Cache for Redis	Cloud Memorystore (Redis, Memcached)

Table 10.5 – Overview of the various NoSQL database offerings from cloud providers

Each cloud provider offers unique features and capabilities, and the best choice depends on the specific requirements of the application, such as data model, scalability needs, global distribution, and specific features such as in-memory data storage or real-time data synchronization.

Rule engine

In the context of IoT, a **rule engine** is a software component in the cloud IoT platform that automates actions based on predefined conditions. Each event in the data stream is processed by the rule engine, and if a condition matches, a defined action is triggered.

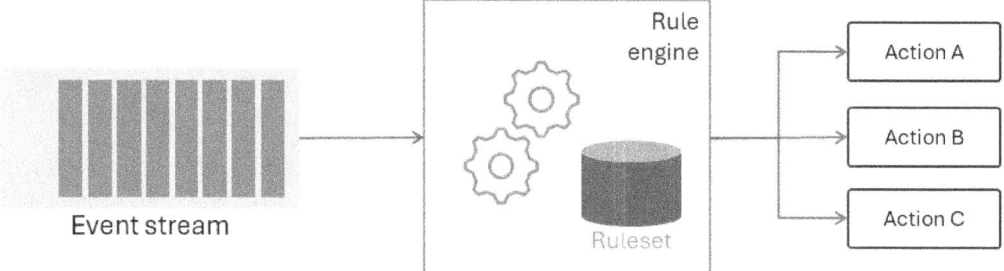

Figure 10.11 – A simple rule engine flow

For simple ruleset processing, all the cloud providers offer out-of-the-box solutions. AWS provides a built-in feature of AWS IoT Core for filtering, processing, and routing device data in real time; Azure provides a similar solution as part of Azure IoT Hub, and in addition, it provides a visual rule builder and supports complex expressions; and GCP provides simple *if-then* rules via Cloud IoT Core.

When it comes to complex logic rule processing, then a custom service needs to be created as a server or serverless service, depending on the data volume and speed. For example, if a vehicle exceeds a specific speed limit or violates a geo-fence, the rule engine will trigger a notification to the fleet owner.

Application Programming Interface (API) gateway

API gateways play a crucial role in modern application development, especially in microservices architectures, promoting security, flexibility, scalability, and efficiency. They are essential for building robust and secure API-driven architectures. An API gateway is a component that sits between the clients and the services and provides a centralized handling of API communication between them. *Figure 10.12* illustrates the API gateway functionality.

IoT application architecture

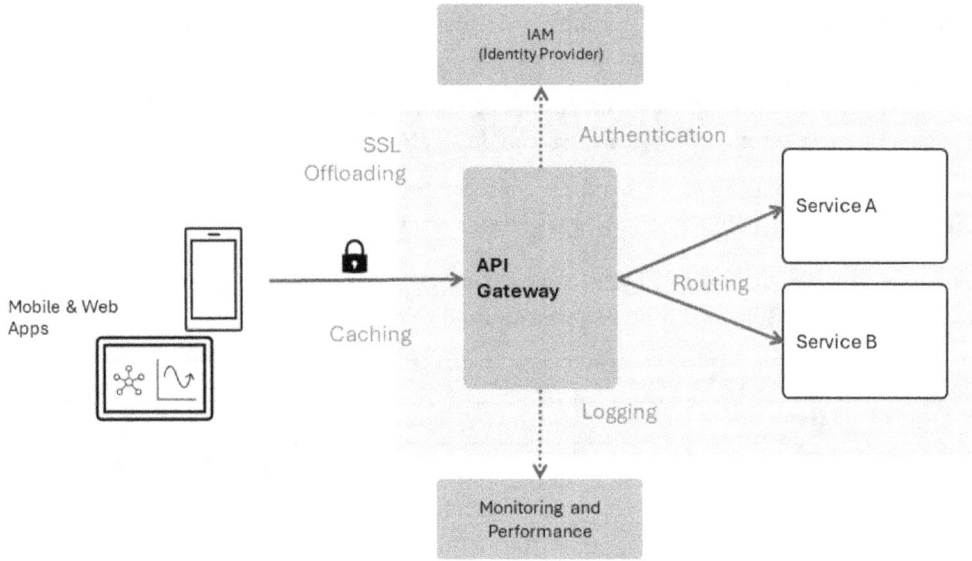

Figure 10.12 – Overview of API gateway functionality

Its key functionalities include the following:

- **Routing** incoming API requests to the appropriate microservice or backend service.

- **Caching** requests and responses to improve the performance of API calls, reducing load on backend services.

- **Request and response transformation** can modify requests and responses as they pass through, such as converting **eXtensible Markup Language** (**XML**) to JSON, adding, or removing headers, or other transformations to meet the requirements of both the client and the service.

- **Service aggregation** can aggregate multiple requests and responses, reducing the number of round trips between the client and the microservices.

- **Authentication and authorization** can handle user identification and ensure that the user has permission to access the requested resources. It often integrates with security protocols such as **Open Authorization** (**OAuth**), **JSON Web Tokens** (**JWT**), and so on.

- **Security**: Beyond authentication and authorization, it can also include features such as **Internet Protocol** (**IP**) whitelisting/blacklisting, encryption, and protection against attacks such as **Structured Query Language** (**SQL**) injection or **distributed denial of service** (**DDoS**).

- **Load balancing** distributes incoming API traffic across multiple backend services to ensure HA and optimal resource utilization.

- **Rate limiting and throttling** can control the number of requests a user is allowed to make in each period to prevent abuse and manage traffic.

- **Logging and monitoring** records data about the traffic and transactions passing through it, which is crucial for monitoring, tracing, and debugging purposes.

- **API version management** can handle requests to different versions of an API, useful during upgrades or when maintaining multiple versions of a service.

All cloud providers offer managed API gateway solutions. Table 10.6 provides a comparative overview of the API management solutions from AWS, Azure, and GCP.

Feature	Amazon API Gateway	Azure API Management	Apigee
Provider	AWS	Microsoft Azure	GCP
Type	Fully managed service	Comprehensive platform	Full life-cycle platform
API types	REST, WebSocket, HTTP	REST, **Simple Object Access Protocol** (**SOAP**), GraphQL	REST, SOAP, GraphQL
Integrations	AWS services	Azure services	GCP services, multi-cloud
Deployment	Cloud-based	Cloud-based, hybrid	Cloud-based, hybrid, on-premises
Security	Authentication, authorization, throttling, IP filtering, **web application firewall** (**WAF**)	Authentication, authorization, rate limiting, IP filtering, policy-based routing	Authentication, authorization, rate limiting, API key management, OAuth 2.0
Analytics	Basic usage metrics	Detailed usage analytics	Comprehensive analytics
Pricing	Based on API calls, data transfer	Based on tiers, usage	Based on tiers, usage
Ease of use	Relatively easy	Moderately easy	More complex

Table 10.6 – Comparing API management solutions from cloud providers

Open source API gateways are a viable alternative to proprietary solutions; they offer flexibility for customization and can be integrated into a variety of environments. Some popular open source API gateways include Kong, Tyk, and KrakenD. This can be a perfect fit if you are looking to avoid vendor lock-in while still leveraging the full suite of API management features.

Connectivity management

The vehicle telematics gateway needs to connect to the cloud backend via Wi-Fi or cellular network. In the case of cellular connectivity, the gateways need to be provisioned with a SIM card, and their life cycle (activation, deactivation, and troubleshooting) needs to be managed along with data plans. Most **mobile network operators (MNOs)/mobile network virtual operators (MVNOs)** offer a **Connectivity Management Platform (CMP)**, which we can integrate via CMP APIs to automate workflows or use their portal to manually manage them. The typical functionality of a CMP is as follows:

- Connectivity management involves managing the network connections of devices, including selecting and switching between different connectivity options (such as cellular, Wi-Fi, and satellite) based on availability and cost-effectiveness.

- Device management enables users to monitor and manage the status of connected devices remotely. This includes tasks such as activation, deactivation, suspension, and troubleshooting of devices.

- Security management provides security features to protect the network and data from unauthorized access and cyber threats, which is crucial in IoT environments, such as providing private **Access Point Names (APNs)**.

- Integration with other systems and platforms, enabling data flow and coordinated actions across different business systems.

- Billing, which includes tools for monitoring usage and managing costs.

Some of the CMPs provided by MNOs to their enterprise customers are AT&T Control Center, Vodafone Global IoT Platform, Verizon ThingSpace, T-Mobile IoT Connectivity Management, Telstra IoT Connectivity Management, and Deutsche Telekom IoT Connectivity Management, and those provided by MVNOs are Aeris Communications, Soracom, KORE Wireless and Sierra Wireless.

IAM

Identity and access management (IAM) is a framework of policies and technologies that enables organizations to securely manage user access to resources such as applications, data, and systems. It essentially ensures that only authorized users have access to the right things, at the right time, and for the right reasons.

Identity management (**IdM**) involves creating and managing user accounts, including assigning usernames, passwords, and other authentication methods. **Access control** determines what permissions users have and what resources they can access. Permission levels can be granular, allowing fine-grained control over who can do what. **Authentication** verifies that users are who they claim to be, typically through login credentials or other verification methods. **Authorization** grants or denies access to resources based on a user's identity and permissions. *Figure 10.13* illustrates the authentication and authorization flow between an API gateway and IAM.

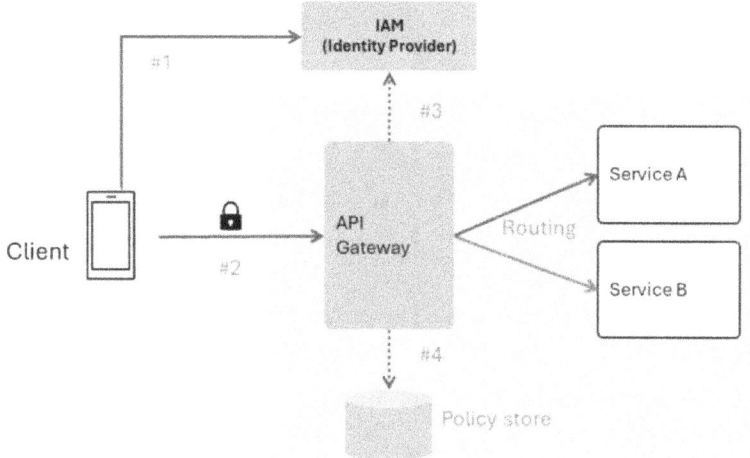

Figure 10.13 – Authentication and authorization flow between API gateway and IAM

The client logs in with the IAM (*#1*) and receives an access token. The client then passes the token (*#2*) along with the request to the API gateway. Once the token is authenticated (*#3*), then the API gateway evaluates against the authorization policy (*#4*) and allows or denies the request.

Vehicle telematics gateway

A vehicle telematics gateway, also known as a **Telematics Control Unit** (**TCU**) interfaces with the rest of the vehicle domain controllers via **Controller Area Network** (**CAN**), automotive Ethernet, or other vehicle network protocols, as shown in Figure 10.14. The TCU hardware block diagram, as shown in the previous chapter in *Figure 9.9*, has an **Application Processor** (**AP**) and a **Vehicle Interface Processor** (**VIP**).

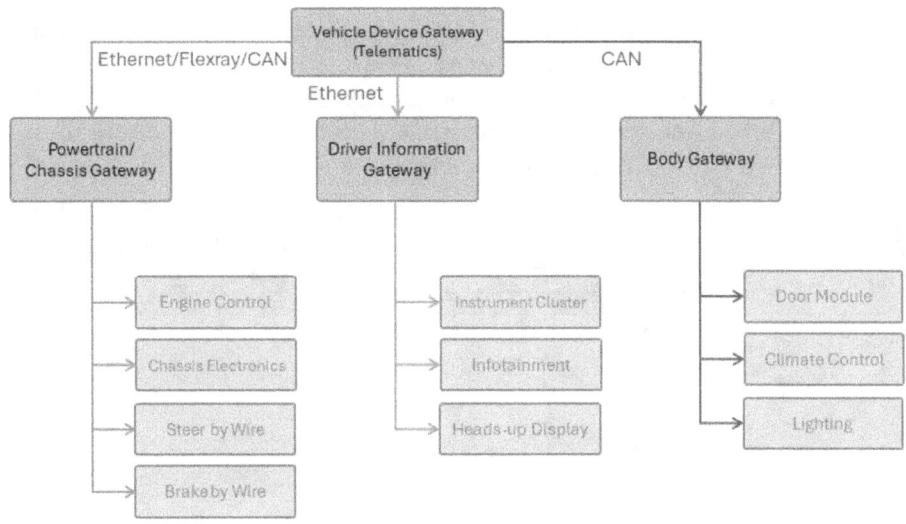

Figure 10.14 – Vehicle gateway interfacing with other domain controllers in the vehicle

The VIP is a microcontroller and in this context interfaces with the vehicle CAN and the AP. Typical APs don't have a built-in CAN transceiver and hence there is a need for a VIP. In the automotive context, most microcontrollers support the Classic AUTOSAR software stack. CAN traffic is routed to the AP via a **Serial Peripheral Interface (SPI)** or **Universal Asynchronous Receiver-Transmitter (UART)** interface. In addition, the VIP can also interface with other sensors such as accelerometers and gyro sensors. *Figure 10.15* shows the various software blocks on a VIP.

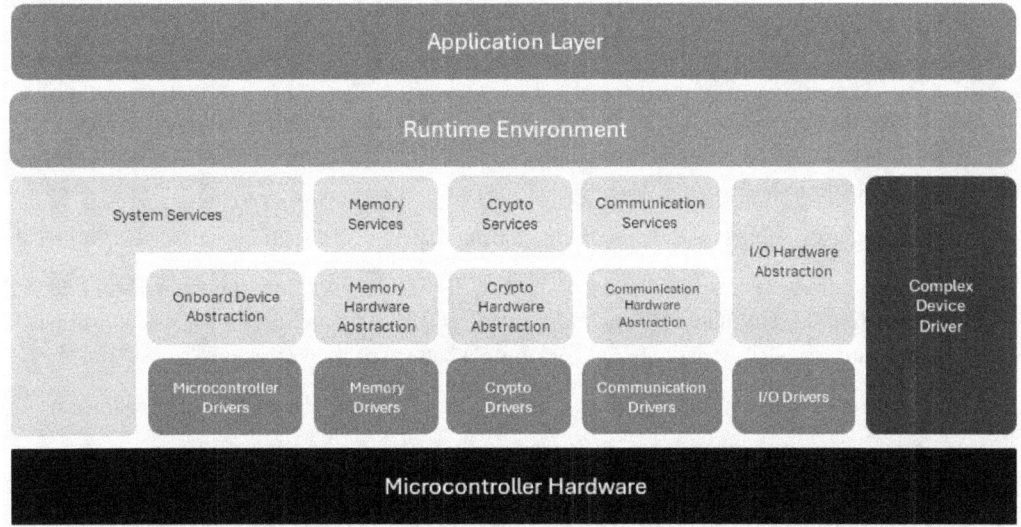

Figure 10.15 – Microcontroller software block diagram based on Classic AUTOSAR

The AP is responsible for all the connectivity and edge functions and interfaces with the vehicle via automotive Ethernet. The AP can run on Adaptive AUTOSAR or Linux, which is supported by the **System-on-Chip (SoC)** vendor. The AP is a multi-core processor where one core is dedicated to the cellular modem software and the rest is used for other edge functions. *Figure 10.16* shows the various software blocks required to develop edge applications on an AP.

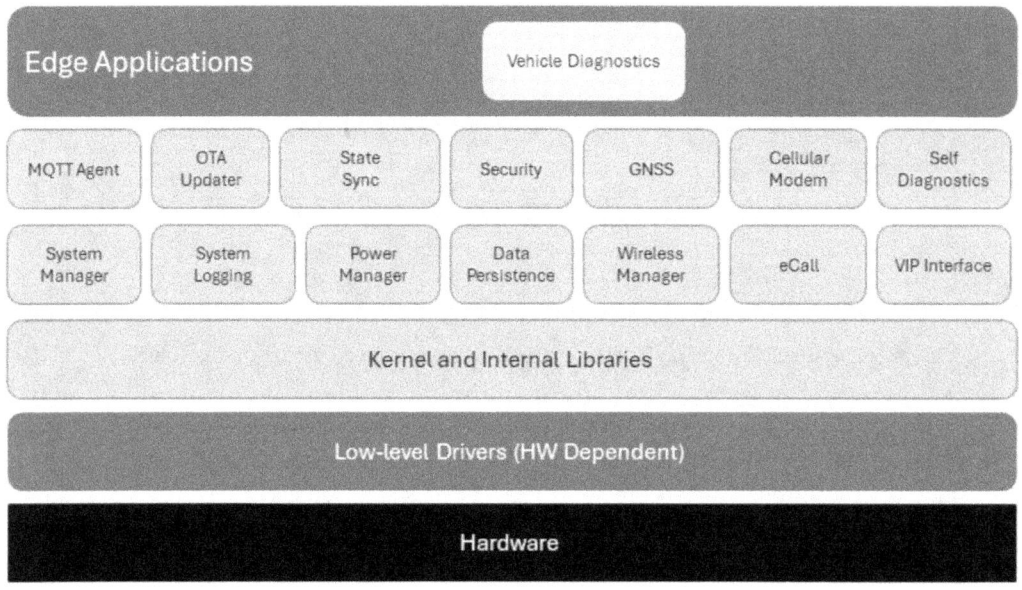

Figure 10.16 – AP software block diagram

The following are the basic functional blocks required to build a secure edge application on a TCU:

- The **system manager** is responsible for managing the life cycle of services and applications; that is, launching, shutting down, and relaunching on a failure. It's also responsible for synchronizing the states between the VIP and AP.
- The **system logger** captures all system information and logs to store locally or remotely as required.
- The **power manager** is responsible for managing the TCU power state transitions from deep sleep to system sleep to full run. Depending on the TCU system architecture, it will coordinate with the system power manager to fall back onto a battery backup as required.
- **Data persistence** is responsible for storing data in a transaction-safe filesystem or external memory.
- The **global navigation satellite system (GNSS) module** is responsible for providing continuous location data. Accelerometer and gyro information, along with this location information, can be used to implement dead reckoning as required for edge applications.

- The **cellular modem** module is responsible for providing connectivity over **Long Term Evolution** (**LTE**) or 5G. Typically, AT commands are used to interface with the cellular modem to configure network connections.

- **TCU diagnostics** are responsible for checking periodically the status of their own subsystems and ensuring they are functioning correctly; they leverage the system manager and logger to perform corrective actions and report issues to the cloud backend as required.

- **State sync** encapsulates the mechanism of synchronizing the state of the TCU or the vehicle with the cloud backend and vice versa. This is an enabler to represent the physical object, system, or process into the virtual model, aka digital twin.

- The **MQTT agent** connects to the TCU IoT applications to the cloud backend (MQTT broker), for efficient communication.

- The **OTA updater** is responsible for updating the firmware on the AP and VIP. It periodically checks for any new software updates with the cloud backend, validates the downloaded packages, and performs the update of the VIP and AP in a specific order to ensure a rollback in case of failure.

- **Security** provides a secure cloud connection via a **Transport Layer Security** (**TLS**) library that implements the TLS protocol between the communicating applications; that is, the cloud backend and TCU. It also provides a means to manage cryptographic objects and store them in dedicated secure storage on the device.

With a comprehensive understanding of the vehicle telematics gateway and its essential components that form the bedrock of in-vehicle connectivity and management, we will now shift our focus to the application layer. In the next section, we will delve into creating a remote diagnostics application.

Remote diagnostics application

Now that we understand the cloud backend and the vehicle telematics gateway technologies and components, let us start building a remote diagnostics application. To do that, we need to represent the physical asset (vehicle) as a virtual asset in the cloud backend; that is, create a digital twin in the cloud. Let us name this service as a **vehicle manager** where we can define vehicle lines and onboard vehicles. We will have the ability to define the vehicle network topology and all the ECUs associated with the vehicle, as shown in *Figure 10.17*. This will be a custom-built microservice.

We will then onboard all the TCU devices to the **device manager**, which is responsible for provisioning the TCU with all the required certificates on the first connection and establishing an association with a specific vehicle. This will ensure that all telemetry data from the TCU is associated with the vehicle. It's important to note that for the sake of brevity, this book will not cover the methodology of connecting end users to vehicles or the mechanisms for access control. When constructing the device manager service, we have the option to either develop it from the ground up, taking the OMA-DM LwM2M standard as a reference and incorporating it with a **Public Key Infrastructure** (**PKI**), or to utilize the ready-made IoT device management service offered by various cloud providers. The following diagram shows the diagnostic service command-and-control flow between the cloud platform and the vehicle telematics gateway.

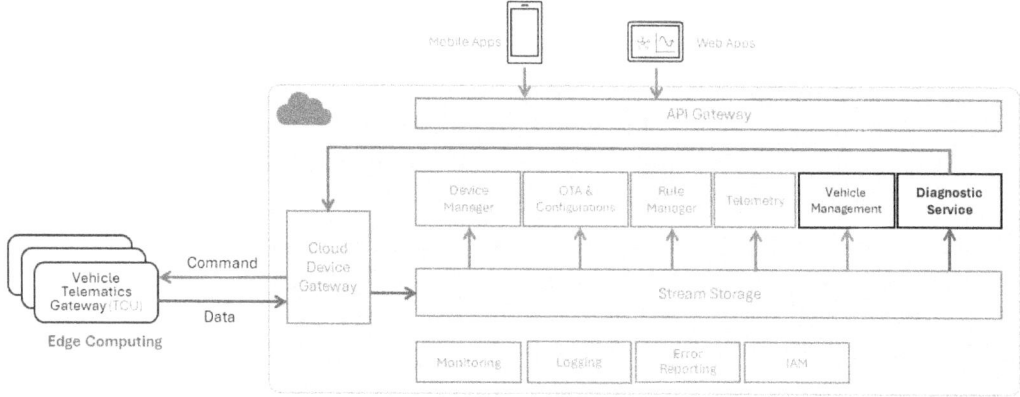

Figure 10.17 – IoT cloud backend with diagnostic service

Now, let us explore how we can implement the **remote diagnostics** service. As the automotive industry transitions from domain controller vehicle architecture to a consolidated zonal architecture, we will consider both the classic diagnostics and **service-oriented vehicle diagnostics** (**SOVD**) scenarios.

In the case of the classic ECU diagnostics scenario, which is based on **Unified Diagnostic Services** (**UDS**) using the **Open Diagnostic eXchange** (**ODX**) and **Open Test Sequence eXchange** (**OTX**) standards, the deployment options are shown in *Figure 9.13*. Option A, where the complete diagnostic stack is part of the TCU and the diagnostic service from the cloud only, sends the OTX and ODX files to the TCU via the cloud device gateway. Option B is where the diagnostic stack is hosted in the cloud, and the TCU implements the **diagnostic protocol data unit** (**D-PDU**) APIs. Every time a technician starts a remote diagnostics session, a container with the diagnostic stack is instantiated in the cloud, and a session is established from the cloud to the vehicle.

In the case of the SOVD scenario, the SOVD server is running on a **high-performance computing** (**HPC**) device, as shown in *Figure 9.14*. The SOVD client can establish a session via the cloud for a remote use case and connect directly to the vehicle via Wi-Fi for a proximity use case.

Building an automotive IoT backend on a public cloud is best approached by maximizing the use of managed services. Cloud providers' IoT Core solutions provide a suite of essential services, including an MQTT broker, device management, and OTA updates for gateway devices (TCUs). However, the OTA update service within IoT Core may not fully address the requirements of comprehensive vehicle ECU updating. Therefore, it is advisable to either utilize a reputable third-party vehicle updater service or to create a bespoke one from scratch. Additionally, other managed services to consider incorporating are an API gateway for interfacing, Kafka for event streaming, and databases for data management and storage. *Figure 10.18* details the options for using managed and unmanaged services.

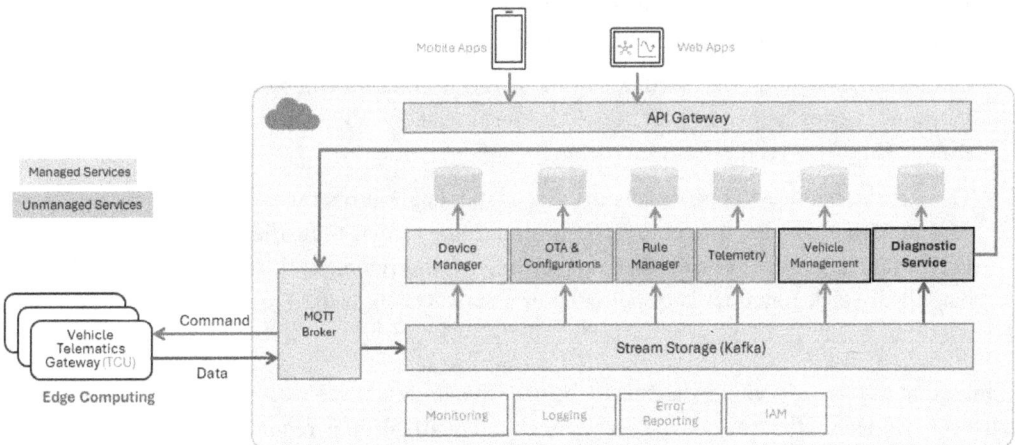

Figure 10.18 – Managed and unmanaged services

The telemetry service is a custom-built microservice used to collect vehicle performance and diagnostics data. The incoming data from vehicles is transformed by the telemetry service and stored in a NoSQL database, and its raw data is stored in an object data store.

Now that we have established how to realize all the building blocks for the automotive IoT backend, let us examine how this system can help us with the remote diagnostics use case detailed in *Figure 10.19*.

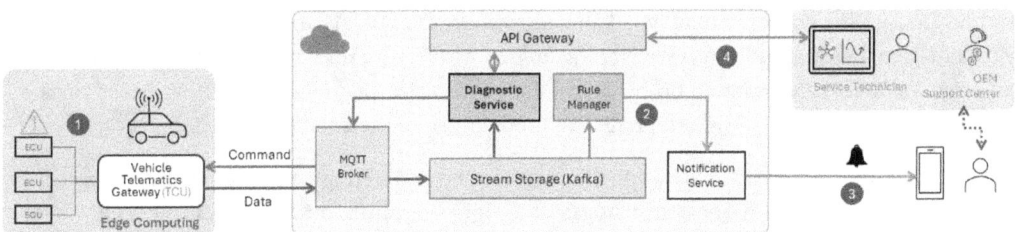

Figure 10.19 – Remote diagnostics flow diagram via the IoT backend

The preceding flow diagram details how the diagnostics data flows from the edge to the cloud, triggering an action by the end user. Here is a step-by-step breakdown of the remote diagnostics use-case implementation:

1. Let us say one of the ECUs in the vehicle throws a diagnostic error that is observed and logged by the TCU; this is then relayed to the cloud via the MQTT broker.

2. The diagnostic error is then placed in the Kafka event stream, and it's available for all the applications that are listening to the specific diagnostic error topic. The rule manager, in this case, processes the diagnostic error event and invokes the action associated with the rule, which is notifying the end user/owner of that vehicle.

3. Once the notification is received by the end user, the end user will review it and can schedule a remote diagnostic session with the service technician by talking to the OEM support center and providing the required authorization.

4. The service technician connects via the diagnostics application, reviews the diagnostics error code, and then invokes the remote connection to the vehicle to further diagnose the specific ECU in question. Under the hood, depending on the vehicle architecture, either a classic diagnostic stack container is instantiated or a SOVD client is used to establish a connection with the vehicle TCU.

The preceding use case for remote diagnostics begins when a certain vehicle's ECU issues a diagnostic trouble code signaling that immediate service technician attention is required. However, it would be significantly more beneficial for the vehicle owner to receive an early warning about a potential failure of a vehicle component or the ECU itself. In the following section, we will explore methods for achieving such proactive notifications.

Predictive maintenance

The goal of predictive maintenance is to forecast potential failures or issues before they occur, allowing for maintenance to be scheduled at a convenient time, thereby minimizing downtime and reducing costs.

To accomplish this goal, we need to collect data from the component or ECU that is being monitored and then analyze the collected data by advanced analytics or ML algorithms to identify patterns, trends, or anomalies that indicate a developing failure. Models are then developed to predict future component or ECU failures based on historical data and real-time monitoring. Now, based on the insights gained from the analysis and predictive modeling, maintenance can be scheduled.

The same framework can be utilized during the development and testing phases of vehicle components to predict failures and improve design. This assists **Original Equipment Manufacturers** (**OEMs**) in preventing design failures in the field, thus reducing recalls.

Let us look at how the IoT architecture can be extended to build the preceding predictive maintenance use case. In the *Stream processing* section, we learned about Lambda architecture (*Figure 10.6*), where the speed layer corresponds to the telemetry service and the rule manager, which delivers real-time views and triggers actions. To analyze the data by advanced or ML algorithms, we need to store the data in a data lake or object store. The data is then batch-processed by the analytics service and insights are generated at a scheduled frequency. These insights can then be routed through the rule manager to trigger predictive notifications. *Figure 10.20* depicts an IoT backend architecture that is extended to support a predictive maintenance use case.

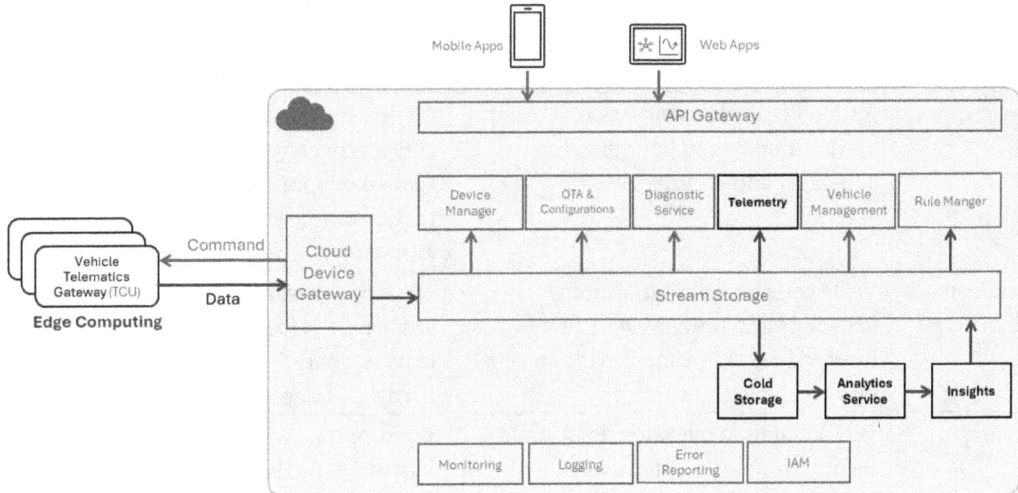

Figure 10.20 – IoT backend architecture extended to support ML and insights generation

With a comprehensive understanding of the cloud backend and the vehicle telematics gateway technologies, as well as their respective components, it's time to turn our attention to the development process. In the next section, we will understand how software is developed for cloud applications and gateway devices.

Development process

Developing software for cloud applications and embedded devices involves distinct processes, methodologies, and considerations, but both share common stages in their development life cycle. Table 10.7 shows the different stages of cloud application and embedded software development.

Stage	Cloud Application Development	Embedded Software Development
Requirements analysis	Understand needs, objectives, scalability, security, and compliance requirements	Understand functional and non-functional requirements, focusing on hardware constraints such as memory and power consumption
Design	Architect the solution, selecting cloud service models and designing for scalability and security	Design software architecture considering hardware limitations and **real-time operating system** (RTOS) requirements. Select microcontrollers and sensors.
Development	Write code using programming languages and frameworks suitable for the cloud. Implement APIs and business logic.	Develop software in languages such as C or C++, focusing on device drivers, control algorithms, and hardware integration
Testing	Perform unit, integration, load, and security testing. Test for scalability, performance under load, and useability.	Perform unit, integration, and **hardware-in-the-loop** (HIL) testing. Focus on real-time performance and resource consumption.
Deployment	Deploy to the cloud environment using **continuous integration/continuous deployment** (CI/CD) practices. Configure infrastructure and services.	Flash the software onto the device's memory, configure initial settings, and calibrate.
Monitoring and maintenance	Monitor for performance, security, and reliability. Perform regular updates and scale resources.	Perform remote diagnostics and firmware updates. Focus on bug fixes and accommodating new hardware components.
Security and compliance	Ensure compliance with regulations. Implement encryption, authentication, and authorization measures.	Ensure compliance with industry standards and possibly obtain certification for regulated industries.

Table 10.7 – Comparing development stages of cloud applications and embedded software

Cloud applications emphasize scalability, security, and integration with other web services, while embedded device software development focuses on hardware constraints, real-time performance, and energy efficiency. As the vehicle architectures start using HPCs, some cloud development technology stacks, tools, and processes are being leveraged by embedded software development teams.

Summary

Software design and development is a well-defined process that starts with requirement analysis and software design. In this chapter, we explored the software design for an automotive IoT application, focusing on remote diagnostics. We introduced the cloud backend deployment (public, private, and hybrid clouds) and service models (IaaS, PaaS, and SaaS). The IoT application architecture was examined through components such as cloud device gateways, edge computing, stream processing, and device management. OTA solutions for vehicle ECU maintenance, telemetry data stores, and NoSQL databases were discussed, followed by the critical role of rule engines and API gateways in IoT infrastructure. The vehicle telematics gateway was detailed, noting its software components. Finally, the development process was compared between cloud applications and embedded software.

In the next chapter, we will deep dive into the deployment and maintenance of an automotive IoT application and understand cloud operations.

References

- [1] https://mqtt.org/
- [2] https://www.openmobilealliance.org/release/LightweightM2M/V1_2-20201110-A/HTML-Version/OMA-TS-LightweightM2M_Core-V1_2-20201110-A.html
- [3] https://kafka.apache.org/
- [4] https://uptane.org/
- [5] https://theupdateframework.io/
- https://aws.amazon.com/blogs/architecture/software-defined-edgearchitecture-for-connected-vehicles/

Get This Book's PDF Version and Exclusive Extras

Scan the QR code (or go to `packtpub.com/unlock`). Search for this book by name, confirm the edition, and then follow the steps on the page.

Note: Keep your invoice handy. Purchases made directly from Packt don't require one.

11
Deploying and Maintaining an Automotive IoT Application

In the previous chapter, we delved into the software design of an automotive **Internet of Things** (**IoT**) application, focusing on remote diagnostics. We will now turn our attention to the deployment and maintenance aspects of the application, in both the cloud and the vehicle.

In this chapter, we will concentrate on the collaborative efforts between the software engineering teams and other key stakeholders to deploy the software application. We will employ the DevSecOps life cycle as our guiding framework, detailing specific activities, tools, and stakeholder interactions throughout the process. Additionally, we will explore how deployment pipelines are established and managed across all the stages of the DevSecOps life cycle.

In this chapter, we're going to cover the following main topics:

- The DevSecOps life cycle – an overview of integrating security at every phase of the software life cycle
- **Continuous integration** (**CI**) – strategies for coding, building, and testing to facilitate seamless integration
- **Continuous deployment/delivery** (**CD**) – methods for releasing and deploying updates efficiently
- Operation and monitoring – techniques for managing and monitoring applications to ensure performance and reliability

The DevSecOps life cycle

The DevOps life cycle is a continuous process that integrates **development** (**Dev**) and **operations** (**Ops**) teams to deliver software applications and services more efficiently and reliably. DevSecOps integrates security practices (refer to *Chapter 6, Exploring Secure Development Processes for Automotive IoT*) into the traditional DevOps life cycle, ensuring that security is prioritized throughout the software development process. Here's an overview of the stages in the DevSecOps life cycle.

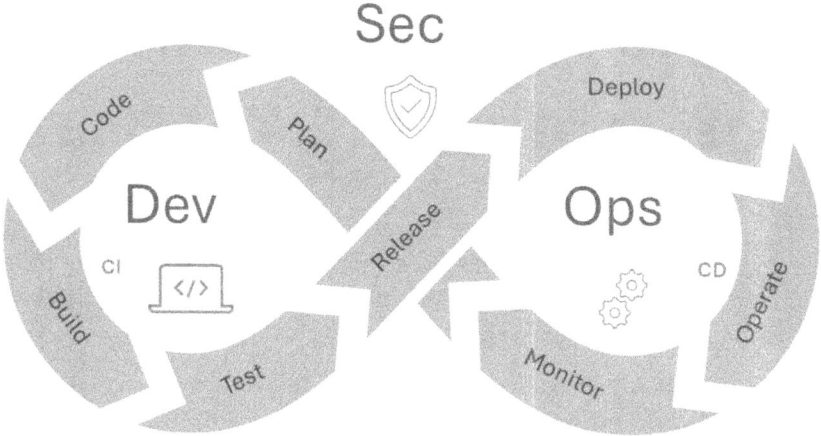

Figure 11.1 – The DevSecOps life cycle

The DevSecOps life cycle ensures rapid deployment and high-quality output through continuous collaboration and integration across various stages. Here are the high-level details of each phase:

- **Plan/Planning** is the initial stage. In this stage, teams define the objectives and requirements for the software development process. This involves gathering user stories, prioritizing tasks, and creating a roadmap for development. Collaboration between development, operations, and other stakeholders is crucial to ensure alignment of goals and expectations.

- **Code/Coding** is the coding phase, where developers write and commit code changes to version control repositories. They follow coding standards and best practices, and collaborate on code reviews to maintain code quality and consistency. CI practices are often employed to automatically build and test code changes as they are committed.

- **Build/Building** is the stage that involves compiling the source code, packaging dependencies, and generating executable artifacts. The goal is to create deployable artifacts that are ready for testing and deployment.
- **Test/Testing** is the stage where testing is performed at different stages (unit, integration, system, and acceptance testing) to validate the functionality, performance, security, and usability of the application. Automated testing frameworks and tools detailed in the upcoming Test section are used to streamline testing and provide rapid feedback to developers.
- **Release/Releasing** is the stage where validated code changes are deployed to production or staging environments. CD practices enable automated deployment pipelines to push changes to production with minimal manual intervention.
- **Deploy/Deploying** is the stage where the application is provisioned, configured, and launched in the target environment. **Infrastructure as code** (**IaC**) tools detailed in the upcoming deploy section are used to automate infrastructure provisioning, ensuring consistency and repeatability across environments. Configuration management tools are used to automate the configuration of servers and services.
- **Operate/Operating** is the stage where the application that is deployed enters the operations phase where it is monitored, managed, and maintained. DevOps teams use monitoring tools detailed in the Operate and Monitor sections to track performance metrics, detect anomalies, and troubleshoot issues in real time. Automated alerts and notifications help ensure timely responses to incidents, minimizing downtime and impact on users.
- **Monitor/Monitoring** is the stage where continuous monitoring is essential to gain insights into application performance, user behavior, and system health. Monitoring tools detailed in the Monitor section collect and analyze metrics, logs, and traces to identify trends, patterns, and potential bottlenecks such as resources, data volume, scalability, and so on. This feedback loop informs future iterations of the development process, driving continuous improvement and optimization.

As we navigate through the intricacies of the DevSecOps life cycle, it's clear that the seamless integration of development, security, and operations is pivotal to achieving rapid deployment and maintaining high-quality standards. This life cycle is characterized by a continuous loop of feedback and improvement across various stages.

Now, let us dive deep into each of these stages, starting with the planning stage.

The plan stage

The **Plan** stage sets the foundation for the entire development and deployment process. Let us now go through the list of activities performed in the planning stage:

1. **Objectives and requirements gathering**:

    I. Engage with stakeholders, including business owners, product managers, customers, and end users, to understand their needs, expectations, and objectives for the project

    II. Gather and document functional and non-functional requirements [1], capturing both technical and business requirements that need to be addressed during development

2. **Architecture and design**:

    I. Define the high-level solution architecture and design of the system, considering scalability, performance, security, and other quality attributes

    II. Determine the technologies, frameworks, and platforms to be used

    III. Design the system architecture, components, modules, and interfaces

    IV. Define the data models, **Application Programming Interface** (**APIs**), and integration points

    V. Consider factors such as modularity, extensibility, and maintainability

3. **Resource planning**:

    I. Determine the computing, storage, networking, and other infrastructure resources needed to support the development, testing, and deployment of the application or service

    II. Identify and select the development tools, frameworks, libraries, and platforms required for the project

    III. Consider factors such as compatibility, scalability, and ease of use

4. **Security and compliance**:

    I. Define security requirements and controls to protect the confidentiality, integrity, and availability of data and resources

    II. Recognize and assess potential security risks and vulnerabilities

    III. Maintain adherence to regulatory mandates, industry norms, and company policies

    IV. Address legal, privacy, and data protection requirements as applicable

5. **Roadmap and prioritization:**

    I. Develop a roadmap or project plan outlining the schedule, milestones, and deliverables for the project

    II. Define iterations, sprints, or release cycles based on Agile or iterative development methodologies

    III. Prioritize features, user stories, and tasks based on business value, customer feedback, and project goals

    IV. Determine the order of implementation to maximize value and minimize risk

6. **Collaboration and communication:**

    I. Ensure alignment and shared understanding among team members, including developers, testers, operations, and other stakeholders

    II. Establish communication channels and workflows for sharing information, updates, and feedback

    III. Use collaboration tools such as chat platforms, project management tools, and version control systems

7. **Risk management:**

    I. Identify potential risks, uncertainties, and dependencies that may impact success

    II. Develop strategies and contingency plans to mitigate risks and address uncertainties

The aforementioned detailed planning process steps are largely identical between the cloud and embedded software development teams. The automotive embedded software development teams are accustomed to the discipline of maintaining comprehensive traceability from requirements, through the design and coding phases to testing. *Table 11.1* details the tools used for various stages of project management and development:

Aspect	Description	Tools
Project management	Facilitate task management, project planning, and progress tracking. Enable teams to create and manage project timelines, milestones, and deliverables.	Jira, Asana, Wrike, Microsoft Project
Requirement management	Capture, document, and track requirements, user stories, and acceptance criteria. Facilitate collaboration among stakeholders and ensure alignment with project objectives.	Jama Connect, IBM DOORS Next, Confluence

Aspect	Description	Tools
Architecture design	Create visual diagrams, flowcharts, and architecture models to represent system components and interfaces. Aid in designing and communicating system architecture effectively.	Enterprise Architect, Lucid chart, `draw.io`, PlantUML
Collaboration and communication	Enable real-time communication, file sharing, and collaboration among team members and stakeholders. Facilitate discussions, decision-making, and knowledge sharing.	Slack, Microsoft Teams, Zoom

Table 11.1 – Tools used in the plan stage

By investing time and effort in the plan stage, teams can ensure that objectives are well-defined, aligned with business goals, and set up for success. Clear objectives and requirements and a solid roadmap lay the groundwork for effective development, testing, and deployment in subsequent stages of the DevOps life cycle.

Let us now delve into CI and the stages that it is involved in.

CI

CI is a process that encompasses the code, build, and test stages, and automates the building and testing of code to quickly identify and address integration issues, enhancing software quality and development speed. Let us dive deep into the coding stage.

The code stage

The code stage involves writing, reviewing, and versioning the source code for the application or system. Code reviews are conducted to ensure code quality, maintainability, and adherence to coding standards. *Table 11.2* details the various tools used in this code stage.

Aspect	Description	Tools
Version control systems	Used to manage and track changes to the code base. They provide features for branching, merging, tagging, and versioning of code.	Git, **Subversion (SVN)**, Mercurial, AzureRepos
Code reviews	Facilitate peer code reviews, allowing developers to provide feedback, discuss changes, and ensure code quality before merging changes into the main code base.	GitHub pull requests, Bitbucket code reviews, Gerrit

Aspect	Description	Tools
Static code analysis	Used to analyze source code without executing it. They scan code for potential defects, security vulnerabilities, coding rule violations, and other issues. These tools help developers identify and fix problems early in the development process.	SonarQube, PMD, ESLint, Cppcheck, Coverity, CodeSonar
Unit testing	Unit-testing frameworks, libraries, and utilities that enable developers to write, automate, and analyze unit tests efficiently.	JUnit, Pytest, Mocha, CppUTest, **Google Test (gtest)**

Table 11.2 – Tools used in the code stage

Git operates as a distributed version control system, while SVN functions as a centralized version control system.

In a version control system, the branching strategy is a set of rules and practices that govern how branches are created, used, and managed. It outlines the structure of branches, their purpose, and the workflow for integrating changes into the code base. A well-defined branching strategy helps streamline development processes, improve collaboration, and maintain code base integrity, as shown in *Figure 11.2*.

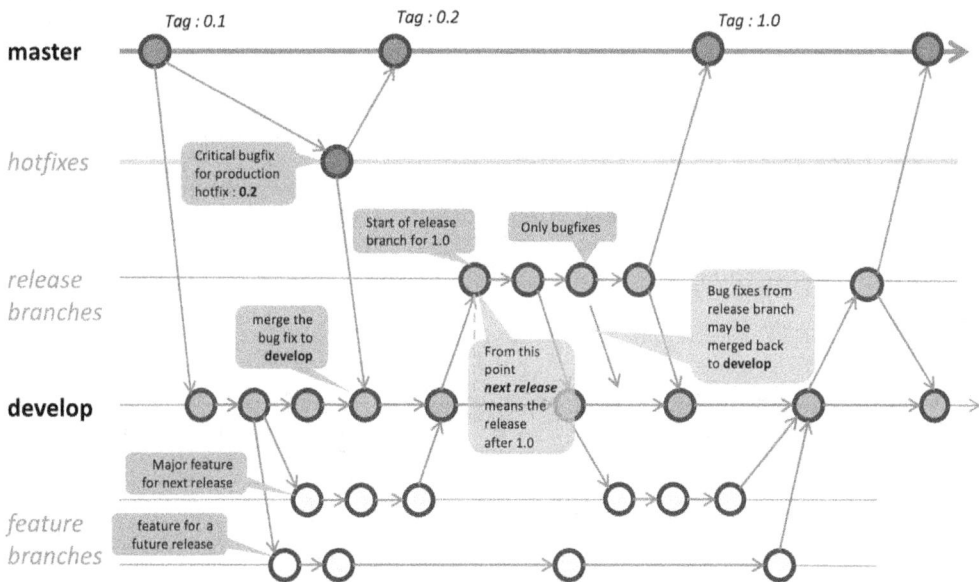

Figure 11.2 – Git's feature branching strategy

So, in the coding stage, developers, through code reviews, ensure the quality and maintainability of the code base. Additionally, employing static code analysis tools aids in identifying and rectifying potential defects and security vulnerabilities early in the development process. Unit testing verifies individual components for correctness, catching errors early, and ensuring each part functions as expected before integration.

With this groundwork laid, we can now explore the build stage's pivotal role in transforming code into deployable artifacts.

The build stage

The **build** stage involves compiling the source code, running automated tests, and generating deployable artifacts. Builds refer to the process of compiling and linking code into a runnable application or system. CI servers automatically trigger builds whenever changes are committed to the version control repository. **Build scripts** or **configuration files** define the build process, specifying dependencies, compilation steps, testing procedures, and artifact generation. Automated tests, including unit tests, integration tests, and functional tests, are executed during the build process to verify code correctness and quality.

In large software projects, building everything from scratch can take hours. This is too slow for a CI workflow, where we aim for frequent updates and fast feedback. To address this, we use **incremental builds**. Incremental builds only compile the specific parts of the code that have changed and any code that relies on those changes. This significantly reduces build times and makes CI much more efficient.

Table 11.3 outlines the different tools employed in the building stage.

Aspect	Description	Tools
Build automation	Automates the build process, orchestrating compilation, testing, and artifact generation	Jenkins, TeamCity, CircleCI
Build scripts/ configuration files	Define the build process and dependencies	Shell scripts, Gradle scripts, YAML files
Artifact management	A repository that stores the build artifacts, such as images, executables, and libraries	Artifactory, Nexus Repository, AWS CodeArtifact, Azure Artifacts, Google Cloud Artifact Registry

Table 11.3 – Tools used in the build stage

There are different types of builds that serve various purposes at different stages of the development process. Here are some common build types:

- **Development builds** are created during the development phase to test new features, changes, or bug fixes. These builds are typically created frequently and may be less stable than other types of builds. Developers use development builds for testing, debugging, and validating changes before merging them into the main code base. Development builds are digitally signed and packaged (prepared for distribution) using development certificates.
- **Nightly builds** are scheduled builds that automatically run daily, typically during off-peak hours. These builds compile the latest code changes and run automated tests to validate the stability and quality of the code base. Nightly builds provide feedback to developers and stakeholders on the state of the code base and help identify issues that may have been introduced during the day.
- **Release builds** are created when the software is ready for deployment to production or release to end users. These builds undergo rigorous testing, including regression testing, performance testing, and user acceptance testing, to ensure that they meet quality standards and are suitable for release. Release builds are often tagged or labeled with a version number, production signed, and documentation generated.

These are some of the common types of builds used in the software development process. Each type serves a specific purpose and plays a crucial role in ensuring the quality, stability, and reliability of the software throughout its life cycle.

The generated artifacts from the build such as images, executables, and libraries need to be stored in a reliable and accessible repository of deployable assets. Several tools specialize in artifact management, providing versioning, storage, and distribution capabilities. All cloud providers offer a managed solution to manage the artifacts, such as **Azure Artifacts**, **Google Cloud Artifact Repository**, and **AWS CodeArtifact**.

In the case of embedded environments, the software image generated is flashed to the target hardware directly, and if it's an **over-the-air** (**OTA**) update, then a delta image is generated. A **delta image** is a partial image that is a difference between the version of the software on the target hardware and the new version of the software. In the case of a cloud environment, the images generated are full Docker images that will be deployed as microservices in a virtual environment.

Overall, the build stage orchestrates the compilation of source code, generation of target images, and execution of automated tests. Now, let us explore the different types of testing employed in both embedded and cloud developments by the **quality assurance** (**QA**) teams.

The test stage

Testing for embedded and cloud deployments involves various approaches to ensure the reliability, security, performance, and scalability of applications and services running in their respective environments. *Figure 11.3* shows some of the different types of testing commonly used for cloud deployments.

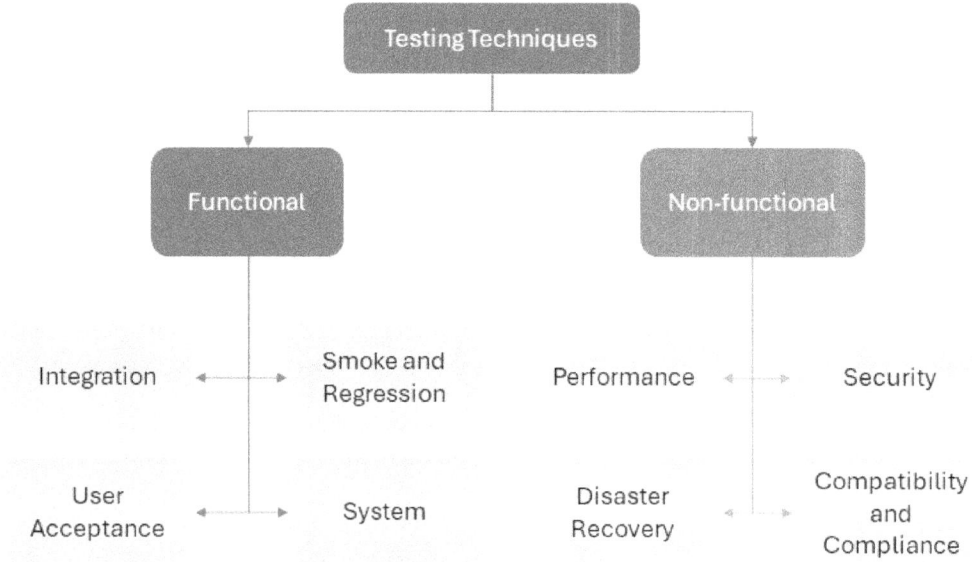

Figure 11.3 — Testing techniques

Functional testing

Functional testing involves testing the functionalities of the application to ensure that it behaves as expected in the cloud environment. It includes the following:

- **Smoke testing**: This quickly verifies basic functionalities to ensure that critical components work. This is done to save time and catch failures early, as running a full regression test suite is both resource-intensive and time-consuming.
- **Regression testing**: This ensures that new changes do not negatively impact existing functionalities.
- **Integration testing**: This validates that different modules or components of the system work together as expected.
- **System testing**: This ensures verifying that the systems perform all the functions as specified in the requirements documentation.
- **User acceptance testing**: This includes tests conducted by end users to verify whether the system meets their requirements.

These functional testing techniques are the same for both the embedded application and the cloud application. Table 11.4 provides a sample of the various tools and frameworks used in functional testing.

Aspect	Description	Tools/Frameworks
Cloud applications	For automating functional testing across web, mobile, and API applications, there are various tools and frameworks that offer a variety of features and capabilities	Selenium, Cypress, Appium, Robot Framework, Katalon Studio, Postman
Embedded automotive	This lists some of the tools used in the automation of functional testing in the vehicle ECUs	VectorCAST, CANoe/CANanlyzer, National Instruments LabVIEW
Testcase management	All the tests and respective test cases are managed using test management tools	TestRail, Zephyr, qTest

Table 11.4 – Functional testing tools and frameworks

There are other testing methods employed to ensure the functionality, safety, and reliability of the vehicle systems. The following are three key testing methodologies:

- **Software-in-the-loop (SIL) testing** involves testing the software component of a system in a simulated environment. Here, the software runs on a virtual model or simulation of the vehicle, rather than on the actual hardware. This allows for early validation and verification of software functionalities without the need for physical prototypes.

- **Hardware-in-the-loop (HIL) testing** involves testing the interaction between the hardware components of a system and the embedded software. In this method, the actual **Electronic Control Units** (**ECUs**) and other hardware components are connected to a simulation environment that simulates real-world conditions. This allows for comprehensive testing of the hardware-software interaction without the need for a physical vehicle.

- **Vehicle-in-the-loop (VIL) testing** integrates both physical vehicles and virtual simulations to evaluate the performance of vehicle systems under realistic operating conditions. In VIL testing, the vehicle is connected to a simulation environment that can simulate various driving scenarios, road conditions, and environmental factors. This allows for comprehensive testing of vehicle systems, including interactions between different subsystems and components, in a controlled and repeatable manner.

These testing methodologies are essential for identifying and addressing issues early in the development process, thereby reducing development time and costs and ultimately improving the quality and reliability of automotive systems.

Non-functional testing

Non-functional testing evaluates aspects of the system beyond the functional behavior, such as performance, reliability, scalability, and security.

Performance testing is crucial to evaluate the speed, responsiveness, and stability of the application under different workload conditions. It includes the following:

- **Load testing**: Measures system behavior under anticipated load conditions
- **Stress testing**: Assesses system performance beyond its operational limits
- **Scalability testing**: Determines how the application performs as usage traffic increases, which includes horizontal and vertical scale testing
- **Endurance testing**: Evaluates how the system handles increasing workloads

Some of the tools used for performance testing cloud applications include **Apache JMeter**, **LoadRunner**, **Gatling**, and **BlazeMeter**.

Security testing focuses on identifying vulnerabilities and ensuring that data and resources are protected in the cloud and vehicle environments, as detailed in *Chapter 7, Establishing a Secure Software Development Platform*.

Disaster recovery testing verifies the integrity and effectiveness of backup mechanisms and recovery procedures in the cloud environment to ensure data integrity and business continuity. It includes the following:

- **Data backup and recovery testing**: Verifies the effectiveness of data backup and recovery mechanisms
- **Data consistency testing**: Ensures that data remains consistent and accurate across distributed cloud environments
- **Fault injection testing**: Intentionally introduces faults/failures into the system to assess its resilience and fault tolerance

Compatibility testing ensures that the application functions correctly across different cloud platforms, browsers, devices, and operating systems.

Compliance testing ensures that the application meets regulatory and industry-specific compliance requirements in the cloud environment, such as **General Data Protection Regulation (GDPR)** or **Service Organization Control 2 (SOC 2)**.

By employing a combination of these testing types, we can ensure that applications and services deployed in the cloud and vehicle environments are robust, secure, and performant, ultimately delivering value to end users and stakeholders.

CD

CD covers the release and deploy stages and ensures seamless and swift transitions from development to live use, minimizing manual intervention and accelerating delivery cycles.

The release stage

The **release** stage involves preparing the software artifact for deployment into production or operational environments. It involves activities such as packaging, versioning, tagging, and documenting software releases. This stage can also be considered as a gating process to determine whether the software artifact can be deployed to the target environment.

The embedded application software release goes to the ECU manufacturing plant or is handed over to the operations team to schedule an OTA campaign or a dealer-managed update. All this coordination, decision-making, and communication among various stakeholders is part of the release stage.

In the case of cloud applications, the software is released to the respective deployment environment such as staging, pre-production, or production via a deployment pipeline. In the next section, we will do a deep dive into various deployment strategies.

The deploy stage

In the case of an embedded application, the hardware manufacturing facility receives the software package via a secure channel. Typically, if the software image size is large, then it is pre-flashed so that the manufacturing takt time is optimized. Unique security keys and certificates are pre-provisioned in the factory's manufacturing line, software is flashed, and diagnostic tests are run at the **end of line** (**EOL**) before they are shipped to the vehicle integration plant. In the case of any issues identified post-production that need a software fix, this will result in a stop-shipment of the vehicles from the manufacturing plant, depending on the criticality of the issue identified.

In the case of cloud applications, the deployment stage is all about securely and reliably releasing applications to different environments. It emphasizes automation, security best practices, and a smooth transition from development to production.

Single-tenant and multi-tenant architectures represent two different approaches to deploying cloud services, each with distinct advantages. A **single-tenant architecture** provides an isolated environment where one customer's data and applications are hosted on dedicated servers. This model offers enhanced security, greater customization options, and stable performance, making it suitable for businesses with stringent data security and regulatory compliance requirements. However, it tends to be more costly due to the dedicated resources required.

On the other hand, a **multi-tenant architecture** hosts data and applications for multiple customers on a shared infrastructure. This approach is more cost-effective, as expenses for software, infrastructure, and maintenance are spread across several users. Multi-tenant solutions also allow for easier scalability and faster deployment of updates, benefiting all tenants simultaneously. However, they might pose challenges in terms of limited customization possibilities and potential performance interferences, known as the *noisy neighbor effect*, where the activity of one tenant could impact the performance experienced by others.

Figure 11.4 illustrates these concepts concretely.

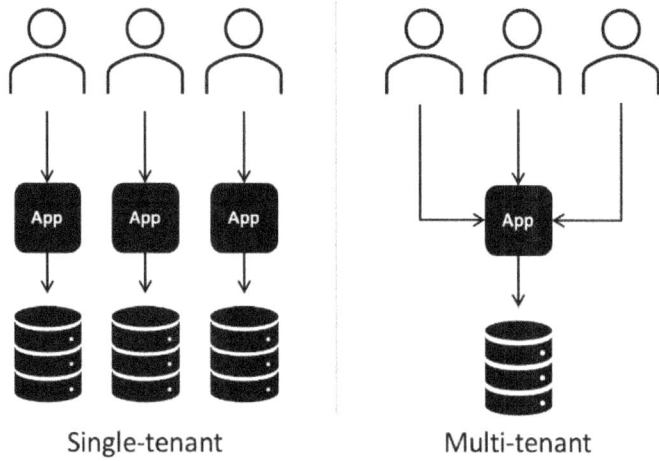

Figure 11.4 – Single-tenant versus multi-tenant deployments

Overall, the goals of a deployment stage are the following:

- **Secure delivery** of the application to its target environment (testing, staging, or production) while adhering to security best practices. This may involve measures such as secure configuration management, access control restrictions, and vulnerability scanning of deployed code.

- **Reliability and consistency** by ensuring a smooth and consistent deployment process across different environments. This minimizes errors and prevents downtime during deployments.

- **Repeatability** by automating as much of the deployment process as possible to ensure repeatable and reliable deployments. This allows for faster rollouts and reduces the risk of human error.

Achieving the goals listed sets a solid foundation for the deployment phase. With these pillars established, we will now focus on selecting the most suitable deployment strategies.

Deployment strategies

Deployment strategies are methodologies used to release software changes into production environments while minimizing risk and disruption to users. Different deployment strategies offer various trade-offs between deployment speed, reliability, and user impact.

Figure 11.5 illustrates some deployment strategies.

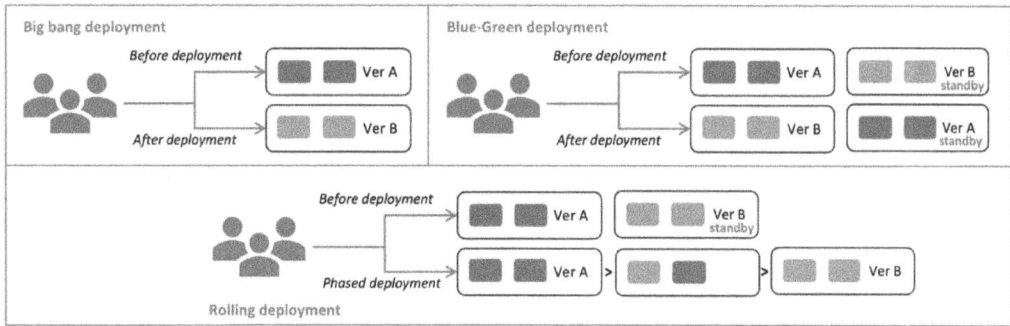

Figure 11.5 – Big bang, blue-green, and rolling deployment strategies

These common deployment strategies are discussed here:

- **Big bang deployment** is when all changes are deployed simultaneously to the entire production environment in one go. It's characterized by a single, comprehensive release of all updates, features, or fixes.

    These deployments are relatively straightforward to execute since they involve deploying a single version of the software or configuration changes across the entire environment. This simplicity can lead to faster deployment times compared to more complex deployment strategies.

    This strategy is typically suitable for smaller projects or updates, where the risk of failure is relatively low and the changes are well-understood and thoroughly tested.

- **Blue-green deployment** has two identical production environments (blue and green) that are maintained. At any given time, only one environment serves live traffic (e.g., blue).

    When deploying a new version, the changes are first deployed to the inactive environment (green). After the deployment is successful and validated, the traffic is switched from the active environment (blue) to the newly deployed environment (green) by updating **Domain Name System** (**DNS**) records or load balancer configurations.

    This strategy allows for zero-downtime deployments and easy rollback if issues are discovered.

- **Rolling deployment** involves gradually updating instances or servers with the new version while maintaining application availability.

A rolling deployment typically involves replacing or updating a subset of instances one at a time, ensuring that there are always enough healthy instances to handle incoming traffic. This strategy allows for the gradual rollout of changes with minimal impact on application performance and user experience.

Rolling deployments are often automated using deployment orchestration tools to streamline the process and minimize manual intervention.

In addition to the aforementioned deployment strategies, we also have the following:

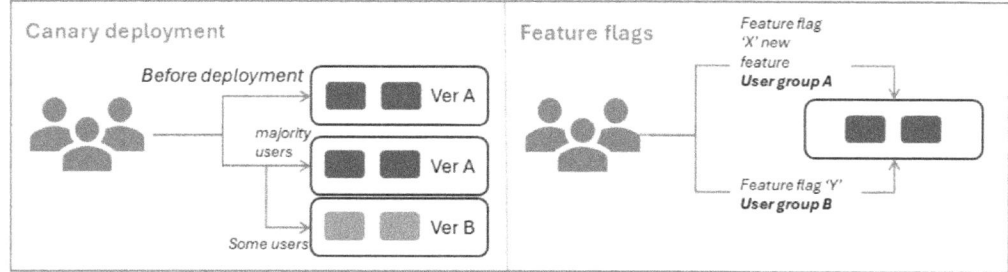

Figure 11.6 – Canary and feature flagging deployment strategies

- **Canary deployment** involves rolling out new changes to a small subset of users or servers before making them available to the entire user base.

  A small percentage of traffic is redirected to the new version (canary), while most of the traffic continues to be served by the stable version. User feedback and metrics are closely monitored during the canary deployment to detect any anomalies or performance issues.

  If the canary deployment is successful and stable, the rollout is gradually expanded to a larger audience. Otherwise, the deployment can be aborted or rolled back.

- **Feature flagging** (also known as **feature toggles**) involves deploying new features or changes to production but enabling them conditionally, based on feature flags.

  Feature flags allow developers to control the visibility and activation of features at runtime without deploying new code. This strategy enables the decoupling of deployment and release, as features can be deployed in an inactive state and gradually activated for different user segments.

  Feature flagging provides the flexibility to roll back changes quickly by simply toggling the feature flag off if issues arise.

These deployment strategies offer different approaches to releasing software changes into production environments, each with its own benefits and considerations. The choice of deployment strategy depends on factors such as the application architecture, risk tolerance, user impact, and organizational requirements.

Regions and availability zones

To ensure reliability, performance, and disaster recovery, there are multiple isolated physical locations [2][3][4] known as **availability zones** (**AZs**). Each AZ is one or more distinct data centers, each equipped with redundant power, networking, and connectivity. Each AZ is set up to be an isolation zone to contain failures within a single zone while remaining operationally independent from other zones in the same region.

The following diagram demonstrates the structure of AZs within a region:

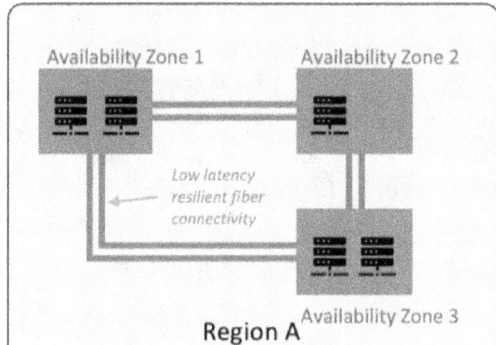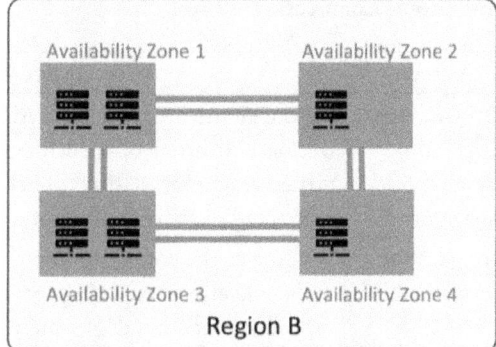

Figure 11.7 – Regions and AZs

Using AZs allows us to make the following improvements:

- **Improve fault tolerance** by deploying applications across multiple AZs so we can ensure that if one zone goes down, the other can continue to function, thereby minimizing downtime
- **Facilitate high availability** by designing systems that automatically failover to backup systems in other AZs without any service interruption
- **Manage data replication** by easily replicating data across zones to ensure quick data recovery in case of hardware failure

A **region** in cloud computing is a specific geographic location where cloud service providers operate data centers. Each region represents a separate geographic area that has been selected to host data center infrastructure. These regions are typically far apart to reduce the risk of a regional outage affecting all of them simultaneously. Typically, each region has multiple AZs; by using multiple regions, we can achieve the following:

- **Reduce latency** by serving users from data centers closer to them, which reduces the time it takes for data to travel, thereby speeding up response times

- **Adhere to compliance** by storing and processing data in specific locations in accordance with local laws and regulations regarding data sovereignty

- **Increase resilience** by spreading their resources across multiple regions to protect against regional failures due to natural disasters, political instability, or network issues

All cloud providers offer multiple regions, with a typical 2-3 AZs per region.

Now let us explore the various tools used in the deployment stage when provisioning and managing the infrastructure, managing configurations, and automating the deployments. The following table provides some examples:

Aspect	Description	Tools/Frameworks
Infrastructure provisioning and management	Automate the provisioning and configuration of cloud infrastructure, such as creating virtual machines, setting up networking, and managing security configurations	Terraform, AWS CloudFormation, Azure Resource Manager
Configuration management	Automates the configuration of infrastructure and applications, ensuring consistency and repeatability across deployments	Ansible, Chef, Puppet
Containerization technologies	Package applications into portable containers for consistent deployments across environments	Docker, Kubernetes
CI/CD pipelines	Integrate development, testing, and deployment processes, enabling CI and CD of applications	Jenkins, GitLab CI/CD, Azure DevOps

Table 11.5 – Tools used in the deploy stage

By implementing a secure and well-defined deployment stage within the DevSecOps life cycle, we can ensure faster, more reliable, and more secure delivery of applications.

Once applications are securely deployed to production, the focus shifts to operating and maintaining them effectively. This Operate stage ensures the applications continue to function as intended, meeting user needs and delivering business value.

The operate stage

In the case of **embedded applications**, if the vehicle is connected and approved to collect diagnostic information, then that information is aggregated in the cloud, which will help engineering teams understand the scale of the issue and work toward rolling out a fix as an OTA software update. If the vehicle is not connected, then this will result in a dealer visit, and the service technician's **diagnosis** will result in a software update or a part replacement. The removed part might be scrapped or sent to the respective manufacturer's service facility for further root cause analysis.

In the case of **cloud applications**, the operating stage involves managing and maintaining the deployed applications or systems in the production environment. It involves ensuring that the application is available, is performing as expected, and meets the user's needs.

The operate and monitor stages are interrelated but are distinct phases. Monitoring focuses on gathering data and insights about the deployed application while operating focuses on maintaining and managing the deployed application in production. So, the key activities in this operate stage are the following:

- **Resource management** involves ensuring that all resources are efficiently utilized and scaled appropriately to meet demand
- **Incident management** involves responding to and resolving issues that affect the normal operation of the application
- **Compliance and governance** involve ensuring operations adhere to legal, regulatory, and policy requirements
- **Performance optimization** involves continuously optimizing the application for better performance based on operational data and feedback
- **Disaster recovery and continuity** involve managing backup processes and disaster recovery strategies to maintain business continuity
- **Security management** involves implementing security measures to protect the application from threats and vulnerabilities in real time

Security management during this stage plays a vital role in protecting applications and data from vulnerabilities and threats. Security monitoring tools are deployed to continuously scan for vulnerabilities, suspicious activity, and potential breaches. These tools can be integrated with CI/CD pipelines to trigger alerts and automated responses when anomalies are detected.

The following table details various security service tools provided by various cloud providers:

Cloud Provider	Service Name	Description
AWS	Amazon GuardDuty	Leverages machine learning, anomaly detection, and consolidated threat intelligence to detect and prioritize potential threats across AWS accounts
	AWS Security Hub	Aggregates security alerts and findings from AWS services and partner tools to provide a comprehensive view of the security status in AWS
	Amazon Inspector	An automated security assessment service that assesses applications for vulnerabilities, exposure, and deviations from best practices
	AWS Macie	Uses machine learning to discover, classify, and protect sensitive data in AWS – particularly effective for identifying and protecting **Personally Identifiable Information** (**PII**) or intellectual property
Azure	Azure Security Center	Offers comprehensive security management and enhanced threat protection across hybrid cloud workloads, utilizing advanced analytics and global threat intelligence
	Azure Sentinel	A scalable, cloud-native SIEM and SOAR solution that uses AI to analyze data across the enterprise for security insights and automated responses
	Microsoft Defender for Cloud	Integrates with Azure Security Center to provide extended threat detection and response features across hybrid cloud environments
GCP	Security Command Center	A comprehensive security and data risk platform that helps identify and respond to threats in GCP, including vulnerability scanning and security health analytics
	Event Threat Detection	Part of Security Command Center and a tool that scans GCP logs for suspicious activity and potential security threats such as malware, spyware, and unusual patterns
	Google Cloud Armor	Provides **Distributed Denial-of-Service** (**DDoS**) protection and **web application firewall** (**WAF**) capabilities to safeguard applications on Google Cloud

Table 11.6 – Security services provided by various cloud providers

A microservices architecture entails building an application as a collection of small, independent services, each operating in its own process and interacting through lightweight methods. Managing these microservices can be complex due to their distributed nature.

Each service is responsible for implementing its own security, monitoring, and data transformation policies, which can lead to inconsistent implementations. Service configurations such as retries, circuit breakers, and timeouts are often hard-coded into each service, requiring redeployments to update them. Developers are tasked with not only business logic but also network resiliency, security, and communication concerns. Monitoring and logging are set up per service, which may require additional tooling and can result in varied levels of observability across services.

This is where the concept of a **service mesh** [5] comes in. The following diagram illustrates the responsibilities of the developer with a service mesh and without a service mesh:

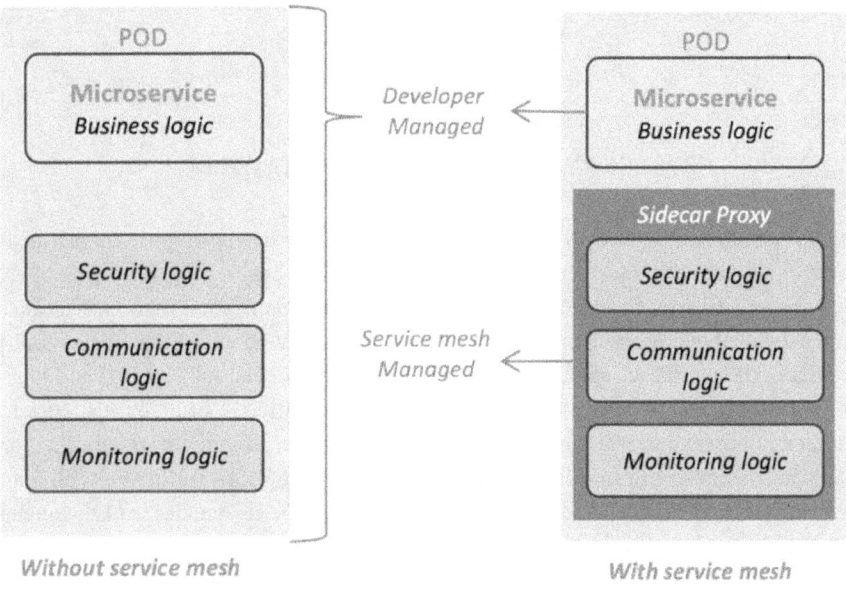

Figure 11.8 – Developer responsibilities with and without a service mesh

A service mesh introduces a sidecar proxy that sits alongside each service instance in the **data plane**. This proxy handles inter-service communication, abstracting the complexity away from the service itself. The service mesh via a **control plane** application provides a consistent way to apply security policies, access control, and observability features across all services without additional code in the services themselves.

The following diagram shows the control and data planes of a service mesh:

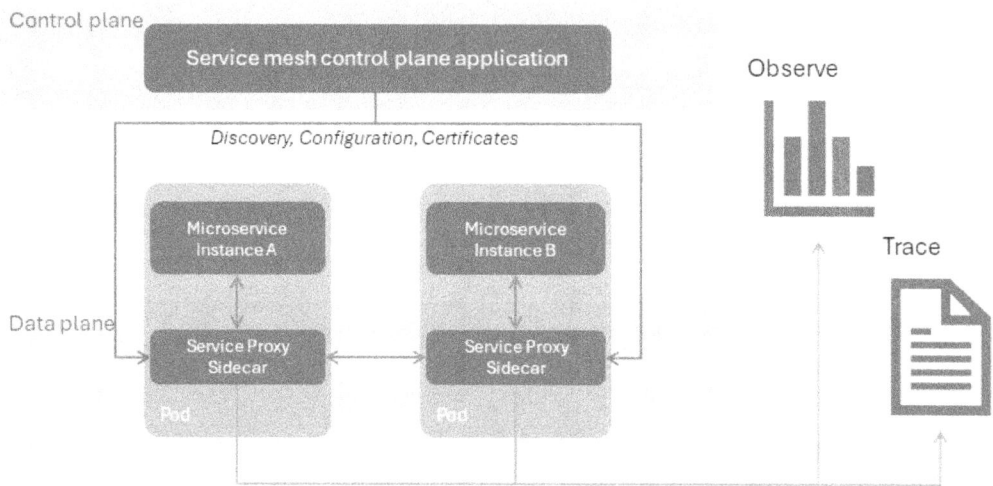

Figure 11.9 – The service mesh control and data planes

Service behavior can be configured dynamically through the service mesh without requiring code changes or redeployments of the services. Developers can focus on business logic, leaving operational concerns such as service discovery, load balancing, and secure communication to the service mesh. The mesh generates detailed telemetry for metrics, logs, and traces, giving a uniform observability layer across all services. The service mesh can facilitate canary deployments, A/B testing, and blue-green deployments with more sophisticated traffic routing rules and patterns. Built-in resilience features such as retries, circuit breaking, rate limiting, and fault injection are provided by the mesh, increasing system reliability. The mesh can handle secure service-to-service communication with automatic mutual TLS, ensuring encrypted and authenticated traffic within the cluster. Some of the open source service mesh options are Istio, Linkerd, and Consul Connect, while managed service mesh on AWS is AWS App Mesh, and on GCP, it is Google Anthos Service Mesh.

Incident management for these applications is a critical component of maintaining service reliability, customer satisfaction, and operational continuity. The process involves identifying, managing, and resolving incidents that disrupt the normal operation of these applications. An effective incident management strategy ensures that any issues are addressed swiftly and efficiently to minimize their impact on users.

The following diagram depicts an incident management workflow:

Incident management workflow

Identification → Logging → Categorization & Prioritization → Response → Investigation & Diagnosis → Resolution & Recovery

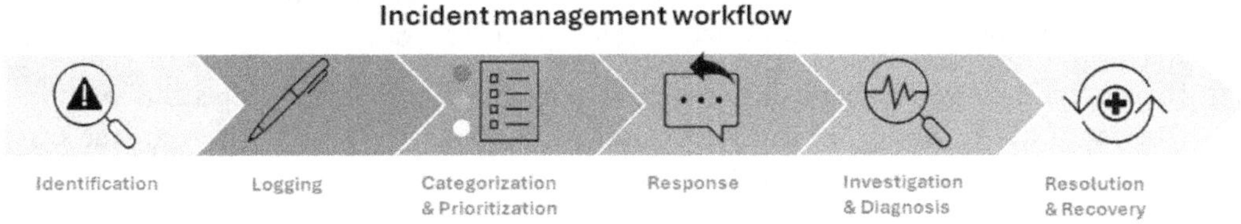

Figure 11.10 – An incident management workflow

Ticketing systems such as Zendesk and ServiceNow are used to log and track incidents. Alerts and notifications are configured to notify the relevant teams immediately when potential incidents are detected. This is usually done via incident management platforms such as PagerDuty and OpsGenie.

When an incident occurs, the **network operations center** (**NOC**) typically handles immediate incident responses and initial diagnostics before escalating complex issues. If the NOC can't resolve it, it then escalates it to the next level. **L1 Support** (**Level 1**) is the first point of contact for users experiencing IT issues. **L2 Support** (**Level 2**) addresses more complicated issues that L1 Support is not equipped to handle. **L3 Support** (**Level 3**) is the most advanced technical level, dealing with the most complex issues that often require debugging of code or deep systemic analysis. This is where the development teams get involved in supporting incidents.

The monitor stage

In the case of cloud applications, the monitor stage involves continuously tracking and reviewing the performance, health, and security of applications to ensure they operate within the desired parameters. It is critical for identifying potential issues before they become actual problems. The key activities include the following:

- **Real-time monitoring** involves using tools to monitor the system's health, performance, and anomalies in real time.
- **Alerting and notifications** involve setting up alerts based on predefined thresholds to notify staff of potential issues.
- **Log management** involves collecting, storing, and analyzing log files to help in troubleshooting and understanding application behavior over time.
- **Performance metrics** involves collecting data on various performance metrics to evaluate the application's efficiency and responsiveness.
- **Security monitoring** involves continuously scanning for security threats and vulnerabilities.
- **Compliance monitoring** involves regular checks to ensure ongoing compliance with regulatory standards.

- **Observability** refers to the ability to infer the internal state of a system based solely on its external outputs. In this context, observability is about being able to understand the system's health and performance and why it's behaving in a particular way.

  The concept encompasses more than just monitoring; it's about gaining rich insights into the behavior of systems and being able to diagnose and resolve issues quickly. Here are the key components that make up observability:

  - **Logs** that provide a chronological record of events or transactions within a system
  - **Traces** that help track the flow of requests through various services and components of a system
  - **Metrics** are quantitative measurements performance over time via quantitative measurements

There are four signals that provide a framework for what aspects of a system should be monitored to maintain its reliability and efficiency. These are called the **four golden signals**, which are the following:

- **Latency**: This is the time it takes for a request to travel from the client to the server and back
- **Traffic**: This is the number of requests a system receives over a specific period
- **Errors**: This is a percentage of requests resulting in errors
- **Saturation**: This measures resource utilization, including **Central Processing Unit (CPU)**, memory, and disk space

They are considered *golden* because they are the most important metrics, as they can tell us about the system's health with the least amount of effort.

Let us look at various monitoring and logging tools in the following table:

Tools	Description	Tools/Frameworks
Monitoring	Provides comprehensive views of application health, resource utilization, and performance metrics	Prometheus, Grafana, Datadog
Logging	Facilitates log collection, aggregation, analysis, and visualization, enabling teams to identify and troubleshoot issues	ELK Stack (Elasticsearch, Logstash, Kibana), Splunk

Table 11.7 – Various monitoring and logging tools

The following diagram shows a Grafana toolchain that can be used for aggregating logs, metrics, and traces, providing a comprehensive observability stack for cloud-native environments:

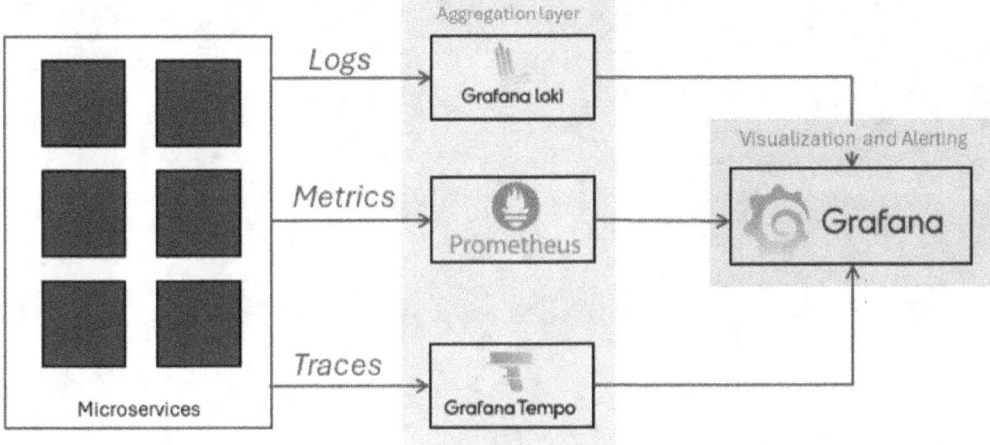

Figure 11.11 – A Grafana-based toolchain for log, metric, and trace aggregation

- **Prometheus** [6] acts as the collector in this stack. It scrapes metrics data at regular intervals from the applications and infrastructure.
- **Grafana Loki** functions as a log storage and retrieval system. It stores and indexes logs, which are detailed records of events happening within your systems. Logs can provide valuable insights into application behavior and potential errors.
- **Grafana Tempo** takes care of trace storage and retrieval. Traces map the entire journey of a request through your system, pinpointing bottlenecks and performance issues.
- **Grafana** is the visualization layer. It ties everything together by providing a unified interface to query, analyze, and visualize the metrics, logs, and traces collected by Prometheus, Loki, and Tempo. We can create dashboards to monitor **key performance indicators** (**KPIs**) and gain insights into the health and performance of the applications.

The following screenshot displays a Grafana dashboard:

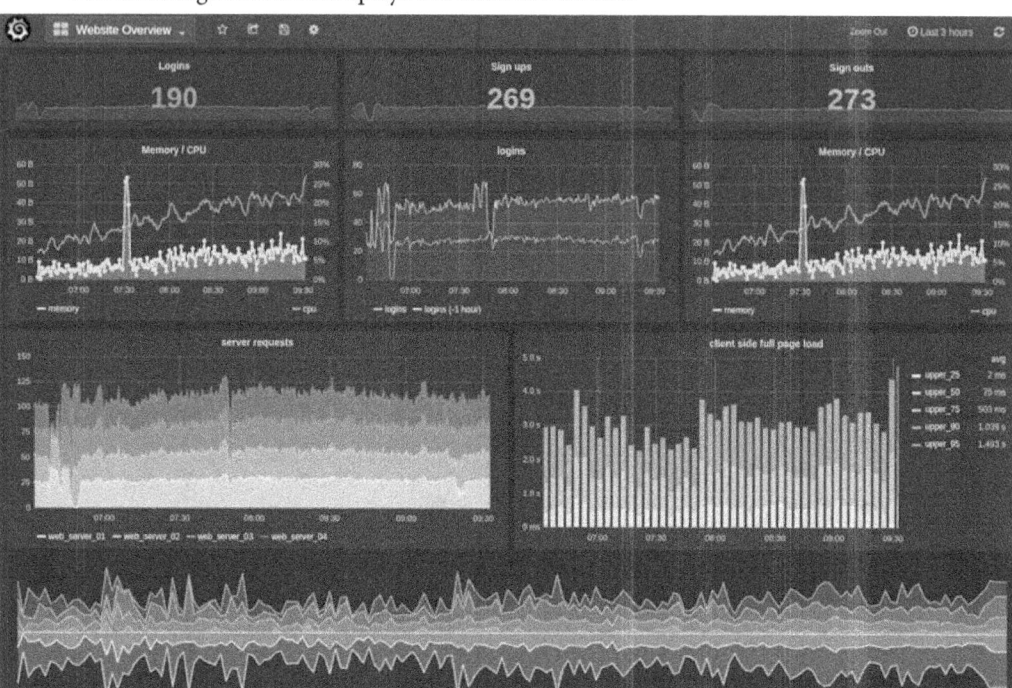

Figure 11.12 – A Grafana dashboard [7]

The **Elastic Stack** (**ELK Stack**) is also widely used for observability of software systems. The ELK Stack consists of Elasticsearch, Logstash, Kibana, and Beats. Most cloud providers support various managed services for monitoring and logging the cloud services and applications. The following table provides the details:

Cloud Provider	Tool/Services	Description
AWS	Amazon CloudWatch	Provides monitoring for AWS cloud resources and applications, tracking metrics and logs
	AWS CloudTrail	Provides governance, compliance, and operational and risk auditing of AWS accounts
	AWS X-Ray	Analyzes and debugs production applications, especially in microservices architectures
	Amazon Elasticsearch Service	Enables real-time analysis of log data using Elasticsearch

Cloud Provider	Tool/Services	Description
Azure	Azure Monitor	Provides comprehensive resource and application monitoring across Azure and on-prem environments
	Azure Log Analytics	Collects and analyzes telemetry data using queries
	Azure Application Insights	Provides a performance management service that detects anomalies, provides analytics, and helps us understand app usage
	Azure Event Hubs	Provides a big data streaming platform and event ingestion service
GCP	Google Cloud Monitoring	Monitors the health and performance of applications across cloud and on-prem resources
	Google Cloud Logging	Manages log data and events from Google Cloud and AWS
	Google Cloud Trace	Collects latency data from applications and displays it for performance optimization
	Google Cloud Pub/Sub	Provides a messaging service for exchanging messages between applications

Table 11.8 – Managed services provided by cloud providers for monitoring and logging

Overall, the DevSecOps life cycle is characterized by collaboration, automation, and feedback loops at each stage, enabling organizations to deliver high-quality software applications and services rapidly, reliably, and efficiently. By embracing DevSecOps principles and practices, teams can accelerate innovation, reduce time to market, and enhance the overall software delivery process.

Summary

The DevSecOps life cycle emphasizes a shift-left approach to security, as a fundamental and proactive part of the development process rather than as an afterthought. The life cycle is depicted as a continuous, iterative process aiming for rapid, reliable software delivery. We explored each stage of the life cycle and identified the specific activities, tools, and stakeholders involved. During the Code, Build, and Test stages of CI, we observed that most of the tools and practices are similar between cloud and embedded application development. The Release, Deploy, Operate, and Monitor stages of CD are where the process and tools are different between cloud and embedded applications.

In the next chapter, we will be exploring processes and practices for automotive IoT software development.

References

- [1] https://www.jamasoftware.com/requirements-management-guide/writing-requirements/functional-vs-non-functional-requirements
- [2] AWS: https://aws.amazon.com/about-aws/global-infrastructure/regions_az/
- [3] Azure: https://azure.microsoft.com/en-us/explore/global-infrastructure/products-by-region/
- [4] GCP: https://cloud.google.com/about/locations
- [5] https://istio.io/latest/about/service-mesh/
- [6] https://prometheus.io/docs/introduction/overview/
- [7] https://www.flickr.com/photos/xmodulo/24311604930/in/photostream/
- https://uptime.is/

Part 5: Automotive Software Insights

Automotive software development in many ways is like software development in other areas, but there are several differences. This section teaches you about these differences and how to develop automotive software.

This part has the following chapters:

- *Chapter 12, Processes and Practices*
- *Chapter 13, Embedded Automotive IoT Development*
- *Chapter 14, Final Thoughts*

12
Processes and Practices

In this chapter, you will learn about the processes and practices that are used in automotive **Internet of Things (IoT)** software development. Often software, engineers bristle at the idea of process. Reasons for this include lack of time, less fun compared to coding, and it being viewed as busy work. Is this you? It is common to hear software engineers say that they would like someone else to take care of the process stuff while they focus on doing design and writing code. It is also common to hear software engineers complain about having to follow processes. Upon hearing this, my typical response is something like, "OK, let's stop doing things that aren't useful and modify the process. Please come back to me with the steps or activities that we should stop doing." So far, I have never had anyone take me up on this. This is not to say that processes cannot and should not get easier to follow. This is the big challenge of creating a process. How do you create a process that improves software quality but doesn't add unnecessary activities?

This chapter will try to address that important question and explain how the typical processes and practices that are used in automotive software development provide benefits. This chapter will cover the following topics:

- Introduction to processes and practices
- **Automotive SPICE (ASPICE)**
- Functional safety
- Other key processes and practices

Let's dive in!

Introduction to processes and practices

The three major standards that are used (or at least talked about) across the industry are *ASPICE*, *functional safety (ISO26262)*, and the newest of the three, *Road Vehicles – Cybersecurity Engineering (ISO/SAE21434)*. Previous chapters covered cybersecurity; this chapter will cover the first two and then cover several other practices that are common in automotive software engineering.

Thomas Jefferson is noted as saying, "*If you want something you have never had, you must be willing to do something you have never done.*" This is a great quote, and it applies to following good processes and practices. If you want to have high-quality software, you may have to follow processes that you have never followed before. The processes you follow may be dictated to you. If not, you and your team may decide as a team, project, or department. *Table 12.1* shows a quick way to help make choices about how many processes to follow.

	More Process	**Less Process**
Smaller team		X
More mature team		X
Less turnover in the team		X
More co-located team		X

Table 12.1 – Process selection guideline

Smaller teams, more mature teams, teams with less turnover, and teams that are co-located can follow lighter-weight processes. According to *Creating a Software Engineering Culture*, by Karl E. Wiegers, there are three major points that enable a healthy software engineering culture:

- Personal commitment by all to create quality products by systematically applying effective software engineering practices

- Commitment by managers at all levels to provide an environment in which software quality is a fundamental success driver, and which enables each developer to achieve this goal

- Commitment by all to continuously improve the processes, thereby continuously improving the products

Everyone effectively applying systematic processes leads to a higher quality culture and thereby higher quality software and a higher quality product.

> **Note**
> Higher quality software and higher product quality are tightly coupled but not the same. You can create a quality product based on design and code that is not high quality.
>
> For example, you can take a poor design and poor coding through significant amounts of testing and applying patches and workarounds and end up with a product that mostly works. However, the design and code are not high quality, which is not easy to extend or maintain.

Effective is a keyword in this statement. Good processes, when applied ineffectively, could be worse for a project compared to bad or no processes applied effectively. *Systematic* is also an important word. Systematically applying processes enables continuous improvement. By systematically applying a process, activities and their results will show what parts need to be changed, where gaps are, and what parts need to be eliminated.

The best part of the book by Karl Weigers is a quote that was attributed anonymously: "*The bitterness of poor quality remains long after the sweetness of meeting the schedule has been forgotten.*"

ASPICE

The first time I remember hearing about *ASPICE* was during a job interview. At that time, I hadn't worked in the automotive industry and the term was new to me. I used my understanding of other process frameworks to answer the question. This was a phone interview with several interviewers on the other end. One of them asked if I had done a quick internet search to answer the question (which I hadn't). I didn't take the job with that company but ended up in the automotive industry with a different company. Now, ASPICE is part of my daily vocabulary.

ASPICE stands for **Automotive Software Process Improvement Capability Determination**. Wait, isn't that ASPICD? Fortunately, the group who created it determined that using the "*E*" from determination is much better, so we have ASPICE. It is common to refer to ASPICE since it's a process assessment model that's used to assess a process or a project to determine the likelihood of higher software quality. Referring to a process as an ASPICE process is not technically correct, but it is probably an OK practice.

ASPICE was created by a group of car manufacturers (European companies) with the intent of improving software quality. It focuses on software, but as it includes system requirements, architecture, and system-level testing, electrical and mechanical engineering work can be in scope. Also, project management is an important aspect. In an earlier version, ASPICE had process groups named ENG.2 to ENG.10. (Some people still use this out-of-date terminology.) ASPICE v3.1 (current at the time of writing this book) includes 32 process groups. However, ASPICE assessments typically focus on four system engineering process groups and six software engineering process groups, as well as supplier monitoring, quality assurance, configuration management, problem resolution management, change request management, and project management. These 16 process groups are deemed as most important.

In this book, we will focus on the following six software engineering process areasas shown in Table 12.2.

Short Name	Full Name
SWE.1	Software Requirements Analysis
SWE.2	Software Architecture Design
SWE.3	Software Detailed Design and Unit Construction
SWE.4	Software Unit Verification
SWE.5	Software Integration and Integration Test
SWE.6	Software Qualification Test

Table 12.2 – The six software engineering ASPICE processes

Each process area is assessed independently. The capability of each process area is assessed to a level between 0 and 5. *Table 12.3* shows levels 0 to 3. It is a typical expectation to assess to level 3. Levels 4 and 5 are available, but they haven't been listed as they are not likely to be considered by projects or departments; they are more focused on departments and processes for managing processes and less focused on projects, which is why these are less important to specific projects and your work as a software developer.

Level	Achievement
0	The process does not achieve its purpose or does not exist.
1	The process achieves its intended purpose.
2	The process achieves its intended purpose based on an approach that is planned, monitored, and adjusted as needed. Level 1 meets the intended purpose but may do so in a non-repeatable way or with some luck. For level 2, activities are planned and monitored during the project, and adjustments are made to ensure that the purpose is achieved.
3	Adds a managed approach to achieve level 2. A simple way to look at this is that a project uses a standard process instead of creating a process as it goes along.

Table 12.3 – ASPICE process levels 0 to 3

The assessment approach is to check each process against a set of **base practices** (**BPs**) and a set of **generic practices** (**GPs**). A project can be assessed to level 1 by meeting all the BPs. To move to level 2 or level 3 assessments, the project must meet the GPs defined for levels 2 and 3. We will narrow our focus to the SWE.1 to SWE.6 BPs.

The software engineering process group is organized in a V-model approach, as shown in *Figure 12.1*.

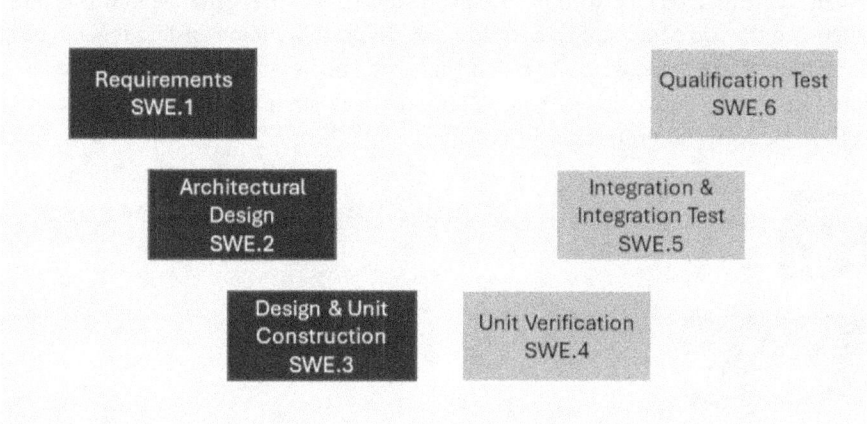

Figure 12.1 – V-model for the ASPICE SWE group

The V-model is a simple way to see that there is a corresponding validation activity for every development activity:

- Software requirements created in SWE.1 are validated by software qualification tests in SWE.6
- Software designs created in SWE.2 are validated by integration tests in SWE.5
- Software units constructed in SWE.3 are validated by unit verification in SWE.4

Bidirectional **traceability** is an important part of how the ASPICE model improves software quality. *Figure 12.2* shows a visual representation of the traceability among the processes. Two systems engineering processes have been added to show how the software processes trace to these: system requirements and system architecture. These are inputs to the software activities and are the basis for software requirements and software architecture; further details are outside the scope of this book. Bidirectional traceability is expected in two dimensions. One dimension was mentioned previously, where traceability across the V-model is expected. The other dimension is that the process must be traceable to an upstream activity. For example, software units that are created must be traced back to a software requirement.

As you read through this section, you will find that traceability is an important part of ASPICE. Many software tools will help automate traceability. It is possible to implement "home-grown" methods. For example, you can create macros in a word processor or spreadsheet that generate new IDs that can then be used between the two connecting parts. However, you may want to consider tools to support you. Rational DOORS Next Generation, Atlassian's Jira, and Jama Connect® are tools that you may want to explore further. Of course, these are just a few and there are many others to consider based on your team's size, budget, and project needs.

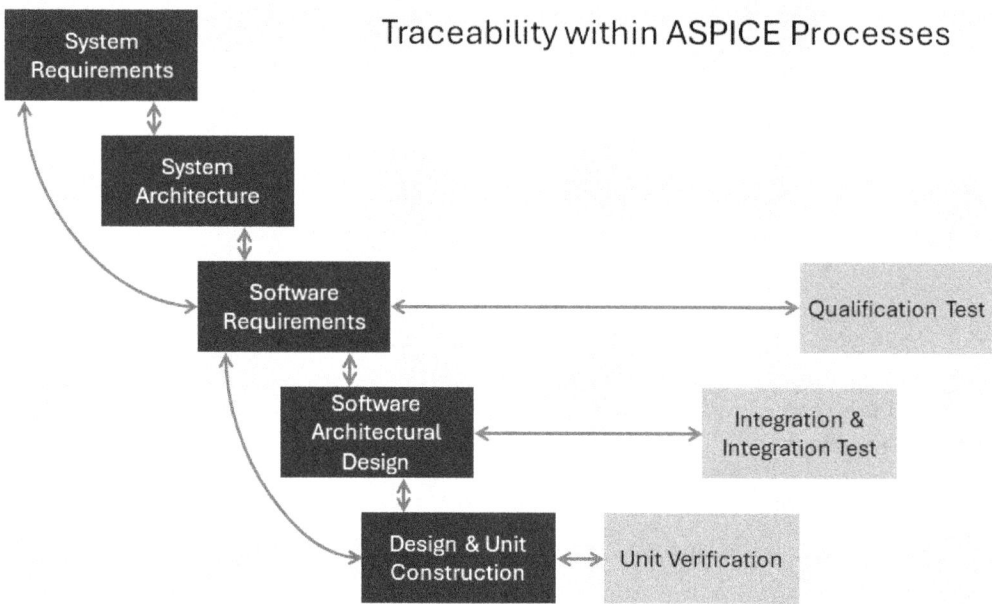

Figure 12.2 – Bidirectional traceability shown for SWE processes and SYS processes

Each has a process ID (for example, SWE.2), a process name (for example, Software Architectural Design), a purpose, process outcomes, and BPs. The process outcomes are the important things to achieve with the process, but the process assessment is done using the BPs. There is a strong correlation between the outcomes and the BPs but not quite a one-on-one mapping. The BPs will give further details about how to determine if a process outcome is met. This makes the assessing part a little easier.

This section risks diving too deep into the ASPICE model, but we intend to focus on the activities that a software engineer is either required to do or recommended to do, which leads to higher software quality. With that in mind, we will look at the process purpose and outcomes for SWE.1 to SWE.6.

SWE.1 through SWE.6 follows the software development flow. It starts with requirements and moves to architecture work, detailed design, and unit construction. Unit verification is next, followed by software integration and integration testing, and finally software qualification. Don't try to do all software requirements work at the beginning of the project followed by all the architecture work before doing detailed design or coding. Begin projects with enough software requirements defined and architecture work completed to create a good foundation and overall structure. From there, proceed iteratively by filling out the requirements and architecture while conducting detailed designs, coding, and testing activities to keep the requirements and architecture in a realistic and usable form. Note that despite the layout of the V-model and the number of processes, ASPICE does not require that each activity be done before the next is started.

SWE.1 – Software Requirements Analysis

The Software Requirements Analysis process has eight expected outcomes to support the creation of software requirements.

Let's focus on the most critical outcomes. The first outcome is the main activity – create the requirement. This is done by analyzing the system requirements and system architecture. (If system-level information is not available, you may need to get customer specifications or go through a requirements elicitation activity to define the software requirements.) You may also find that the system requirement adequately covers the software requirements without further specification. Aim to not replicate information.

The next outcome is setting the expectation that the software requirements are written correctly and can be verified (or tested). Among the many good mentors I have had during my career, one had a keen focus on well-written software requirements. I was able to learn how to write software requirements from him. If you're tasked with writing or reviewing software requirements (and you should be at some point), you need to spend some time learning to write a proper requirement. It isn't difficult, but it is something that must be learned.

Two more expected outcomes on software requirements. One, they must be reviewed, and two, an agreement must be made by the relevant stakeholders. The review is extremely important to find defects early and to make sure those implementing and testing the requirements understand them. The final point is traceability. All software requirements need to be traced back to a system requirement and all system requirements must be traced to a software requirement. This helps ensure that the software is implementing all the expected behaviors and nothing more.

SWE.2 – Software Architectural Design

The Software Architectural Design process has five process outcomes that support the software architecture activities. These help you end up with a good architecture based on the requirements.

ASPICE expects the software architecture to be based on the software requirements. By doing this, the software architecture covers what it needs to without supporting additional capabilities. This is checked as part of the traceability expectation. Often, the software architecture work is done by one main person or a small team, with others supporting it in select areas. The recommended approach is to have a seasoned software architect who has significant experience in software development.

SWE.3 – Software Detailed Design and Unit Construction

The Software Detailed Design and Unit Construction process has six outcomes that support it.

With this, we come to the fun part: writing code. Well, almost. In this process, a detailed design based on the requirements and architecture is created. Software units are defined, and the units are written. Determining what a software unit is can be an interesting exercise. If you're defining your standard process so that it meets ASPICE expectations, depending on your assessor, you may have a few rounds of back and forth on how to describe the steps to define and produce a unit.

It's a great idea for your software engineering department to have templates for the architecture and design work. These templates will help in several ways. Their consistency will help when people move from one project to another and when reviews are conducted. They will also ensure that the work considers all the relevant topics. For example, the template can have a section on **Read-Only Memory** (**ROM**) life cycle management. This will be a good reminder to consider and plan for this important work. Reviews continue to be critical to improve software quality and to find defects earlier and fix them at a lower cost.

SWE.4 – Software Unit Verification

The Software Unit Verification process has five outcomes that help verify the detailed design and non-functional requirements.

At this point, when looking at the V-model, we've finished going down the left-hand side of the V and are now starting to work our way up the right. The right-hand side is about verifying the work that has been done as part of the left-hand side activities. Each of the remaining three processes verifies the work that was done on the opposite side of the V, beginning with unit verification.

The main point of this process is to verify the software units, but ASPICE requires that this verification be done in a structured manner to increase software quality. The process expects the verification to be done based on a strategy that includes regression testing. Unit verification is done based on defined criteria (such as a static analysis target or code coverage). The defined criteria are used to see whether the verification is adequate and whether the units are passing or not. This is not an ad hoc activity but is structured with a planned intent. As a software developer, someone else on the team is likely responsible for defining the strategy and criteria and you will follow these things. Make sure these things are in place so that you are following them and not simply creating unit tests solely based on your judgment or even worse, not doing any unit testing.

SWE.5 – Software Integration and Integration Test

The Software Integration and Integration Test process guides the integration and integration testing activities. There are eight outcomes to support these activities.

This process is the most difficult to grasp and put into practice since it's about integrating all the software units into the complete integrated software before testing that the integration has been done correctly. The correctness of the integration is based on the architectural design.

An automotive application project that involves IoT is likely to be a project that's big enough to have a dedicated build engineer. This may not be a full-time role, but most software developers will not be creating the integration strategy or creating the scripts to do the integration. As such, this won't be considered in this book. Similarly, when it comes to integration testing, there will likely be a person or small team that has this responsibility.

SWE.6 – Software Qualification Test

The Software Qualification Test process has six expected outcomes that help with creating a test strategy, selecting test cases, and ensuring traceability for software requirements.

This process is really important in that it is the last step in the software activities. It is the last opportunity to catch any defects, non-compliance, or usability issues with the software. After this step is completed, the software will be turned over to a system test team, a factory for production, or a customer.

The main intent of this process is to ensure that the software implements the software requirements. The process expects that this activity be done based on a defined test strategy. Test cases are created based on the software requirements and are traceable between the requirement and test case. Generally, it is expected that every software requirement has a test case, and every test case traces back to a requirement. This bidirectional traceability ensures all software requirements get tested and that there are no extraneous test cases. Having extra test cases may seem like a good idea, but if the test case is not based on a requirement, it isn't known whether the test case results indicate an actual issue. Also, extra test cases take time to create, execute, and maintain.

With that, we've covered the software-specific ASPICE processes. Depending on what part of the automotive IoT application you're working on, you may be involved in many of the activities described previously. These processes define how you should be accomplishing the software development activities so that you get a higher quality product and software. *Chapter 14, Final Thoughts* will come back to ASPICE and Agile and describe when ASPICE is and isn't used in more detail.

There is another ASPICE process called **ACQ.4 Supplier Kickoff and Monitoring** that is likely to be important for you in automotive software development. It will be discussed in the next chapter. You can learn more about ASPICE at reference [1]. Next, we will learn about functional safety.

Functional safety

Functional safety is rapidly growing in importance. It is more formally known as ISO26262: Road Vehicles – Functional Safety. This ISO standard was first released in 2011 and the current version was released in 2018. ISO 26262 is an adaption from an earlier standard named IEC 61508, which was a functional safety standard that applied to all industries. These standards have many similarities, but ISO 26262 was derived to only focus on road vehicles.

The purpose of functional safety is to ensure that automotive software (and hardware) operates as intended or defaults to some known safe state. With increasing complexity in automotive software and hardware and with increasing reliance on technology (for example, **Advanced Driver Assistance Systems** (**ADAS**) and electronic controls), mission-critical software needs to operate as intended. Functional safety is about reducing the risk of failures that could lead to accidents. The functional safety standard accomplishes this through development processes and expects failure detection, failure recovery, and mechanisms to be included to establish a safe state if failure recovery isn't possible. So far, this standard isn't required by law, but more and more people in the automotive industry are choosing to follow the standard. Also, while laws do not require that functional safety be followed, failures in software could have legal ramifications.

A department or project that's going to follow the functional safety standard will have experts in this area. These experts will establish processes and practices that meet the expectations of the standard and work through the course of development projects to ensure that the standard is being followed. Note that functional safety is not just about software development or development across disciplines; it covers the life cycle of an automotive component and includes aspects such as management, production, and service. This book will focus on software developers, not functional safety experts. This section will provide a solid background of this understanding and focus on what software engineers will need to do when working on a project that requires functional safety.

There are three ways to describe functional safety:

- Vocabulary
- Risk classification system
- Development process

Let's take a closer look.

Vocabulary

The functional safety standard defines over 100 terms (the full list is part of the ISO 26262-1:2018 standard) to support the standard's goals. The common vocabulary supports precise communication, which, in turn, supports achieving the standard's goals. Here are a few key terms to help you better understand functional safety. You don't need to be an expert in functional safety, but if you are involved in a project with functional safety requirements, these will be quite helpful:

- **Functional safety**: This is the absence of unreasonable risk due to potential sources of physical injury or health damage caused by **Electrical/Electronic** (**E/E**) systems malfunctioning
- **Hazard Analysis and Risk Assessment** (**HARA**): This is a risk assessment that focuses on events that could cause physical injury or health damage
- **Functional safety concept**: This refers to functional safety requirements and architecture
- **Technical safety concept**: This refers to technical safety requirements and system design
- **Technical safety requirement**: This is a requirement that's derived from functional safety requirements
- Several types of faults are defined, including latent faults (not detected or noticed), detected faults, permanent faults, and residual faults
- **Safety mechanism**: This refers to fault detection or failure control to support operating without unreasonable risk

Next, we will look at the risk classification system.

Risk classification system

The risk classification system is called **Automotive Safety Integrity Levels** (**ASIL**). It has various levels: A, B, C, and D, along with **Quality Management** (**QM**). QM means that risks are low enough (from a safety perspective) that functional safety standards don't need to be applied. ASIL A has the lowest rating while ASIL D has the highest rating, with the risk of hazards increasing from A to D. *Figure 12.3* shows a few examples of potential ASIL ratings.

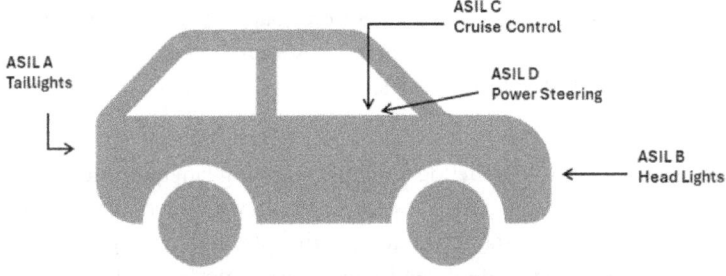

Figure 12.3 – Examples of ASIL ratings

As shown in *Figure 12.4*, the ASIL rating is determined by considering the severity of a fault, its likelihood of occurring, and the ability to control the hazards. You probably won't be determining the ASIL as this is done by functional safety experts. However, you need to understand these ratings as they apply to the components that you work on or interface with.

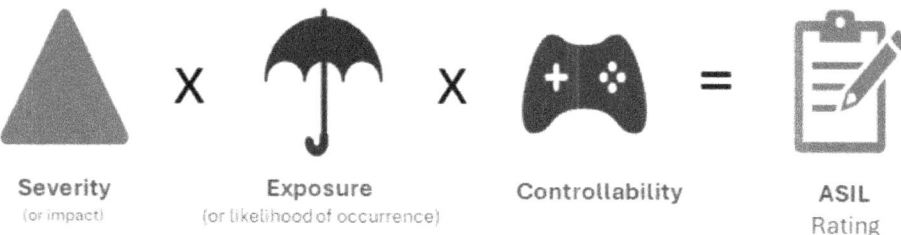

Figure 12.4 – Determining an ASIL rating

Now that we've looked at the various vocabulary and the risk classification system, we will complete this introduction to functional safety by looking at the development process.

Development process

As mentioned when we covered ASPICE, some of the most value-added parts of your development processes (whether they're related to ASPICE, functional safety, or other standards) are checklists and templates. These help you standardize activities and remind you of good practices and things to consider. For example, a template can help you consider memory wear leveling for memory parts, while a code review checklist can help you remember important things to check during your review. If your processes do not have checklists or templates, consider developing these as part of your project development work.

The functional safety development process leverages the ASPICE model. It has similar steps. These will be discussed next.

Software project initiation

Software project planning is required to ensure that the functional safety activities will be completed as part of the development process. This is important for a couple of reasons. First, the activities are included in the project's schedule or work list so that they aren't missed or forgotten. Second, the activities are planned with the rest of the activities, so there is less chance that they will be skipped due to schedule or budget pressures.

As part of the planning process, the modeling and programming languages are selected based on specified criteria. Design and coding guidelines are required, which mandate things such as low complexity, strong typing, and using established design principles and naming conventions. If you're working on a project with functional safety requirements, it's important to know about and use or follow the specific programming languages, design guidelines, and coding guidelines.

Specification of software safety requirements

Next, a HARA is done, and the functional safety concept is created during the project's concept phase. (You likely won't be involved in these activities, depending on the size of the project, but based on your experience and seniority, you may be asked to contribute.) The HARA will be used to determine ASILs for the events/components of the system. It will also be used to create safety goals which lead to requirements as shown in *Figure 12.5*.

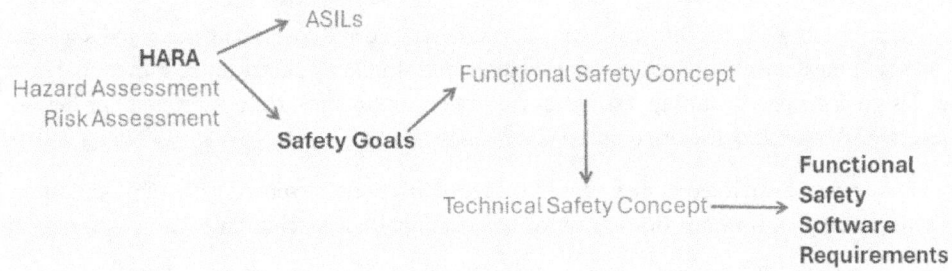

Figure 12.5 – Steps to create the functional safety software requirements

The functional safety concept takes the safety goals, derives the functional safety requirements, and allocates them to the architecture. It does the following:

- Detects faults
- Mitigates failures and goes into a safe state when needed
- Provides mechanisms to allow faults, so long as they don't violate safety goals
- Sends warnings to the driver to reduce risks

The technical safety requirements are written based on the functional safety concept. The technical safety requirements must specify fault detection, a safe state, and details regarding warnings and degradation.

Safety mechanisms with safe states define the following things:

- Transition
- Time interval allowed for a fault
- An emergency fallback if the transition to a safe state does not happen
- Staying in a safe state

The software safety requirements follow the technical safety requirements and address software-based functions that must support the technical safety requirements.

The hardware-software interface (which is jointly owned by systems, hardware, and software groups) that's created during the system design activities may need to be updated with further details that specify the correct control and usage of the hardware system.

Software architectural design

After the requirements work, the software architecture is developed and covers both safety and non-safety requirements. It must have modularity, encapsulation, and simplicity as attributes. Functional safety specifies design principles including restricting the size of components, restricting coupling between components, and restricting the use of interrupts.

All components must be marked as new, reused with modifications, or reused without modifications. Components in the first two categories must follow functional safety requirements.

By default, embedded software must be completely developed to the highest ASIL level of any component. However, the functional safety standard allows the system to be decomposed so that components can be treated differently if they meet certain criteria. Functional safety experts need to identify the components and ASILs in the system. Once they've done this, you need to implement components to the right ASIL (or QM) as identified.

The software architecture must support error detection, including range checks for the input and output data, plausibility checks, detection of data errors, external monitoring facilities, control flow monitoring, and a diverse software design.

The architecture must support error handling, including a static recovery mechanism, graceful degradation, independent redundancy, and correcting codes for data.

Finally, a verification step is required to check the architecture for various properties, including compliance with the software safety requirements, compatibility with the hardware, and adherence to design guidelines. There are different ways to verify these properties, including inspecting the design, simulation, prototyping, and formal verification.

Software unit design and implementation

After the architectural work, the software units are specified, implemented, and verified. Functional safety calls out design principles and properties to support good designs. Design principles include things such as variable initialization, functions with one entry and exit point, no recursions, limited use of pointers, and no implicit type conversions. Design principles include readability, simplicity, and robustness. They also include correct order of execution and consistency in interfaces. Design work needs to follow these principles and the resulting design should have the aforementioned properties. The detailed design and implementation are statically verified before the software unit testing phase is conducted.

Verifying the software unit design and implementation process demonstrates the following:

- Compliance with the hardware-software interface specification
- The software safety requirements have been fulfilled
- The code complies with the design
- Compliance with the coding guidelines
- Compatibility with the target hardware

You can use the following methods to verify the unit design and implementation process:

- Walk-through
- Inspection
- Semi-formal verification
- Formal verification
- Control flow analysis
- Data flow analysis
- Static code analysis
- Semantic code analysis

Next, we'll cover various testing methods.

Software unit testing

The following methods can be used for software unit testing:

- Requirements-based test
- Interface test
- Fault injection test
- Resource usage test
- Comparison testing between the model and code, if applicable

After unit testing comes integration testing (although these activities are not always done serially or by the same person).

Software integration and testing

Integration testing verifies that the design is implemented as intended. The following methods can be used for software integration testing:

- Requirements-based test
- Interface test
- Fault injection test
- Resource usage test

Functional safety specifies that production releases of software only include specified functions. Unspecified functions are only allowed if they do not impair the software safety requirements. Integration testing is used to verify compliance with this specification.

Verifying software safety requirements

During the software safety requirements verification step, testing must be done in a hardware-in-the-loop environment, electronic control unit network environment, or vehicle environment so that more real-world scenarios can be covered.

OK – that was like a 70 MPH drive through functional safety. This section introduced this topic at a level deep enough that you should have some grasp on what needs to be done and understand the important terminologies. This will act as a jumping-off point as you work on functional safety projects.

So far, we've covered ASPICE and functional safety. These two standards overlap quite a bit. Functional safety uses ASPICE as a base, adding the necessary practices and expectations that are required to achieve its goals. If your department is using separate processes for ASPICE and functional safety, discuss with your management team ways to merge these. If done well, the merged processes will reduce confusion and improve efficiency as the team works on various projects or parts of a project. Next, we will look at several other processes and practices that are helpful in automotive development.

Additional automotive processes and practices

ASPICE and functional safety are the biggest two processes that are used in automotive software development. However, many other smaller processes and practices can be used to aid in software development. This section will cover four smaller tools that focus on problem prevention and problem-solving.

DFMEA

Design Failure Mode and Effects Analysis (**DFMEA**) is a process, tool, practice, or approach that's used to reduce and prevent design failures. It is a specialization of FMEA and doesn't just apply to software engineering – it is an excellent tool that can be used by software designers. It's a structured approach that considers what failure modes may exist in a design, prioritizes them, and makes design updates as appropriate based on how they've been identified and prioritized. It's similar to risk management in terms of how it identifies, prioritizes, and specifies actions to be taken.

DFMEA supports finding defects earlier in the process, where they are cheaper to address, and can find more latent defects that may not show up until software is in wide use or has been changed to accommodate new requirements or through maintenance.

The DFMEA tool is pretty basic and can be a simple spreadsheet with columns to help with identification, impact assessment, prioritization, and specifying actions to be taken for each failure mode. Here, each row of the DFMEA is a specific failure mode.

When you're using a DFMEA on an IoT application, you will likely be splitting the entire application into many different ones. You could approach this hierarchically, where a higher-level DFMEA is split into lower levels. By following either of these approaches, you can design camera software within a central controller. *Figure 12.6* shows an example of a DFMEA spreadsheet with a **rear-view camera** (**RVC**) failure mode analysis.

Potential Failure Mode	Effect	Causes	Likelihood of Occurrence	Priority	Preventative Actions
RVC not displayed	Black screen instead of camera image	Serdes fails to send ack command	Medium	High	Add a timeout for ack command; restart serdes as needed
RVC frozen	Camera image frozen	Camera module stops sending data	Medium	High	Add a frame checker to make sure image is updating; restart camera as needed
RVC doesn't go away	Camera image continues to be displayed	When changing gears display statement machine misses gear change signal	Low	Medium	Ensure that each state can accept a gear change signal

Figure 12.6 – Example DFMEA

Figure 12.6 is a bit simplified but covers the main components. The failure mode (that is, the potential defect) is given in the first column, and the effect in case the failure mode occurs is described in the next column (this is like the severity and detectability). Next, the causes of the failure mode are listed. This helps determine the preventative actions that can be taken. Your department or project will likely have a specific format for DFMEA. If not, a template can quickly be put together that you can use. Then, the likelihood of occurrence is given. These are important factors in prioritizing the failure mode and determining actions.

Using the priority, you can determine which of the failure modes to address and how they should be addressed. The *Preventative Actions* column is filled in with the actions that will be taken.

The form isn't hard to understand or fill in. However, the challenge and importance are in thinking through your design and identifying the failure modes. This will come partly from past experiences and from reviewing the design documents. This is where having a design documented (instead of just in the heads of one or more developers) is helpful. If the design is documented, it is visual and can be reviewed methodically. The second hardest part of the DFMEA tool is the discipline it takes to use it to reduce failures. You may not be required to use this tool, but it may be something you should do. Will you? It may be required, but will you just fill in enough of the blanks to get past an audit or management review?

There is a cost to doing DFMEA and the amount of time spent is often a judgment call by the software/project management team but more often made by software developers. The choice of how much effort to put into a DFMEA does not typically boil up to the level of a manager to decide. Processes and reviews are critical to help make the right calls, but the experience, wisdom, and maturity of individual software engineers are critical to successfully using DFMEA. Every software engineer ought to believe that the DFMEA tool is useful when applied appropriately. Every software engineer ought to know how to use the DFMEA tool as they mature from a junior engineer.

Doing a good job at DFMEA will lead to lower development costs (by finding defects earlier), higher customer satisfaction (fewer defects found by the customer), and lower life cycle costs (resulting from software that is more extensible and easier to maintain).

5 Whys root cause analysis

We will now move from problem detection to problem-solving. The most famous and most used problem-solving tool is the **5 Whys root cause analysis** tool. This tool comes from lean manufacturing and the Toyota Production System. The 5 Whys was invented by Sakichi Toyoda and is a methodical approach to problem-solving. The methods are designed to get to real root causes and apply countermeasures that truly fix the problem and prevent it from recurring. Too often a symptom is seen and addressed while the root problem continues to exist and will show up again. Taiichi Ohono, who is considered the father of the Toyota Production System, stated, *"The basis of Toyota's scientific approach is to ask why five times whenever we find a problem … By repeating why five times, the nature of the problem, as well as its solution, becomes clear."* Like DFMEA, the 5 Whys tool is a basic spreadsheet that begins with the problem definition and then uses five columns for each of the 5 Whys before the real root cause is documented. Then, short-term and long-term countermeasures are documented.

Getting the earlier parts of the 5 Whys analysis done correctly is important to end up with a good solution to the problem at hand. Defining the problem properly is critical. This seems simple but often requires time. The better the problem is defined, the easier it will be to get to the root cause and then to solutions/countermeasures.

The following are a couple of things I've seen around problem-solving, specifically with the problem definition:

- There is a rush to get to a solution, so proper time isn't put into defining the problem. This may lead to an incorrect root cause and a countermeasure that doesn't address the problem.
- The problem isn't defined specifically enough, leading to people interpreting the problem statement differently.

Regarding the second issue, problem-solving is often a team sport, meaning several people are involved. If people on the team have a different understanding of the problem, the 5 Whys analysis will not go great. Once the problem has been sufficiently defined and all those working on the problem-solving activity understand and agree, doing the 5 Whys is the next step. But remember, define the problem properly!

The purpose of the 5 Whys is to get to a root in a systematic way that requires you to think carefully about each step toward the root cause. Again, there is often a strong temptation to jump too quickly to the root cause (and then the solution). Let the process help. Each of the 5 Whys should take you a small step closer to the root cause. Taking too big of a step likely depends on assumptions that may not be valid. Also, each successive Why needs to directly follow on from the previous Why. It is a common problem to jump around in the Why sections. For example, you may have the problem that your car stopped in the middle of the road. In this simple example, it is probably appropriate to jump to the cause that the car ran out of gas and more gas is needed. But let's try this approach out.

Problem statement: *The car stopped in the middle of the road.*

By noticing the gas gauge or the behavior of the car, the root cause is that the car is out of gas and needs more gas. But let's use the process. Why did the car stop? Because the engine died. Why did the engine die? Because the car was out of gas. Why was the car out of gas? Because there's a hole in the gas tank. In this made-up example, adding more gas to the car isn't going to help very much as the gas will leak out again and the car will stop again. So, the hole in the gas tank must be fixed. It is possible to also go a step further and ask why there is a hole in the gas tank, in which case you may find that something is sticking up in the road that needs to be removed. Fixing the hole in the gas tank might only be a temporary fix if the car goes along the same road again. It is easy to make up examples to show how the 5 Whys analysis works; you must do the hard work of applying it to software engineering problems.

The power of the 5 Whys is its simplicity, but it requires discipline and patience. Next, we'll cover the fishbone tool.

Fishbone

A **fishbone** is a visual way to represent possible causes of a defect, as shown in *Figure 12.7*. It is a useful tool when a defect is difficult or costly to try to reproduce and when there is limited data available to deduce the root cause of the problem.

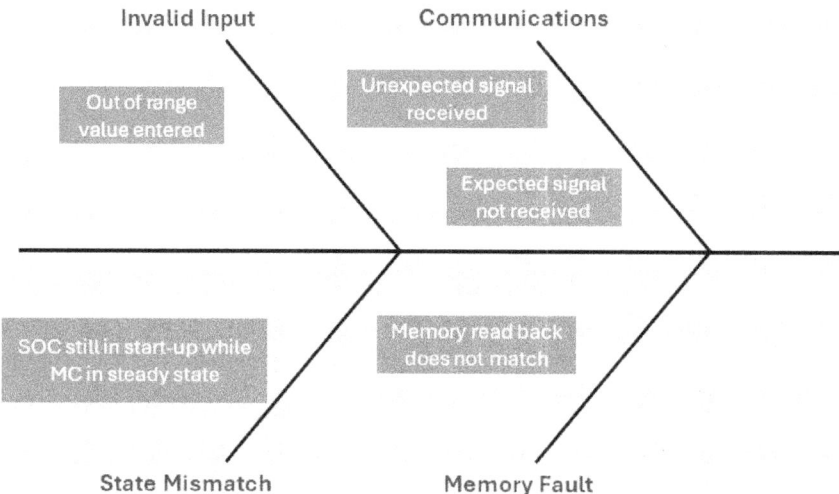

Figure 12.7 – An example fishbone diagram

Figure 12.7 shows an example of a fishbone diagram. Each bone can be a category of causes, with specific causes coming off each bone. Then, through experimentation or reviews, the causes can be eliminated as a cause or suspect of the effect.

You must start a fishbone diagram by identifying the types of causes for the effect. Then, you must identify as many specific causes as you can. This is done through experience, reviewing designs/requirements/code, or through experimentation. After, you must test each of the causes until one cause remains or a small set of causes remains. These become the focus of the rest of the problem-solving activity. You may not be able to 100% identify the root cause in cases where a problem is rare and/or information is limited. You can use the fishbone activity to identify changes that have the best chances of eliminating the effect.

One additional benefit of using the fishbone and 5 Whys processes is that the problem-solving activities are visible. You can show your managers or your customers the progress that is being made. They can also get involved in the activity by offering suggestions and asking questions. This engagement (when structured) can be very helpful. They will have more confidence and probably a little more patience while you are working through a solution.

As you work through the 5 Whys and fishbone processes, you should be able to trace problems back to requirements, design, or coding problems. Especially for design problems, go back to the DFMEA and other design documents. If these were done well, you will be able to find the problem, or you will be able to see where additional details were needed. Take the time to update these documents as this will help with future problems and keep the software in a maintainable state (for example, you aren't changing the code so that it doesn't go out of sync with the design). You should also review test cases to determine why the problem wasn't caught sooner. Add test cases at the unit, integration, or qualification level based on what you've learned from the problem-solving work.

A-B-A testing

A-B-A testing is a method that increases confidence that a change (that is, a fix) addresses a fault. Once a fix has been identified, you can proceed with the A-B-A test. The first test, A, tests without the fix and the fault should be seen. Next, the fix is applied, and the B test is done. The fault should not be seen. Finally, remove the fix and do the second A test. The fault should be seen again.

If you do not see the fault during the second A test, the fix may not be fixing the fault, in which case another variable might be involved. The test circumstances for the B and second A tests may be different from the first A test. For example, you may be faced with a failure that occurs when the software is running in a vehicle. After some analysis, you determine a fix and run the fix in a lab or on your desk. The fault isn't seen, and the solution is good to go. Right? It might be that the fault is only occurring at higher (or lower) temperatures that are present in a vehicle and not in a lab or at your desk. The second A test helps to avoid implementing a "fix" that doesn't fix the problem. That can be costly and quite embarrassing.

Summary

In this chapter, we covered the two biggest automotive software development standards: ASPICE and functional safety. Both standards follow a V-model approach where activities on the right-hand side of the V verify the activities on the left-hand side of the V. Functional safety is a standard that is applied in specific areas, given the additional efforts that are required to follow the standards. ASPICE may be required but most of what's expected in ASPICE is typical good software development practices and is not necessarily adding extra work. Typically, a project or department's development processes will be based on a standard such as ASPICE. For automotive software development, I recommend that you leverage the work that you've put into defining ASPICE as the basis for your software development processes.

Then, we looked at a few tools that can support problem detection and problem-solving. DFMEA is one important tool that should be applied to your designs. More time spent on the design phase is likely going to be less costly and produce higher quality products. The DFMEA is a simple but structured way to help find faults in the design before they end up in your code and your customer's hands.

It might be required that you follow some of the tools, standards, and processes that were covered in this chapter. Even if they are required, software developers make day-to-day decisions on how they spend their time. Managers and software quality assurance leaders can talk about following good processes and can put checks in place, but it is up to you to make good decisions about how much time is spent on things such as requirements reviews, designing, design reviews, code reviews, and test case development. It is recommended that you learn the tools and processes. By doing this, you will be in a much better position to apply them properly. These processes are mostly common-sense approaches to software development. It is often up to you to apply these so that they lead to higher software quality. Start with the culture and quality mindset and use the processes and tools mentioned here to help.

This book covers various important topics regarding vehicle architecture, cybersecurity, and cloud development. Embedded software is also important in automotive IoT applications. We'll cover this in more detail in the next chapter.

Reference

[1] VDA-QMC. Automotive SPICE. Available at: `https://vda-qmc.de/en/automotive-spice/`

Get This Book's PDF Version and Exclusive Extras

Scan the QR code (or go to `packtpub.com/unlock`). Search for this book by name, confirm the edition, and then follow the steps on the page.

Note: Keep your invoice handy. Purchases made directly from Packt don't require one.

13

Embedded Automotive IoT Development

In this chapter, you will learn about embedded development. Some readers will be in the process of learning. Other readers will be new to embedded development. You may be coming from a cloud computing or database-intensive software engineering background and are transitioning into automotive software development due to the growing need for cloud and database experience in automotive **Internet of Things (IoT)**. Much of this chapter will be for the second and third types of readers. If you are in the first group, you might choose to skip this chapter except for the last section on supplier management.

This chapter will dive into five topics related to automotive software development:

- Embedded software development
- Power management
- Operating systems
- Hypervisors
- Software development ecosystem

Embedded software development

Woodrow Wilson was noted as saying, *"If I am to speak ten minutes, I need a week for preparation... if an hour, I am ready now."* The quote reminded me about the differences between software development and embedded software development. Writing a program to run in a typical computing environment can be much easier and involves the consideration of fewer constraints. Writing an embedded program is likely to take longer.

Becoming a good, embedded software developer is difficult and takes a combination of software engineering skills and electronics hardware knowledge. Software developers working in embedded systems often come from an electrical engineering or electronics background and are not as strong in core computer science principles. Or they come from a computer science background and need to learn more about electronics and hardware.

Figure 13.1 shows topics that need to be considered when developing embedded software. They are grouped into themes to help organize them. Real-time **Operating Systems (OSs)** and related topics will be covered in the next section.

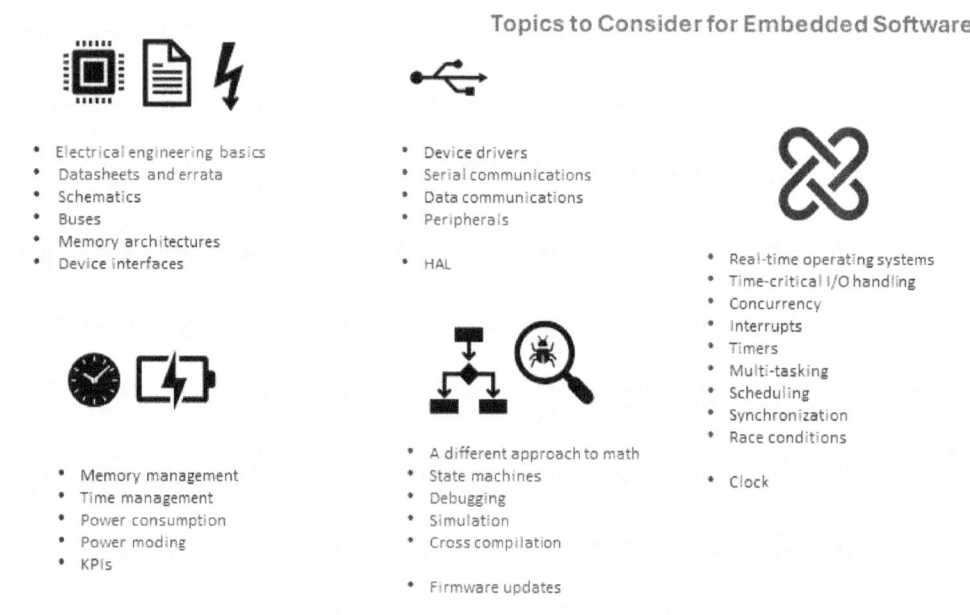

Figure 13.1 – Embedded software engineering topics

This section will cover electrical engineering-related topics, devices and drivers, and memory management topics.

Electrical engineering

Whether you are writing applications, **Human-Machine Interface** (**HMI**) software, or low-level drivers, you will need to know some electrical engineering to be a successful embedded software developer. The lower down in the software stack that you are working on, the more you will need to know. At the end of the application, you will need to know how to read system block diagrams and understand the capabilities of the hardware. You will likely be optimizing your code in terms of memory usage and performance. Your understanding of how the hardware works will increase your ability to make your code work great. If you are working at the bottom of the software stack, you will need to know about schematics, programming control pins, and data lines.

In this section, we will look at system block diagrams, data buses, schematics, datasheets, memory architecture, and device interfaces. *Figure 13.2* shows a sample system block diagram of an **in-vehicle infotainment** system (**IVI**). The IVI is centered around a **System on a Chip (SoC)**.

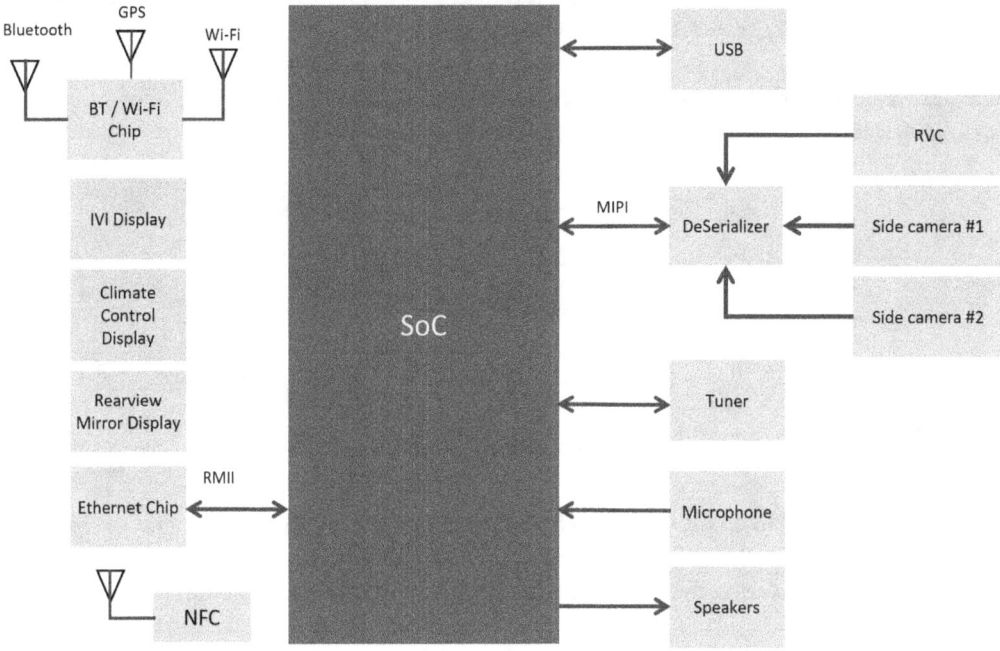

Figure 13.2 – Infotainment system block diagram

This diagram shows important components and their interfaces. The diagram supports project planning and the next level of software design. The software design is informed of things such as the number of displays and audio outputs and the number of camera and microphone inputs. If you are a system designer or architect, it is your responsibility to create this diagram based on requirements and intended future uses.

Understanding data buses is important for automotive software engineers. Common buses are Ethernet, **Controller Area Network (CAN)**, **Local Interconnect Network (LIN)**, FlexRay, **Media Oriented Systems Transport (MOST)**, **Integrated Interchip Sound (I2S)**, **Inter-Integrated Circuit (I²C)**, and **Serial Peripheral Interface (SPI)**. Communications with displays may use **High-Definition Multimedia Interface (HDMI)**, **Low-Voltage Differential Signaling (LVDS)**, or **Mobile Industry Processor Interface Display Serial Interface (MIPI DSI)**. Communications with cameras may use **Mobile Industry Processor Interface Camera Serial Interface (MIPI CSI)** or FPD-Link. Ethernet seems to be the future for vehicle communications because of its higher bandwidth capability. Ethernet may displace CAN, LIN, FlexRay, and MOST in many instances. However, I²C and SPI will likely remain as necessary data buses due to their lower hardware costs and ease of implementation.

You may be a decision-maker or influencer in choosing what serial data bus to use in your design. If so, you will get to make trade-offs in cost, implementation time, complexity, and data usage requirements. We will dive into the SPI serial bus in this chapter. SPI is often used to communicate between a controller (for example, the **Vehicle Interface Processor** (**VIP**)) and peripherals such as **Electrically Erasable Programmable Read-Only Memory** (**EEPROM**). SPI supports full-duplex communication at speeds of 10 Mbps or higher. It works well when communicating with peripherals that require relatively large amounts of data, such as writing to EEPROM. *Figure 13.3* shows a design of an SPI interface.

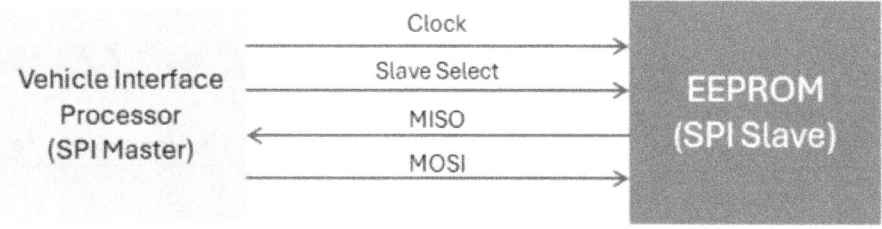

Figure 13.3 – Diagram of an SPI interface between a VIP and EEPROM

With SPI, data is shifted in on the **Master In Slave Out** (**MISO**) line and shifted out on the **Master Out Slave In** (**MOSI**) line. Several devices can be connected to the same SPI lines so the slave select (or chip select) is used to select the device and then the data is shifted based on the clock edges.

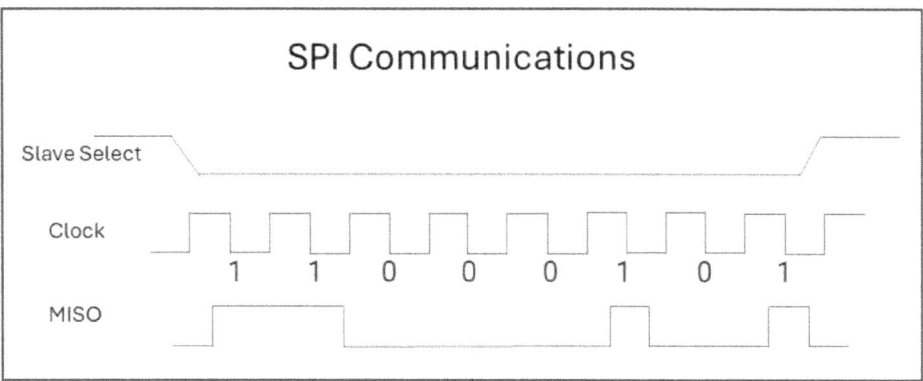

Figure 13.4 – Signal diagram for SPI communications

Figure 13.4 shows a signal diagram of SPI communications. 0xC5 is being passed from a peripheral (the slave) to the control (master). The master sets the slave select line low, indicating to the specific peripheral that it should be accepting data from the master. Data from the MISO line is sampled based on the clock. SPI is flexible to allow the **Least Significant Bit** (**LSB**) or **Most Significant Bit** (**MSB**) first. In this example, the MSB is first.

Schematics/block diagrams

You may need to read schematics, but you will almost certainly need to read detailed block diagrams. Hopefully, schematics reading will only be required for tough debugging scenarios. For example, you can find a reference design for the NXP S32K3 microcontroller at reference [1]. It will be helpful if you open this web page as you read the next couple of paragraphs.

The NXP S32K3 microcontroller is the focal point of that reference design for a telematics box. Start by learning what each of the major components in the design will do. Table 13.1 lists the components.

Part	Description
S32K344	Microcontroller
SJA1110B	Ethernet switch
AG550Q	5G module
SGTL5000	Low-power stereo codec for audio I/O
PF5020	Multi-channel PMIC (for power management)
TJA1XXX	Transceivers for CAN, LIN communications

Table 13.1 – List of major components in the reference block diagram

The next thing to consider is the types of communication between the microcontroller and the peripherals. SPI, I²C, **Universal Asynchronous Receiver/Transmitter** (**UART**), FlexCAN, I2S, and **Reduced Media-Independent Interface** (**RMII**) are some of the key buses. The software will need drivers to support all of these and then a **Hardware Abstraction Layer** (**HAL**) to know which communication bus to use to control and interface with the peripherals. *Figure 13.5* shows a simplified software stack view for quick reference.

Figure 13.5 – A simplified software stack view

From this base, you can start software development activities.

Datasheets, errata, and application notes

If the software you are writing is directly interfacing with hardware, you will need to live and die by the datasheet. While you should trust the datasheet, you must also acknowledge that it could be wrong. Regularly check for updates or errata in the datasheet. As other poor, helpless software engineers find problems with a datasheet, the part supplier will make revisions or publish errata to correct or add important details. You could be stumbling over a problem that has already been solved and the solution published.

Let's look at the *PF5020* as an example of using a datasheet. The datasheet is available from the NXP website [2] and can be found on other websites. *Rev. 3 - 20 September 2021* is the latest available version at the time of writing this book. Noting the revision number/date is important to know whether you are using the most recent version or not. The PF5020 has a capability for frequency tuning. This capability is described in *Section 15.7.3*. This section indicates that the *CLK_FREQ[3:0]* bits configure the manual frequency tuning. *Table 63* gives the specific settings. *Section 16.2* gives the register map, and the address for the *CLK_FREQ* settings is *0x3A Bits 0 - 3*.

Application notes are another source of useful information in learning how to program and control a device. For example, Texas Instruments has an application note for its *DS90UB913/4* part. It is titled *I²C over DS90UB913/4 FPD-Link III with Bidirectional Control Channel. Version SNLA222-May 2013* and it is the version being referred to next. The purpose of the app note is to describe I²C communication between *DS90U913/914* devices. The app note starts with an introduction to the scenario.

Section 2 provides details of the control registers used for the application. There is a very important note near the beginning of *Section 2* telling the reader that *bits[7:1]* are used and bit 0 is not used or serves a different purpose. For example, if you are loading *0x42* into a register, you would not load *1000010b*. You would shift left one bit and load *10000100b (0x84)*. Missing this could lead to unnecessary frustration.

Section 3 describes two scenarios of operations for these parts. *Section 3.1* describes how to communicate from the host controller to a camera sensor. *Figure 13.6* shows the layout for this application.

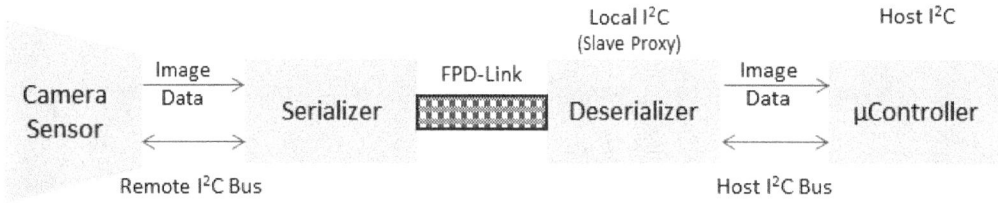

Figure 13.6 – Connection diagram for the application being described

The host and sensor are connected via a deserializer and serializer (a typical configuration). The host and sensor communicate via I²C but the deserializer and serializer communicate via FPD-Link III. This allows control of the sensor via I²C but doesn't require a direct I²C link between the sensor and the host. In three relatively basic steps, the app note describes how the host controller communicates with the deserializer, communicates with the serializer, and communicates with the camera sensor. The app note goes on to describe more complex scenarios. This app note can save you time and some agony of having to dive deep into the datasheets. The app note pulls together the relevant information in a more succinct format.

Get into the good practice of understanding datasheets, spotting errata, and reading application notes.

Device drivers

Along with the OS (discussed in the next section), device drivers are a fundamental building block. You may be working on an IoT application, but getting data from the vehicle (e.g., camera, sensor, or vehicle speed) is essential for the application.

Devices will be connected physically to an SoC or microcontroller using some means from a basic **General-Purpose Input/Output** (**GPIO**) line, an SPI or I²C, **Universal Serial Bus** (**USB**), MIPI, HDMI, FPD-Link, or many other mechanisms. A device driver is a piece of software that knows the physical connections to a device and handles the low-level interaction with the device. Device drivers do several things to support good software design. Among these are encapsulating details about physical connection and allowing software components to be more independent of the specific hardware connection, supporting simplicity and reusability.

The job of device drivers includes the following:

- Recognize the device.
- Initialize the device.
- Control the device.
- Access status of the device.
- Pass data to, from, or both to and from the device.

You may be faced with writing a device driver from scratch, but try to avoid this when possible. It is likely that a similar driver has already been written. Depending on the device, the device manufacturing may provide device drivers that you can adapt for your specific needs and hardware implementation. Many OSs, such as Linux, provide libraries and drivers. Also, the open source community will have many drivers available. A warning, though: if you are choosing to use drivers from the open source community, take care of how this code is integrated and how it is brought into your build and configuration management. Without proper handling, you will risk your company's intellectual property due to open source licenses. Purchasing device drivers is also an approach to be considered.

Bosch offers an automotive device driver library. They claim to have a *"fully managed cloud service that enables development teams to obtain the necessary device driver software for sensors and actuators in the form of modules and to configure it as necessary, all via a web platform ... available for the majority of sensors and actuators on the market."*

Hardware Abstraction Layer (HAL)

Continuing with exploring embedded development, HAL is a very commonly used term. But it likely means something different to different people. HAL, **Board Support Package** (**BSP**), and device drivers are terms used interchangeably. The purpose of the HAL is pretty obvious in that it abstracts the hardware from the middleware and application layers. Whether a device driver is part of the HAL or a layer below, and similarly whether the BSP is part of the HAL or not, is not too important. But aim for consistency in terminology. For example, if using Android, there are specific definitions for various HALs. Stick to these.

Your design needs to abstract the details of the hardware from the upper layers of the software. Device drivers help to do this, but the HAL provides more abstraction. A device driver such as an I²C driver encapsulates specifics about how to communicate via I²C, but the next layer above needs to know that when talking to the I²C driver, the connection is via I²C. If the hardware changes to use SPI, the higher levels of software will have to know this. The HAL creates an interface such as a camera controller. The camera controller can support one or more communication buses with the camera, which the upper layer of software will not know or care about. The middleware layers focus on business logic and there's no complication around the type of communication.

Additional aspects of embedded development

This book is not teaching you to become an embedded developer, but it is providing important considerations to help you get started. Some of the most critical topics are introduced and help you know where you need to learn more from other sources.

You will have to consider interrupt handling, multi-tasking, concurrency, and race conditions. Your design will likely need to be very specific about what tasks are running, where they are running (for example, if you have a multi-core processor), and the task priority at which they are running. Specifying these things will ensure that the applications run in a way that the user expects. You must avoid situations where tasks are hogging the CPU (for example, if a hardware device is continually interrupting the system, you can't allow the interrupt processing to prevent other activities from continuing). You will want to make sure that the user interface events can stop or change the processing of other tasks. For example, if map routing calculations are happening but the user wants to change a destination, the map routing needs to stop when a new destination is entered. Automotive applications have many activities going on in parallel (e.g., audio decoding and playing, video decoding and playing, navigation, camera operations, **heating, ventilation, and air conditioning** (**HVAC**) control changes, multiple display outputs, and being ready to take input from the user via speech input).

The computational power of **Electronic Control Units** (**ECUs**) is growing, decreasing the need for this next embedded software development practice, but you may still find yourself in need of it. Floating-point calculations are often avoided in embedded software development. Fixed-point arithmetic can be used instead of floating-point calculations. Fractional numbers are represented as an integer. For example, *38.56* can be represented as *3856*. In general, a scaling factor is selected to represent a fraction number as an integer. The scaling factor will be selected based on what is known about the calculations and the precision needed for the application. Then, addition, subtraction, multiplication, and division operations can be performed on the integer representation of the numbers. (Care must be taken to avoid or deal with overflow cases.) The fixed-point math can typically be done faster than floating-point math when floating-point hardware is not available. *Figure 13.7* shows an example of this.

> 3.12 x 40.8 = ?
> Scale the numbers to: 312 and 4080 with a scaling factor of 100
> 312 x 4080 = 1,272,960
> Revert the scaling factor (100x100) to end up with 127.296

Figure 13.7 – Example of fixed-point arithmetic

This chapter will now turn attention away from general embedded software topics to automotive-focused topics.

Automotive-focused aspects

Embedded and automotive development intersect in areas such as memory management, **key perfomance indicators** (**KPIs**), time management, and power modes.

Memory management will likely be very important as your application will be doing as much as possible with the least amount of memory. Early in the project, memory must be allocated to the **Random Access Memory** (**RAM**) and **Read-Only Memory** (**ROM**) for all subsystems or components. Remember to include a buffer for a couple of reasons. For RAM and ROM, later in the project when you are fixing bugs in time-crunch situations, you may need to use extra memory. If RAM usage is near 100%, your software will almost definitely slow down and exhibit unexpected behaviors. A 20% buffer is a good recommendation to follow. The memory layout is something that must be aligned and agreed to with all contributors. This is especially true when third parties or partners are co-developing and using the same memory. In these cases, it is ultra-critical for agreements to be well documented. These might belong in statements of work, acceptance criteria, or other business agreements. If later the overall memory layout does not all fit, it will be difficult to go to the third-party company and get them to reduce their memory footprint.

After the up-front planning and agreement with all parties is complete, mechanisms need to be built in to monitor/measure the memory usage and to take reasonable actions in case RAM usage grows beyond certain limits. Measuring and monitoring memory usage needs to be part of regular testing (for example, weekly and release testing). Use cases are needed to stress the memory usage so that you avoid false confidence in your software by testing only *happy paths*. (A happy path is, for example, if step 1, step 2, and step 3 are followed, the expected result occurs. However, what does the software do if step 1 and step 3 are followed but step 2 is skipped? Does the software handle this situation in a reasonable way?) Regular memory usage reports (both RAM and ROM) should be captured and monitored. A standard format that is easy to comprehend (even by managers) for these reports is recommended.

KPIs are another critical part of automotive software development. There are expectations or even laws regarding how quickly various parts of the automotive software must start up or react to inputs. For example, in the United States, the **rear-view camera** (**RVC**) must be displayed within a legally specified amount of time. CPU usage, the time audio takes to play, and the time to display a cluster tell-tale are other examples of KPIs. There are dozens of, if not more than 100 KPIs that will be specified for automotive applications. KPIs should be established early in the project. Mechanisms for measuring them should be created early in the project and then they should be measured and reported on at a regular cadence. If possible, include the measurements as part of the build and sanity testing. Like the discussion on memory usage, if various companies are contributing to a project, KPI responsibilities need to be established early in the project and even included in business agreements.

Time synchronization is another area of critical importance. Time synchronization has been a challenge in networks, and with automotive software growing in complexity, that challenge is one that now must be solved in the automotive industry. In automotive applications, data comes from many sources, such as various sensors in a vehicle. Data needs to be ordered and interleaved correctly for it to make sense and proper actions to be taken based on the data. Audio and video data synchronization is another use case that needs to be handled. CAN and FlexRay have some mechanisms that support time synchronization, but with Ethernet moving into automotive applications, Ethernet time synchronization is needed.

It seems obvious enough to say that all components use the same time, but this is an overly simplistic statement. Various processors and controllers in a vehicle can use a real-time clock but these will not likely be synchronized due to factors such as different startup times and clock drift.

A solution is to have a clock master. But even this approach is not as simple as it sounds. When the clock master passes its time through the system, there are different propagation delays and time that a component takes to process the time message so that the time as one part of the system will be different than at another place.

AUTOSAR has a time synchronization protocol specification. The protocol aims to support time-critical and safety-related applications such as deploying airbags and braking, but it supports other applications. It handles the distribution of time information over Ethernet. This specification is based on networking standards including IEEE 1588 and IEEE802.1AS. The AUTOSAR specification claims advantages over these networking standards. In automotive networks, the network is static, and the nodes are known so the claim is that the AUTOSAR specification can work more efficiently.

Embedded software development

Directly following the IEEE802.1AS standard, which is known as the **generic Precision Time Protocol (gPTP)**, may be a better solution if your project is not using AUTOSAR. Ethernet was built for applications that do not have time-critical requirements and is now evolving to support these requirements. Data needs to be presented or used at the same time (for example, audio and video). In automotive software development, a central controller needs to act on data from multiple sensors in a time-coordinated manner. The beginning of the gPTP solution is to select one Grandmaster. The Grandmaster has two primary responsibilities:

- Measure propagation delays from the Grandmaster to the devices
- Measure residence time (time for a node to handle the message)

Synchronization is done by the Grandmaster transmitting special messages to all nodes. Then, each node can know how much time it takes to get a message from the Grandmaster. This allows each node to synchronize its time to within 500 ns of the Grandmaster's time. Data from various sources, such as cameras and RADAR, can be passed to a zone controller or central controller and be synchronized.

Power state management

Power state management is about making sure that the embedded automotive applications (for example, infotainment applications) are running quickly, do not cause the battery to run down, and are able to respond to external sources when the vehicle is not running (for example, the key fob). *Figure 13.8* shows four scenarios for power management.

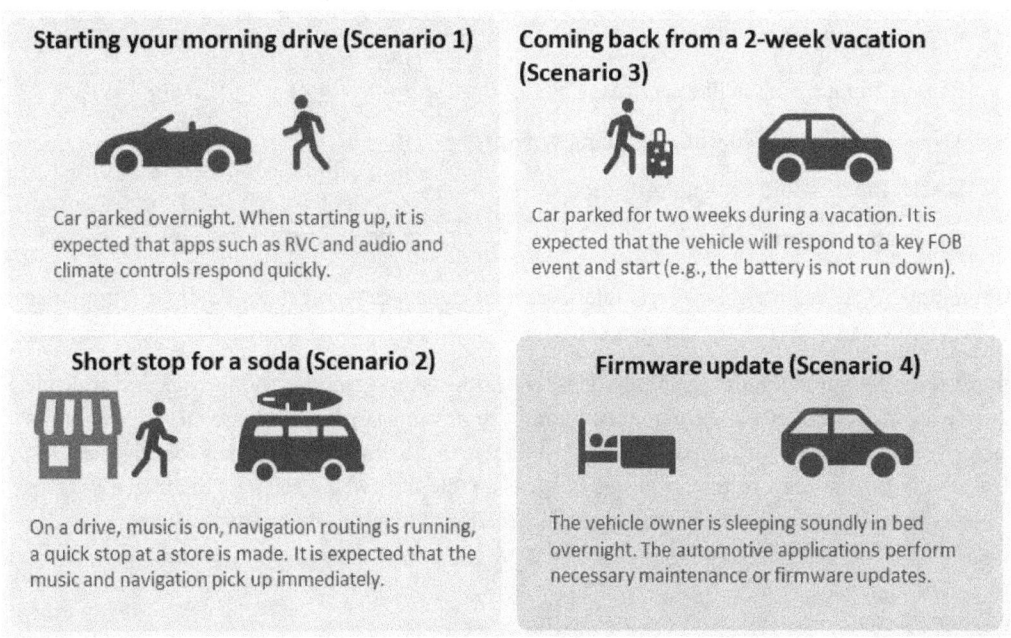

Figure 13.8 – Power state management user scenarios

Let us have a look at these scenarios in detail:

- The first is the most frequent. When starting a morning commute, RVC must be up within two seconds of the ignition turning on. The driver will expect climate controls to work immediately as well as audio playing. These apps need to start up quickly.
- Another scenario is that the driver stops at a convenience store to pick up a drink on the way to work. The car ignition is turned off for a couple of minutes. When the driver returns to the vehicle, it is expected that apps such as navigation and audio pick up from where they stopped and are running immediately.
- The third scenario shown is a driver returning to a car that has been parked for two weeks (or longer). It is expected that the vehicle responds to the key fob and starts. In other words, the software shouldn't run down the car's battery.
- The last scenario is where the car is parked in a garage and the software is able to perform necessary maintenance, including software updates. Then, when the driver is ready to use the car, the software is also ready.

The following lists inputs to which the automotive software must be able to respond:

- Key switch transitions – Off, Accessory, On, and Start
- Vehicle doors opening/closing and door locks engaging
- Seat sensor and seatbelt closure switches indicating occupants are present
- Events from key fobs
- Remote input from satellite services
- Vehicle security enabling and disabling systems
- Special input to initiate diagnostic modes
- Other user input from dashboard and device-based switches and touchscreens

The automotive OSs include libraries and interfaces to help manage power states. Android Automotive OS offers dedicated power state management.

One of the modes it offers is **Garage Mode**. This reflects the growing amount of software and available connectivity. Like cellphones and computers, automotive software needs various maintenance activities to be performed. One of the biggest is software updates to address things such as security patches, defect resolution, and new features. **Garage Mode** is an idle time where software maintenance tasks can be performed. This mode needs to be carefully managed to not drain a vehicle's battery.

As you work on automotive IoT projects, you may be directly involved in designing and implementing the power states, but even if you are not, you need to understand these. You need to understand the sequence your application will be running, when inputs will be available, how to properly shut down your app, and how it manages update activities.

The next section will discuss OSs that are used in automotive applications. These are becoming more similar and standard with non-automotive applications, which is helpful in transitioning between industries. But they also have automotive-specific support.

Operating systems

Your automotive applications will run on an OS that in many ways looks like those running on a phone or computer. Various services and thread management will be the same or similar. Differences will exist. These differences are due to many things already discussed in this book, such as power state management, cybersecurity, and functional safety. The processors selected are also a major factor.

The OS kernel architecture is a key factor. There are two main types of kernels: monolithic and microkernel. Linux is a leading monolithic kernel OS and QNX is a leading microkernel OS. A monolithic kernel includes core functionality and services. A microkernel includes a minimal amount of software and provides the ability to layer on services as needed. This modular approach is often preferable for automotive ECUs.

Functional safety ratings may be a factor in which OS is selected. According to an EE Times article, *Perspectives on Automotive Operating Systems* by Egil Juliussen [3], all AUTOSAR-based OSs, including some from Vector, Elektrobit, and BlackBerry, have functional safety ratings. But as of 2024, many of these are limited in ability to serve larger and more complex software, such as infotainment and ADAS applications. QNX is an exception in that it both has a functional safety rating and can be used in large software applications such as infotainment. It is likely that as functional safety rating requirements grow and zone controllers and central controllers are to be functional safety certified, more OSs will come along as alternatives to QNX. This same article notes that it believes Linux would have a functional safety version and possibly Google's infotainment OS would also.

MarketsandMarkets™ [4] is forecasting Android OS to grow significantly as an automotive OS in the 2022-2030 period and have a significant share by 2028. Android's automotive OS is called **Android Automotive OS**, or **AAOS**. AAOS is open source and provides access to Google's Play Store. Global Market Insights reported that QNX held a 34% industry share and is forecasted to grow through 2032. QNX can provide real-time performance needs in many applications. It is also highly scalable and flexible to be used in many types of applications.

Figure 13.9 shows a brief overview of automotive OSs being used at the time this book is being written. The information was gathered from *Everything You Wanted to Know About Types of Operating Systems in Autonomous Vehicles*, published by the Intellias global technology partner [5].

OS	Brief Description	Used By
QNX	One of the most used OSs. A real-time OS with functional safety ratings and a Hypervisor option.	Partnering with about 40 automakers including Ford, Acura, VW, BMW, and Audi.
Wind River VxWorks	Widely used in embedded applications. A real-time OS with Functional Safety ratings.	Toshiba, Bosch, BMW, Ford, VW, and others.
Green Hills INTEGRITY®	A real-time OS claiming the highest safety and security ratings making it a good candidate for ADAS applications.	Works mostly through partnerships with OEMs and Tier 1/Tier 2 providers.
Linux	There is a project called Automotive Grade Linux (AGL) that is aimed at automotive applications.	BMW, GM, VW, Toyota, Chevrolet, Honda, Mercedes, Tesla, Lyft, and Baidu.
Android Automotive OS	Enables an Android phone-like infotainment user interface and gives access to the Google Play Store.	Volvo and Audi have signed contracts with Google to start using the upgraded version of this automotive OS in 2020. They're followed by the Renault-Nissan-Mitsubishi Alliance.

Figure 13.9 – Brief introduction to some top automotive OSs

Autonomous driving, higher computing performance demands, and AI are disruptors opening opportunities for other OSs to enter the automotive industry. NVIDIA is one example of a company that has leveraged its position as a chip manufacturer, introducing the NVIDIA DRIVE™ OS, which is drawing interest from various automotive **Original Equipment Manufacturers** (**OEMs**) and tier 1s. Mentor Graphics and Microsoft are investing in supporting autonomous vehicles. The Renault-Nissan-Mitsubishi Alliance is working with Microsoft.

Another approach to automotive OSs is for OEMs to develop their own or partner closely with another company for a customized OS. Tesla has demonstrated this is possible and other OEMs are looking at following. Toyota, Volkswagen, GM, and Daimler are all working on proprietary OSs.

So, what does this mean for you? You may need to learn or improve your skills around OSs used in automotive development. A recommendation is to learn either QNX or Linux. There are many opportunities to become competent in working with the Linux OS, including many free courses. Learning QNX may be more challenging as training courses are likely to cost money and be offered by fewer sources. One advantage of QNX is that you can take courses directly from the company that develops and distributes QNX, BlackBerry QNX. BlackBerry QNX offers a course titled *Real-Time Programming for QNX Neutrino RTOS*. At the time of writing this book, this 28-hour course has a listed cost of $2,495. Learning the Android OS is another option. This is a highly valued skill within a narrower band of opportunity as this OS is not as widely accepted across the industry. But like Linux, there are opportunities to learn Android with free options.

As a first step, if you have not already done so, take a college-level course on OSs. The background that you will learn will be helpful. There will likely be theory that is not important but the concepts that you learn are not likely ones you will be able to pick up in other ways. Especially through a local community college, you may be able to take a course for free or at a low cost. Another option to explore is whether your company reimburses for courses. Time will be the main thing you are giving up. Here is a sample of courses (from the U.S.) with similar ones likely available in your country and area:

- **Atlanta Technical College**: CIST 1130 Operating Systems Concepts
- **Portland Community College**: CS 140U. Introduction to UNIX
- **Clackamas Community College**: CS-140 Introduction to Operating Systems

For Linux-specific learning, many options are available, including free options. One to try is Linux Journey [6]. This journey ends with network administration but the first two levels, Grasshopper and Journeyman, are good ways to learn about the Linux OS.

This completes our introduction to automotive OSs. Next up is a discussion on hypervisors, which are ancient in computer industry years, but very new for automotive applications.

Hypervisors

Hypervisors have been around for more than 50 years but have recently become important in automotive software development. The main driver for the use of hypervisors is centralized computer controllers, which reduce the required hardware (for example, one SoC instead of two). A more specific use case is for running cluster software and IVI software on the same SoC. We will come back to explaining this use more after some background on hypervisors.

Hypervisors allow separate programs or operating environment running on the same hardware. It creates **Virtual Machines** (**VMs**). There are two main types of hypervisors, which are referred to by different names, but one convention is to refer to them as bare metal and hosted. Another convention is to call them type 1 and type 2. *Figure 13.10* gives a visual representation of the types of hypervisors:

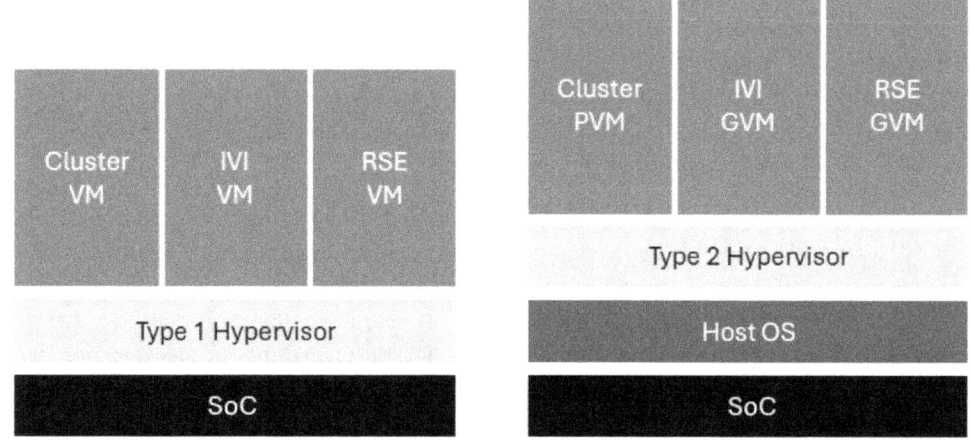

Figure 13.10 – Visualization of type 1 and type 2 hypervisors

Type 1 is also called bare metal as there is no main OS and the VMs run on their own on top of the hypervisor. Type 2 is called a hosted hypervisor as they have a host OS.

Some of the current hypervisor providers are QNX, Continental AG, Renesas Electronics Corporation, NXP Semiconductors N.V., Wind River Systems Inc., Green Hills Software LLC, Sasken Technologies Ltd., OpenSynergy, and Mentor Graphics Corporation. QNX provides a type 1 hypervisor and is one of the most common.

Both types will serve automotive applications. A type 1 hypervisor can be more efficient and secure as the VMs run directly on the hypervisors, compared to type 2 where the **guest virtual machines** (**GVMs**) run on top of the host VM. Type 2 hypervisors can have a lower cost of adoption and maintenance. For a QNX type 1 hypervisor, there could be an additional trust factor in using QNX for the cluster, which has more real-time OS requirements. With QNX, you will likely be using both the QNX OS and the QNX hypervisor. In this setup, the QNX OS is running in the **Primary Virtual Machine** (**PVM**) and then one or more guest VMs are running other OSs (could be Linux or Android). The choice between the hypervisors probably comes down to your company's experience, business choices, and partner relationships.

As mentioned, a hypervisor can reduce the hardware requirements. A hypervisor can also increase security and better support functional safety applications as parts of the software can be separated into different VMs.

A good thing about a hypervisor is that it sets up the execution environment for software so that mostly, you do your software development within a VM without being concerned with what is happening in the other VMs. But let's take a little closer look at implementing a hypervisor based on roles on a project. As a system engineer or system/software architect, you will be making critical choices regarding a hypervisor. In one of these roles, you will be deciding things such as the following:

- Whether to use a hypervisor
- What hypervisor to use
- What VMs are needed
- The allocation of applications and services running in each VM
- The allocation of cores, priorities, and other resources within the SoC

If you are a software engineer working on the lower level of your system, for example, in the BSP, you may be involved in setting the configurations for the hypervisors and with the build systems for the hypervisor and the VMs.

If you are working within a VM, you will be doing most or all your work without thinking about the fact that you are working in a hypervisor system. One service that will need to be implemented is inter-VM communication. A communication stack will be created on each VM. You will also need to understand the software architecture to know whether your software components need to get or pass data or commands between the VMs.

In addition to commercially available hypervisors, there are also open-source hypervisors. Xen is one example. Xen is working to expand to embedded and, more specifically, automotive systems. One of their solutions is pairing Linux with Zephyr, which is a small **Real-Time Operating System** (**RTOS**). Xen is even working toward better support for safety certification. ACRN is another open-source hypervisor coming from the Linux Foundation. These and others are options for your project if you are looking for an open-source or a potentially lower-cost solution.

That completes our quick introduction to hypervisors. The next section continues on the theme of embedded development. It will discuss key development tools typically used in automotive software development.

Development tools

This section introduces some of the most important software development tools used in automotive IoT. This section will cover life cycle management tools and briefly touch upon programming languages, code management tools, and debuggers.

Life cycle management tools

The **Software Development Life Cycle** (**SDLC**) is a process for designing and building software (and is not specific to the automotive industry). Typically, the SDLC is defined by 5-7 stages. The stages may include planning, requirements definition, design, development, testing, deployment, and maintenance. The first five stages line up closely with ASPICE processes and the V-model. SDLC has similarities to **Application Life Cycle Management** (**ALM**) and these terms are often used interchangeably. The SDLC is more focused on development. DevOps is another related term, a shortened form of **development operations**. DevOps is a way of automating and improving the building and delivering software. DevOps will come up again in this section.

There are SDLC tools that help automate software development. These tools support activities including the following:

- Project planning, release planning, and sprint planning
- Assigning and managing daily and weekly activities
- Managing your work items
- Communication of project status
- Process adherence (think ASPICE and functional safety)

Figure 13.11 shows one layout for SDLC tools.

Figure 13.11 – SDLC tools layout

In the example shown, the main tools are as follows:

- Requirements database
- Design documents database
- Test case and execution management
- Activity planning and management tool

Each of these four tools will need to connect to one or more of the other tools to show the workflow and traceability throughout the life cycle. Let's start with the activity planning and management tool. This is where the work begins, at least from a planning point of view. The project backlog or feature list is created with the tool. A release plan is created and then the upcoming release is planned. Planning is done by breaking down backlog items or features into various tasks. Here is a list of potential tasks:

- Requirements authoring task
- Requirements review task
- Requirements approval task
- Design task
- Design review task
- Design approval task
- Unit construction task
- Code review task
- Test case creation
- Test case execution
- Code merge task

Whether all these (or even more) tasks are needed or not depends on the process requirements and department/project choices. The previous chapter shared information and guidance on process selection.

The requirements task will be done using a requirements management tool. The design task might be done with a word processor along with a drawing program. However, a design tool that supports modeling is likely a better choice. Code is stored in a configuration management system. Test cases might also be kept in a word processor or a spreadsheet. A test management system is likely to be helpful. It can support defining what test cases to execute and keeping track of the execution and results.

For each of the parts – project management, requirements, design, code, and tests – traceability is needed. Requirements need to be traceable to design and testing. Especially when your project aiming at ASPICE is level 2 or level 3, requirements work needs to be connected to project management. Design work must be traceable to the corresponding code and integration tests. Additionally, every piece of code must be linked to its respective unit tests.

It is possible to accomplish this with a word processor or spreadsheet. Unique identifiers can be created to allow tracing across the workflow. Macros can even be created to semi-automate this activity. Depending on factors such as the size of your project and process expectations, this may be an acceptable approach. Likely, while working on your project, you will need specialized tools to support your software development. Jira is one of the most popular project management tools. It supports the tracking of the backlog, tasks, bugs, and product requirements and it integrates into other tools such as GitHub. *Figure 13.12* shows one set of tools for managing software development activities and supporting process goals.

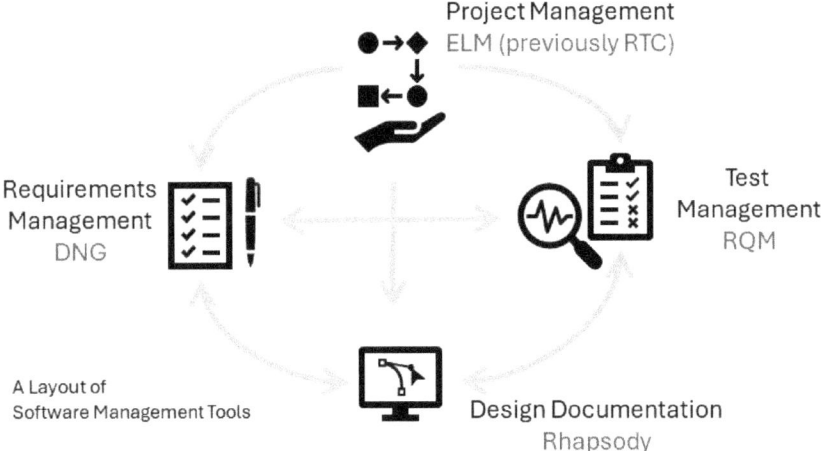

Figure 13.12 – Tools that can support software processes (Reference: `https://softacus.com/blog/articles/jazz/installation-of-ibm-elm-en`)

DOORS Next Generation (**DNG**) is a requirements database that can link to test cases in **Rational Quality Manager** (**RQM**). Rhapsody can be used to trace back to requirements in DNG and test cases in RQM. Then, **Rational Teams Concert** (**RTC**) can connect all the pieces and manage the workflow.

Whatever tools your project or department is using, you will be much better off taking the time to learn to use them. With some investment, this part of your work can go from painful and frustrating to mostly bearable.

The next software development tool that we will discuss, although briefly, is a debugger. In this author's personal experience in writing software as well as a manager, it is clear that debuggers are not used frequently enough. Sometimes the reason is that the debugger is not configured to work in the development environment. Sometimes it is easier to use debug statements. Sometimes it is easier to guess with code changes. If you are working at the application level, you may be using an **Integrated Development Environment** (**IDE**) that has a debugger built in and is straightforward to use. If you are working on lower levels of code, you may not have an IDE.

GNU Debugger (**gdb**) is a debugger that you need to learn and use. Debugging code without a debugger is often done and it works. However, there are shortcomings. There will be more iterations while trying to get to a fix. Also, the fix may only address part of the problem. You may have a test case that is failing, and you can fix that specific test case and miss other errors in the code. Of course, code inspection should be used to find errors. Next, a unit test should be used to verify that the unit construction is per the design. When a test case failure is detected that cannot be resolved through code inspection, consider a debugger as your next tool of choice. With a debugger, you will be able to step through conditional code to properly understand what path is being taken. You will be able to examine the stack to check function call paths and check the variables.

Invest time into getting your debugger set up, learn how to use your debugger, and then use it.

We will spend just a minute mentioning some other automotive software development tools. If you are not already, become an expert in C++. Be an expert on not only the syntax and programming but also the related object-oriented design and implementation. This is an important foundation that seems obvious but too many people working in automotive software are not strong enough with this skill. C programming skills are important, but C++ is critical. Being proficient in scripting languages is also helpful. Python proficiency, for example, will make you a much more effective software engineer. You can use your Python skills in a variety of ways, including for quick prototyping and creating test scripts. Git, Gerrit, and Jenkins are common tools that you will need to use.

Finally, automotive software development involves **Controller Area Network** (**CAN**) signals and, to a lesser degree, **Local Interconnect Network** (**LIN**) signals. Message passing is being done more and more by Ethernet, but CAN will continue to be used in parts of automotive systems. CAN signals are likely to be key inputs and outputs of your software system. To test your software at your bench or even in a lab, you are going to need to simulate CAN signals. Testing and simulation tools can be purchased. Vector, for example, has the CANalyzer and CANoe products. GENIVI/CANdevStudio *"aims to be cost-effective replacement for CAN simulation software." "CANdevStudio enables to simulate CAN signals such as ignition status, doors status or reverse gear by every automotive developer"* [7]. Learning how to use these tools and getting them set up takes time. Take the time to learn how to use these tools when you are not *under the gun* to make a software delivery or fix a critical defect.

Like a good craftsman, being proficient or an expert in your tools will greatly enhance your software development efficiency and effectiveness. Be an expert in C++. Know how to use a debugger. Be proficient in Python. Know the system around your software so that you can test your software. Know how to simulate CAN or other signals.

That wraps up the fourth section of this chapter. The final section before the summary discusses the software development ecosystem. Automotive applications are rarely created by one team or one project. Navigating the ecosystem of development is critical for success.

Software development ecosystem

Automotive IoT development is going to involve many partners. Partners refer to multiple companies coming together to build the overall solution. Traditionally, the ecosystem starts with the automotive OEM. Then, the tier 1 is responsible for the development, manufacturing, and delivery of parts to the OEM. The development is a combination of electrical, mechanical, and software activities. Specific to software development, there are typically tier 2s and other suppliers of components. A tier 2 supplier might be the navigation component supplier, for example.

With the growing value of software, OEMs are taking more control over software development. They are operating more like a tier 1 with respect to software development. This adds more partners to the software development mix. *Figure 13.13* shows an example of a typical ecosystem of an automotive IVI system.

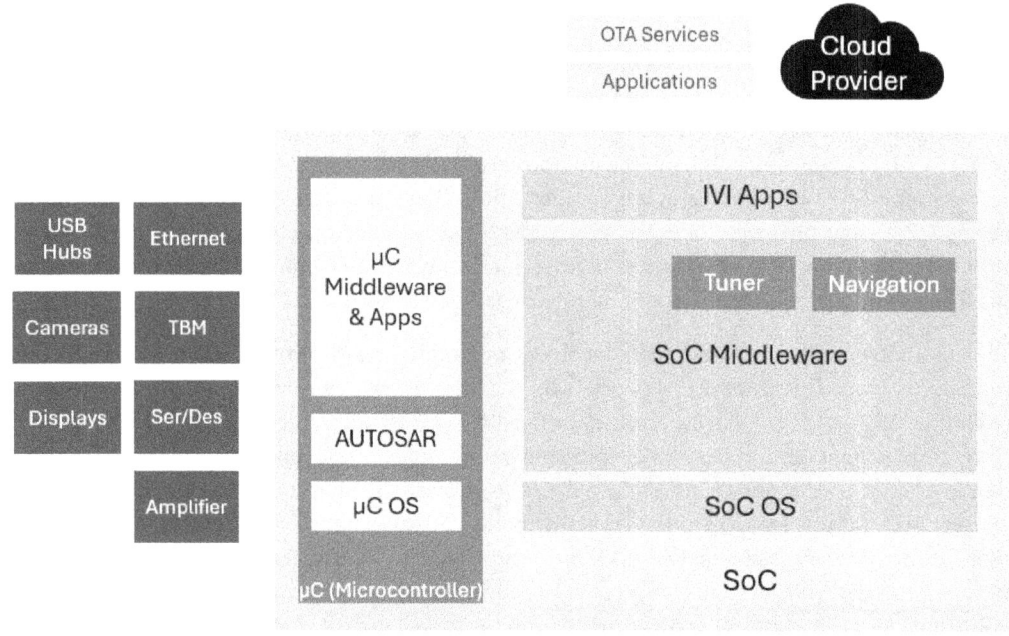

Figure 13.13 – Example of the ecosystem of an automotive IVI system

Each colored block represents a different company. In this figure, there are nine colors/companies. It is possible to have fewer, but it is more likely that this is understating the number of companies providing content.

You may be creating software that is completely within your company and not dependent on another company. But most likely, you will be working in a complicated ecosystem with suppliers, partners, and a customer. The success of a project will be improved with good business agreements, partnerships, communication, established timelines, defined deliverables, and joint working practices. We will look at three types of relationships: the one with your customers, your co-suppliers, and your suppliers.

You and your customers

This is the most important relationship in your development project. As a software engineer, you likely won't be at the forefront of establishing business agreements with your customers. You will play important roles in making the project a success and excel in customer engagement. Some aspects where you will provide support are as follows:

- Understanding and helping to clarify the specifications and scope of the project
- Estimating the project development efforts and timing
- Helping to clarify conditions within the agreement, such as deliverables from the customers to your company (this might include what test properties need to be provided by the customer)
- Continuously helping to manage escalating scope creep when needed
- Ensuring deliveries to your customers are aligned with the project plan/business agreement
- Helping to adjust project plans during the course of the project

During the bidding phase, you will support reviewing specifications and project requirements, helping to get clarification from the customer, and filling in gaps in the specification. Following this step, you will help to estimate the project effort and with the timeline of deliveries and the content of each delivery. Also important is to help identify conditions that are important to project success. Examples of these things are as follows:

- Define when customer specifications need to be delivered
- Define what and when test properties are needed – a test property might be a display that your software is writing to
- Provide clarifications on what is in and out of scope for the project

During the execution phase, you will be helping to manage scope creep. Some amount of scope creep and change in your company's deliverables is expected. (Hopefully, some additional unknown work was accounted for during the bidding phase.) Your approach should be to do a great job at customer engagement and support the success of the project. This means the need for some flexibility. You will be on the frontline of discussions as your customers may be pushing you to implement things that are not in the defined scope. Depending on the direction from your management team, you may decide when to accept the proposed changes and when to request the customer submit a formal change request.

You and your co-suppliers

You are likely to have "co-suppliers" contributing to parts of the customer's overall system. In this case, your customer has other suppliers contributing. In the ideal situation, your company and the other suppliers will develop to specifications and project requirements. We work in the real, not the ideal, world. A couple of challenges that you will face include gaps in the specifications and issue investigation. When there is a gap in the specifications, either you or the other supplier will have more work than planned, and negotiations will be needed to determine which company steps up to fill in the gap. Your customer may want you and others to work things out among yourselves. This is easier for the customer as the problem is solved with less time and effort from them. But likely, the customer will have to arbitrate and make a choice.

Your role in this may be important. You must help with the discussions with your management, the other supplier, and the customer. Is the change or gap the responsibility of the other supplier? From a system architecture point of view, what makes the most sense? Your management or business team may be taking the lead on these types of discussions, but your input is vital to support them. Also, you will likely be taking part in the discussions with the customer or other supplier. Your influence may help to bring resolution with less escalation.

Issue triaging is another area that may involve you and other suppliers. When the customer finds an issue, they will want it to be resolved as quickly as possible. An example is RVC not displaying properly when putting a vehicle into reverse. *Figure 13.14* shows a view of an RVC system.

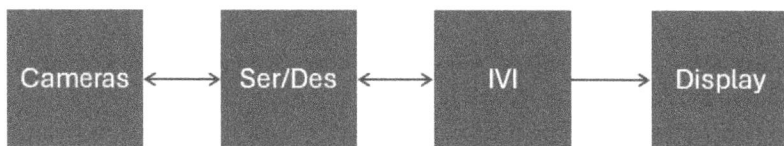

Figure 13.14 – RVC system view to help consider subsystems that might have a fault

Each block comes from a different company. When RVC is not showing, which company takes the lead in figuring out the root cause and resolving the issue? Issue triaging is a time-consuming process, and if you and your company are spending time working through issues that are not your fault, time is wasted. One approach to minimizing this is to include adequate logging so that you can show when issues are not within your subsystem. Being able to quickly triage issues and get the customer to assign the issue to the right company will help you and your company so that you are not wasting time.

You and your suppliers

Managing your suppliers is the area that you will have the most control over and the most responsibility in terms of project success. One view of the landscape of suppliers is shown in *Figure 13.15*:

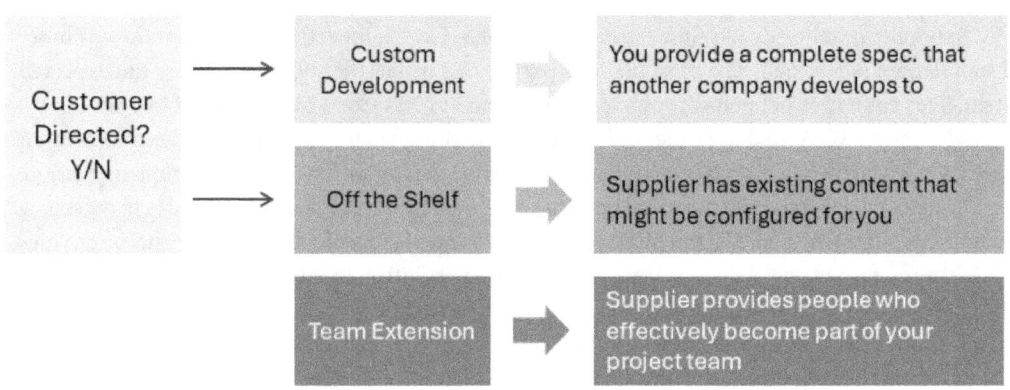

Figure 13.15 – Three types of suppliers you may work with on a project

Team extension is where contractors join your team, and from a project execution perspective, they are part of your team. Off-the-shelf suppliers are ones that have an existing product being provided to your project. It is likely that some configuration is needed as well as support while you are integrating their product. Custom development is when you are outsourcing part of your project. You will create a specification and have a supplier create content based on this.

At the beginning of your project, your team or management will be making buy versus make decisions. Sending out **Requests for Quotes** (**RFQs**) and reviewing responses will determine whether to outsource work and which supplier to choose.

> Note
> If there are functional safety aspects to the work being outsourced, there are functional safety process activities that must be followed.

Once it is determined that you will be outsourcing some or all of your project or purchasing components, supplier management becomes a critical part of project success.

The committee that put together the ASPICE standard recognized the importance of supplier management and created a process named ACQ.4, Supplier Monitoring. The purpose of the process is to manage the performance of the supplier based on the agreed commitments. The software engineering ASPICE processes were covered in the previous chapter. We have come back to ASPICE in this chapter as ACQ.4 fits best here and is also an important aspect of many software development projects. Even if there are no ASPICE requirements for your project, it is highly recommended to incorporate some of these practices into your project if you have suppliers providing content.

ASPICE supplier monitoring has four expected outcomes. First, joint activities between the customer and supplier are performed. The most obvious part of this is that the supplier delivers the expected content. This content is defined through a specification, an RFQ, or a **Statement of Work** (**SOW**). You might use an SOW to define the high-level deliverables, such as monthly releases with main deliverables. Then, there is a specification that provides the details for the design and implementation. The software coding is something that will be outsourced. Whether your company chooses to outsource requirements authoring and design will depend on how much control you want over these activities and how many resources your company has to do these activities. In addition to the main content, there will be other activities that the supplier is expected to perform. These may include status reports, test reports, and software quality audit results. These things can get overlooked when agreements are put in place. Requesting your supplier to add these to their list activities later becomes harder. Your company will have certain activities that must be performed that support the success of the supplier outsourcing. Some things your company may need to do are as follows:

- Provide specifications
- Participate in or lead reviews of things such as requirements, designs, code, and test cases
- Approve things such as requirements, design, code, and test cases
- Integrate deliveries
- Test integrated software and report defects back to the supplier in a timely manner
- Provide test properties

Ideally, these things are spelled out as part of the project deliverables. A good practice is to use a standard SOW or RFQ template. The standard template helps you to remember to include things that are sometimes overlooked.

The second outcome from ASPICE is to ensure regular communication between you and the supplier. The third outcome is to ensure the supplier is sticking to the agreements. Establish regular check-ins or status reports. Have a standard project report format, include it in the initial agreement, and adjust it as the project continues. Often, when a project gets off track, there are escalations involving upper management. A recommendation is to do pre-escalation. This is getting the right level of management engaged early in the project. This helps to get the right level of attention on the project. It helps to reduce escalations, and when an escalation is needed, connections have already been established.

The last outcome is that changes are negotiated. This is another good point to include in the initial agreement. For example, have a section in your standard SOW that describes how changes will be negotiated and agreed to.

The last topic to be covered in this section is regarding directed suppliers. A directed supplier is a supplier delivering content to you to integrate and use. Your customer is requiring you to use a specific supplier and the business agreement may be between your customer and that supplier. This situation presents challenges. Your customer will likely be holding you to your deliverables, which includes content from the supplier, and expect you to hold the supplier to their deliveries. However, since the customer has selected the supplier and the business agreement might be between the supplier and the customer, your company has less control. Leverage your supplier management practices as much as you can with directed suppliers. When sharing the project status with your customer, be sure to give the status specifically related to directed supplier responsibilities. Then, as needed and as practical, encourage tri-party meetings where the direct suppliers give their project status to your company and your customer. These things will help hold the directed suppliers accountable and support smoother project execution.

Summary

That completes a view of some important parts of automotive IoT development. We had an introduction to some embedded development topics, which was helpful especially if you are new to embedded development. We looked at typical OSs and hypervisors used in the automotive industry. We touched on several software development tools and closed the chapter with a look at supplier management. Much of automotive software development is about integrating software components coming from various suppliers. This makes supplier management important. Important takeaways from this chapters include being introduced to aspects of embedded software development and the need to grow in your knowledge of this area, an overview of automotive OSs and hypervisors, with suggestions of where to learn more, software development tools commonly used, and, very importantly, managing projects in complex automotive software development ecosystems.

Next, we will move on to the final chapter of this book. The chapter will share further insights on automotive IoT software development. Each of the authors will be sharing a few final thoughts to help you to grow as a software engineer.

References

[1] NXP S32K3 Automotive Telematics Box Guide, https://www.nxp.com/document/guide/getting-started-with-the-s32k3-automotive-telematicsbox-t-box:GS-S32K3-T-BOX

[2] NXP PF5020 datasheet, https://www.nxp.com/docs/en/data-sheet/PF5020.pdf

[3] Egil Juliussen, "Perspectives on Automotive Operating Systems," EE Times, https://www.eetimes.com/perspectives-on-automotive-operating-systems/

[4] MarketsandMarkets™. Automotive Operating System Market. Available at: https://www.marketsandmarkets.com/Market-Reports/automotive-operating-system-market-257628775.html

[5] Intellias, "Everything You Wanted to Know About Types of Operating Systems in Autonomous Vehicles," accessed at `https://intellias.com/everything-you-wanted-to-know-about-types-of-operatingsystems-in-autonomous-vehicles/`

[6] Linux Journey. Available at: `https://linuxjourney.com/`

[7] GENIVI/CANdevStudio. Overview. Available at: `https://github.com/GENIVI/CANdevStudio#overview`

Get This Book's PDF Version and Exclusive Extras

Scan the QR code (or go to `packtpub.com/unlock`). Search for this book by name, confirm the edition, and then follow the steps on the page.

Note: Keep your invoice handy. Purchases made directly from Packt don't require one.

14
Final Thoughts

You have made it to the last chapter of this book. Drawing upon our combined 80+ years of experience, this chapter provides additional insights that will significantly enhance the success of your project. Before we summarize this chapter and this book, there are a few topics that haven't gotten enough attention so far.

Congratulations on your learning journey through automotive **Internet of Things** (**IoT**) (unless you skipped around or skipped ahead to the end). We've covered a lot in this book, and you have learned about a wide range of important topics and dived deep into key areas. Let's finish strong with this last chapter!

In this chapter, we will cover the following topics:

- Agile, scaled Agile, and ASPICE
- Automotive embedded testing considerations with real examples
- Security

Finally, we will recap this book.

Agile

This section will discuss Agile development, scaled Agile, and how Agile development can work with ASPICE, which was discussed in *Chapter 12, Processes and Practices*. With IoT, cloud computing, and **software-defined vehicles** (**SDVs**) coming into automotive development, Agile development practices are common. **Scaled Agile** is an approach being taken by some.

There are both good and bad reasons to follow Agile development practices. First, we'll look at the wrong reasons. Sometimes, people and teams use Agile as an excuse for poor software development practices. Some examples include not capturing requirements as they're not needed because Agile is being followed, saying that design work will be captured later once coding is done, and doing ad hoc testing of code rather than following a strategy and a plan. Agile might just be an excuse to write code that may end up either being more like demo code or possibly being production code that doesn't have a solid foundation. If these are the reasons for following Agile, the result may be worse.

Enough with the negatives – let's review the positive reasons for choosing Agile development. An Agile approach reduces the time from requirements to design to coding to testing. This reduced cycle time has many benefits. First, the requirements and design will be more just-in-time. They will not be sitting around for weeks or months getting stale. However, they may be out of date and incorrect. When coding happens, they might have been forgotten about and must be refreshed – or worse, they might be ignored. Implementing testing while code is being developed can significantly shorten the time it takes to resolve issues. This is because root cause analysis occurs closer to the time the problem was introduced. A final advantage to mention is that software can be incrementally demonstrated to stakeholders. The feedback time from stakeholders to the software engineers is reduced more easily, allowing changes to be made. This also helps with schedule predictability. The team may think the product is nearly ready for production, but then the stakeholder feedback comes in late. There may not be much time to react to this feedback, meaning that either the feedback isn't considered or the schedule is pushed out.

Additional advantages of Agile approaches are that they promote teamwork and daily progress. Daily scrums or kanban boards require team members to come together and help each other make progress each day. If done right, roadblocks are removed sooner, and better progress can be made. You are almost certainly using Agile in your software development. Make sure that you and your team aren't following Agile as an excuse to avoid solid software development practices in areas including requirements and design.

With Agile development as a given (whether teams are following it in a beneficial way or not), many automotive companies are scaling up in terms of Agile practices. Scaling means that they are taking Agile development beyond software teams and amplifying the impact of Agile across projects, portfolios, and the business. *Figure 14.1* shows a view of how Agile can be scaled.

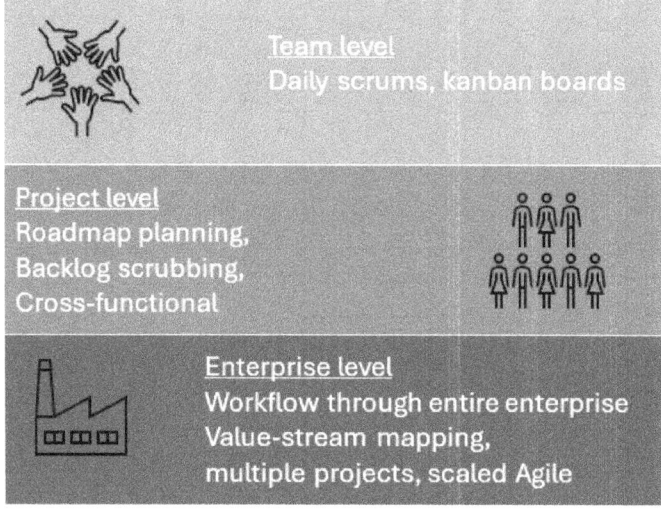

Figure 14.1 – How Agile can be scaled

If just a software team is following Agile, some of the benefits will be limited. One of the biggest limitations is flexibility in features. The feature set that's envisioned for a project is typically too ambitious. Agile helps to focus efforts on the most important features and have them ready to be delivered early. If the business, sales, or product managers are not supportive of Agile development practices such as delivering the most value, they may continue to press for all features while ignoring schedule and quality risks. Project managers who aren't ready to accept Agile development approaches may demand detailed schedules and expect more documentation early in the project. Stakeholders may not buy into and support this incremental development and may not provide valuable feedback during development.

Scaling Agile to the cross-functional project level or even further to the business, enterprise, or company level provides opportunities to magnify the benefits of Agile development. More people are thinking, acting, and deciding in an Agile way. For example, if hardware and software are both following Agile practices, the teams can be more closely aligned in their development cycles. Software will be better prepared to provide feedback to the hardware team on changes to designs and part selections. Then, each team can react/change as needed based on the feedback. This will speed up the overall development of the HW+SW and will better support the customer's intended design (rather than the purely documented design.)

There are several scaled Agile frameworks. A few of the more popular frameworks are Scrum@Scale, Disciplined Agile, **Large-Scrum Scale (LeSS)**, Nexus, and Enterprise Kanban. **Scale Agile Framework (SAFe)** [1] may be the most popular.

SAFe is continuing to evolve from its inception with SAFe 1.0 in 2011. In 2023, it is up to SAFe. 6.0. At the time of writing, there are four levels of SAFe: Essential SAFe, Large Solution, Portfolio, and Full SAFe. Essential SAFe is the initial level and is project-focused; it is typically the Agile development process that's familiar to most people. Large Solution supports multiple sub-projects that coordinate with each other. Portfolio combines value streams into a more coordinated effort. Finally, Full SAFe focuses on complete businesses and driving value for the entire enterprise. GM has been using SAFe since 2011 [2].

Scaled Agile focuses on applying Agile development approaches to a wide scope within a company. This helps magnify the positive impact of Agile. As a software developer, your activities will still be focused on traditional Agile development, but you will see how your day-to-day work connects to large parts of your company.

Agile+ASPICE

A common perception is that if you are following Agile development, you cannot also follow ASPICE. Following Agile might also be used as an excuse to not follow ASPICE. Following ASPICE for all parts of automotive IoT development ensures a high level of quality, reliability, and compliance with industry standards, which is crucial for the safety and security of connected automotive systems. Cloud services and application layers may not need some of the additional rigor expected by ASPICE. Embedded software, especially software that is closer to the vehicle's functionality and is more difficult or expensive to update once deployed, can benefit from following ASPICE.

Let's go back to Agile and ASPICE. You can follow Agile development approaches and also follow ASPICE. ASPICE involves setting expectations around requirements, design, coding, and testing. Agile development doesn't remove the need to do all these things and do them well. The biggest difference may be around the timing when each of these activities is performed and how complete each activity is at various stages of development. ASPICE adds some rigor around things such as traceability and following some defined strategies.

If your company or projects have been following ASPICE processes and you are beginning to adapt to more Agile processes, some changes may be needed to better enable continued ASPICE adherence while taking advantage of Agile development. Your project's or department's defined processes may need to be updated to support Agile. Some types of changes that might be needed are specified here:

- Changing specified development tools so that they're more flexible or ones that better support Agile
- Timing requirements for when activities must be completed (to support iterative completion of work)
- Support more process tailoring to allow scrum teams to tune their development
- Simplify the required ASPICE strategies and plans
- Update and choose process metrics that match Agile development
- Identify completeness criteria
- Aim for ASPICE level 1 or 2 compliance, not level 3

The people who are assessing your projects may also need to adapt their methods. A simple example of this is that the assessor cannot expect that all requirements are complete at earlier stages as the requirements will be created iteratively along the way. The **International Assessor Certification Scheme** (**intacs**) is a group that certifies process assessors, and their certifications are widely accepted. Intacs is working to support the growing need for Agile development in automotive processes and ways to implement ASPICE in an Agile development model. Agile SPICE training is being offered for process assessors.

Learn and follow Agile development. However, when doing this, make sure that Agile is not an excuse to skip good software development practices that are critical to ending up with a high-quality result. ASPICE can work with Agile but should be leveraged for the right parts of the project since it adds effort. Update standard processes so that they support both Agile and ASPICE as needed.

With Agile and ASPICE covered, this chapter will continue to dive into a couple of important parts that often do not get the required attention. Next, we will cover automotive embedded testing.

Automotive embedded testing

Security testing and system testing offer their own challenges that require good planning and execution. *Chapters 7* and *11* cover these forms of testing.

Ideally, the testing phase is about verifying that the software is working properly. It should not be about finding problems. Problems should be found during the requirements, design, and coding activities. But we aren't perfect, and testing is an important part of the software development process.

This section intends to help you plan your testing activities by discussing the various parts of the system that need to be tested. It will help you create your test plans and help keep your testing on schedule. Following the advice provided will support left-shifting, which will, in turn, support finding issues earlier.

If you wait too long to think about testing, you will likely run into problems later in your project. Potential challenges include not having the right test properties available, testing happening too late and leading to late code or design changes, test case development not being planned and ad hoc testing being done instead of using proper test cases, and support that's required from other suppliers or partners not being communicated to them. This section will help reduce the risks of these and other issues from occurring.

Figure 14.2 provides an overview of automotive IoT applications. We will use this figure to help illustrate the types of testing that are required. Of course, depending on your project's specific requirements and location within the overall system, the specific testing will vary. We will look at the whole picture so that you can determine what your project requires.

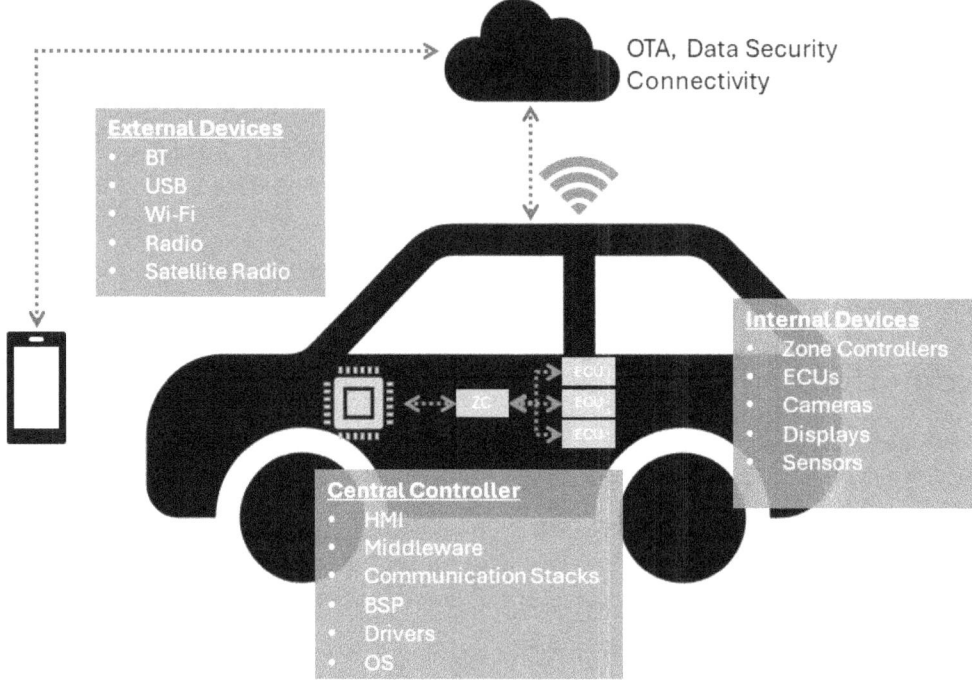

Figure 14.2 – Overview of automotive IoT testing

Figure 14.2 shows a vehicle with a cloud and a phone connected to the vehicle through the cloud. Then, within the vehicle, there's a central controller, typical devices that are connected to the central controller internally, and externally connected devices.

Let's consider testing in terms of the **central controller**. The **Central Controller** box within the preceding figure provides a simple view of the layers that will need attention during the testing phase. Here, the **Human-Machine Interface** (**HMI**) and the lower layers need to be tested. This figure also shows many of the types of connections with the central controller. For example, the **External Devices** box shows external devices that can connect. This includes phones and other devices that can connect through **Bluetooth** (**BT**), **Universal Serial Bus** (**USB**), and Wi-Fi. Radio signals may connect through AM, FM, or satellite. The **Internal Devices** box shows internal devices that may connect to it. There may be other vehicle controllers here, such as zone controllers or **Electronic Control Units** (**ECUs**). There will likely be cameras, displays, and other sensors. Once the central controller has all of these, it will connect to the cloud, which supports things such as over-the-air updates, security verification, connecting to phones, and accessing data (such as for navigation and traffic).

Testing planning needs to consider the complexity of interfaces. Planning is one of the keys to the success of the testing activities. Consider how and when the central controller's camera and display software components and features will be tested. You may need to plan the acquisition of the displays and cameras so that they align with the project and development schedule. You may need to deal with situations where some of the displays or cameras aren't available when needed. This might be because it hasn't been determined what will be used or whether these are being developed in parallel with the work on the central controller. It might also be because of cost or supplier chain factors, in which case you may have limited access to these. So, what can be done about this? Here are some recommendations:

- Do test planning early in the project, determine what displays and cameras are needed, and place a request early to get these.
- Monitor the delivery of these devices to ensure they will arrive on time. Escalate the situation if delays occur.
- Consider creating alternative test setups using different cameras or displays. Even consider creating special test boards where you can verify the interfaces.
- Deal with interfaces to zone controllers, ECUs, and other sensors similarly by planning how testing will be done with these devices or with a surrogate test device. Next, consider the external device connections. Phone connections, both wirelessly (for example, via BT) and wired (for example, USB). will need to be tested. The specific phones to be tested, the availability of the phones, and the cost of the phone (if purchases are needed) all need to be considered. Following standards such as Apple CarPlay makes testing both simpler and harder – it is simpler as there is a standard to follow but harder because the standard is difficult to meet (more on this in the next subsection).

Radio and satellite radio testing is another important consideration. If the system has a remote telematics box, this testing can be more isolated and may be the responsibility of another supplier or partner. But if you and your team are responsible, there are several things you should plan when testing. For example, if you're testing within a building, how will you access signals (for example, does your site have good signal strength or is it equipped with appropriate antennas and repeaters)? How will you test across the regions that the vehicle will operate in (Japan, China, the US, and Europe all differ in terms of the types of signals and in-region testing or validating regional capabilities that are important)? For satellite radio, having the right subscriptions may be necessary for testing.

Another important consideration is where your testing will be done. Specifically, what testing will be done in a vehicle, a bench setup, or a desk setup? A bench setup is a setup that emulates a vehicle in terms of having vehicle-like connections and controls. If in-vehicle testing is needed, this can be a big challenge. Requests to access vehicles need to be placed early (and often). It is common to have to carefully schedule access to a vehicle for testing. It is better to get testing done without a vehicle. Benches can be extremely helpful. These can be costly and takes time to design and acquire the parts to put together a bench.

With that, we've provided an overview of test properties that need to be considered while focusing on testing a central controller. Your part of the system may be much smaller with fewer things to be concerned about. However, you will still need to plan how your testing will be done and what test properties are necessary to perform the testing.

Another recommendation, regardless of what part of the system you are responsible for, is to have a standardized test report. Your customer may have a required format. If one doesn't exist, it may be beneficial to create one and get this aligned with your customers and/or stakeholders. This test report can be used as validation that your portion of the system meets specifications and performs as required. If you're challenged about the behavior of your software (or product), you can refer to the test report to show that it's working as required.

Table 14.1 provides a list of test properties that can help you plan your testing strategy.

Testing Properties Checklist
HMI testing support
Phones
Radio signal connections for all regions supported
Satellite signal connections for all services and regions supported
Antennas, signal repeaters
Other vehicle controllers
Cameras
Displays
Speakers
Microphones
Other sensors
Cloud connectivity
Security test keys/certificates
OTA server access
Navigation and traffic service access
Cables
Vehicle access
Bench setup
Access to partner or other supplier software

Table 14.1 – Test properties checklist

The next subsection will cover another view of the test activities and cover various types of testing.

Types of testing

When thinking about testing your software, testing its features, functionality, or requirements comes to mind first. This makes sense and captures much of the important testing that must be done. There is another broad category of testing that must be done that's based on functional versus non-functional requirements.

Functional requirements cover what the software is supposed to do. They are the features of the software. Non-functional requirements cover important properties of the software (or system) and include performance measures and design qualities such as scalability, compatibility, and maintainability. They may include usability, regulatory, and security aspects not covered by the functional requirements.

Some of these non-functional requirements must be tested alongside the functional requirements. Stability, **key performance indicators** (**KPIs**), and certification or pre-certification are all ways to cover many of the non-functional requirements. Your functional requirements may have explicit requirements around stability, or they might be assumptions that have been made by your customers and stakeholders. Regardless, doing stability testing to ensure that your software runs over an expected period without locking up, resetting, or having other non-recoverable malfunctions is essential. If they don't exist, establish some standards for measuring stability and start the testing early and often through the development cycle. While designing stability tests, consider ways to stress your software so that you aren't just testing simple scenarios that don't uncover problems. Stressing your software with valid and invalid inputs over extended periods will help you end up with a stable product that makes customers and users happy.

KPI testing is another important type of testing to include in your test strategy. Specifically, KPIs that measure the timing of various functionalities (for example, startup and response times to inputs), **Random Access Memory** (**RAM**) and **Read-Only Memory** (**ROM**) usage, and **Central Processing Unit** (**CPU**) usage are all vital. Measuring RAM, ROM, and CPU usage is important to do throughout the development cycle. Limits should be established and measured against, leaving room in case late changes are needed that take up more memory or CPU usage. Like stability testing, the conditions for taking the measurement ought to stress the usage as much as possible to find as many issues and potential problem cases as possible. Having standard reports for the KPIs is a good idea. These can be used as evidence that your software is meeting requirements and expectations.

The third type of non-functional testing is certification or pre-certification testing. When conducting this type of testing you can meet other regulations or standards. If your software has any certifications that need to be given or other regulations that must be met, you should consider doing testing for these earlier when possible. Official testing may be done by another company or group within your company. Even in this case, you should consider pre-certification as part of your testing process. Official testing to meet standards or to be certified will likely happen near your final release and the start of production. You have limited time to address findings at this point in the project timeline. Run through the certification tests early to find and address issues earlier. Doing this will greatly reduce the risk of late schedule delays and quality issues in production.

That wraps up this section. Testing topics are areas that typically have the best opportunities for improving your project execution and product quality. If testing is done well, your project will have a much better chance of staying on track (for example, within budget and schedule), your product quality will be much better, and your code base will be in a much better position to be extended and maintained.

Automation in testing sits in a different dimension from the other types of testing we've covered. For example, KPIs can be done manually or automated. Often, people are too busy doing manual testing to create automated tests. Then, as the project continues, the amount of manual testing grows, leaving even less time to try to automate tests. Automation in testing is such a valuable tool in terms of checking for regressions and getting fast turnaround results. So, invest in automated testing. Vision and discipline are required to do this. Do it and you will be glad you did.

Security

In today's age of rapid development, there's a need for constant updates and new features – think about how often there are new updates for the apps on your phone or how often there is a new update of Android OS or iOS. On top of major or minor releases with new features or bug fixes, there are often many more frequent security patches being released. Automotive IoT follows the same pattern, so there must be a focus on security. This focus on security must be **top-down** so that a cybersecurity culture can be created within the organization and security activities and solutions can be budgeted and planned for. The focus on security must also be driven **bottom-up**, ensuring that security technologies, requirements, and constraints are understood within a project, as well as that appropriate cybersecurity activities and testing approaches are applied to help achieve a certain level of cybersecurity assurance.

Unfortunately, in real-world projects, not everything goes as planned and often, since there are strict deadlines and limited budgets, security activities or security features are the first things to be dropped. While security is often thought of as *nice to have*, actual functionality *is a must*, so project managers, development teams, and organization leaders often consider security as an afterthought.

Even though several useful security standards and frameworks exist, and while they may look good on paper, applying and enforcing such standards and frameworks is often very time-consuming and cumbersome. Security teams and project teams have different objectives and priorities and often need to fight for resources and budgets to ensure that cybersecurity activities are performed within the already strict schedule. A development team that is already fully booked with development activities to meet a deadline will not agree to take on new security tasks. So, how can we all work together to build secure products?

Well, while considering how to apply security solutions within your organization, it is also important to make sure that security teams can work together with project teams, functional safety teams, cloud teams, IT teams, process teams, and others. Internal processes must be established that incorporate cybersecurity activities as part of the typical development processes and security testing must be automated as much as possible. Proper roles and responsibilities must be established and cybersecurity must be budgeted and planned – it must be built-in, not bolted on. Thus, there must be synergy between these different teams, something that's often easier said than done, and making sure everyone in the organization is embracing security. Security is not the security team's job; it is everyone's job!

Summary

In this chapter, we covered some critical pieces that you will need to put your automotive IoT project together. You are likely to follow some Agile process with more and more projects and companies tending toward a scaled Agile approach. ASPICE and functional safety continue to be important processes for some parts of automotive IoT. This chapter discussed how these can be compatible with Agile development.

Since we have reached the end of this book, let's do a quick recap. We started by looking at trends that drive automotive IoT development. *Chapter 1, Automotive Technology Trends* described these, as follows:

- Mobile apps and the cloud
- Modern software development
- Standards and regulations

These trends are creating a need for new applications. *Chapter 2, Introducing Automotive IoT Use Cases* described a few of these, including the following:

- Enhanced driver experience and safety, including **Advanced Driver Assistance Systems** (**ADAS**) and phone as a key
- Optimized fleet management, including predictive maintenance
- Revolutionizing the road with applications such as **vehicle-to-everything** (**V2X**) communication

These trends led to new use cases and applications being required, with significant changes having to be made to automotive software development. Such changes are required in the areas of vehicle architecture, secure development, system design development within the cloud, external connectivity, and external applications. To support their development, new standards are necessary, as well as new development approaches.

Traditional vehicle architectures do not support a variety of data sources, data bandwidth, and data processing speeds that are required. *Chapters 3 to 5* dived into novel approaches to vehicle architectures and vehicle diagnostics (covered in *Chapter 4, Vehicle Diagnostics*). *Chapter 5, Next Wave of Vehicle Diagnostics* provided an in-depth look at **Secure Onboard Vehicle Diagnostics** (**SOVD**), which is the next wave of diagnostic protocols that are required in modern vehicles.

Secure development is an extremely important and challenging part of automotive IoT development. *Chapters 6 to 8* covered this topic broadly and dived deep into key areas. There are new cybersecurity threats abound in automotive IoT that need to be considered. First, *Chapter 6, Exploring Secure Development Processes for Automotive IoT* discussed security processes, including ISO/SAE 21434. Then, *Chapter 7, Establishing a Secure Software Development Platform* provided step-by-step practical guidance on secure software development. Secure software development is not just for security-focused developers. All software developers need to follow secure development practices. Finally, *Chapter 8, Securing the Software Supply Chain* looked beyond your specific development and showed you how to have a secure software supply chain.

The next three chapters showed the system design for automotive IoT applications. *Chapter 9, System Design of an Automotive IoT Application* showed you end-to-end system design by providing a remote vehicle diagnostics use case to teach you. The next chapter, *Chapter 10, Developing an Automotive IoT Application*, explored various development processes, including backend deployment, architectures, and service models. Finally, *Chapter 11, Deploying and Maintaining an Automotive IoT Application* covered how to deploy and maintain applications. Emphasis was put on the DevSecOps life cycle. It also covered security integration, coding, building, testing, releasing, deploying, operating, and monitoring, providing a comprehensive guide to ensuring rapid deployment and maintaining high-quality standards in automotive IoT applications.

The final part of this book covered processes and practices for automotive development, as discussed in *Chapter 12, Processes and Practices*, and focused on some embedded development approaches, as discussed in *Chapter 13, Embedded Automotive IoT Development*. This last chapter, *Chapter 14, Final Thoughts*, added a few more thoughts from the authors' perspective based on real-world experience to help you on your continued journey of automotive IoT development.

That's all we've got for you in this automotive IoT book. Software advances are a significant disrupter in the automotive industry. Companies are undergoing restructuring to adapt and take advantage of the opportunities that are available to them. Partnerships are changing, with new ones forming and existing ones changing. Many new players are entering the industry. These things are happening because of the increase in automotive IoT software.

As a software engineer, the automotive industry is a great place to be. You have opportunities to move across many domains and move up and down the software ecosystem. You can work at low levels, writing code that works directly with hardware. You can work in middleware layers that implement business logic between the human interface layers. You can also work on cloud services and external applications that connect to vehicles. You can even specialize in an area such as secure development and then transition into another specialization area such as HMI development.

Hopefully, this book has greatly increased your understanding of automotive IoT software development. We wish you the best of luck on your current project and best of luck in your career in the automotive industry.

References

- [1] https://scaledagileframework.com/
- [2] *Experts Say Auto Industry's Software Strategy Needs To Get Agile To Complete*, by Ed Garsten: https://www.forbes.com/sites/edgarsten/2021/03/08/experts-say-auto-industrys-software-strategy-needs-to-get-agile-to-compete/?sh=30f996ae3938

Index

Symbols

4G LTE 174
5G 174
5 Whys root cause analysis tool 268, 269

A

A-B-A testing 271
access control 210
Access Point Names (APNs) 209
ACQ.2 98
ACQ.4 Supplier Kickoff and Monitoring 259
activities, SSDLC 112
   code review 113
   design review 113
   dynamic application security testing (DAST) 114
   fuzz testing 114
   interactive application security testing (IAST) 114
   penetration testing 114
   requirements review 113
   static application security testing (SAST) 113
   TARA/threat model 113
   vulnerability scanning 114

Adaptive AUTOSAR 43-46
   components 45, 46
   development tools 46, 47
   diagnostic service management 64, 65
   versus Classic AUTOSAR 47
Advanced Driver-Assistance Systems (ADASs) 16-18, 47, 176
Agile 106, 301
   advantages 302, 303
   limitations 301
agile development 84
Amazon Elastic Compute Cloud (EC2) 194
Amazon Kinesis Data Streams (KDS) 202
Amazon Managed Streaming for Apache Kafka (MSK) 202
Amazon Web Service (AWS) 192
American Institute of Certified Public Accountants (AICPA) 101
Android Automotive OS (AAOS) 285
Anti-lock Braking System (ABS) 4
Apache Cassandra 205
Apache JMeter 232
API gateway 206-208
   API version management 208
   authentication and authorization 207
   caching 207

load balancing 207
logging and monitoring 208
rate limiting and throttling 208
request and response transformation 207
routing 207
security 207
service aggregation 207
API management solutions, from cloud providers
comparing 208
API Security Top 10 104
Application Life Cycle Management (ALM) 290
Application Processor (AP) 210
application programming interface (API)-based attacks 84
Application Programming Interfaces (APIs) 7, 128
application security (AppSec) testing 111
application security posture management (ASPM) 123
AppSec tooling
dynamic application security testing (DAST) 128
software composition analysis (SCA) 126
static application security testing (SAST) 124
architectural and detailed designs 113
artificial intelligence (AI) 53
ASPICE 304, 305
ASPICE for Cybersecurity 10, 83, 92, 96
security activities 97
ASPICE SWE group
V-model 255
assets 87, 88
Association for Standardization of Automation and Measuring Systems (ASAM) 68
URL 68

attack feasibility 90
attack potential 90
augmented reality (AR) 174
authentication 210
authorization 210
Automatic Emergency Braking (AEB) 18
automation 108
automotive development
Agile 106
scrum 107
V-model 105, 106
automotive embedded testing 305, 306
considerations 307
functional requirements 309
non-functional requirements 309
properties checklist 308
recommendations 307
types 309
Automotive Ethernet 176
automotive IoT 11
automotive IoT applications 3, 83
automotive IoT ecosystem
Backend Solution 11
Mobile Device 11
simplified threat model 85-87
Vehicle 11
automotive IoT testing 306
automotive IoT, use cases 12
data management 13
phone as key 12
remote diagnostics 12
vehicle management 12
Automotive Open System Architecture (AUTOSAR) 31, 39
Automotive Safety Integrity Levels (ASIL) 261
Automotive Software Process Improvement Capability Determination 253

Automotive Software Process Improvement Capability dEtermination (ASPICE) 3.1 10
Automotive SPICE (ASPICE) 253
  process levels 254
  Software Architectural Design process 257
  Software Detailed Design and Unit Construction process 258
  software engineering processes 253, 254
  Software Integration and Integration Test process 259
  Software Qualification Test process 259
  Software Requirements Analysis process 257
  Software Unit Verification process 258
automotive systems 98
automotive trends 3, 4
  CASE 4, 5
  cloud 8
  mobile apps 7
  modern software development 8, 9
  SDV 6
  SOA 6, 7
  standards and regulations 9, 10
autonomous driving (AD) 53
autonomy domain controller (ADC) 32
AUTOSAR C++ 125
AUTOSAR Runtime Environment for Adaptive Applications (ARA) 44
AUTOSAR XML (ARXML) 42
availability zones (AZs) 237
  improvements 237
  structure 237
AWS CodeArtifact 229
AWS IoT Core 200
AWS IoT Greengrass 200
Azure Artifacts 229
Azure IoT Edge 200
Azure IoT Hub 200
Azure Kubernetes Service (AKS) 197
Azure Machine Learning (AML) 200

B

backend IoT app 86
base practices (BPs) 254
Basic Software (BSW) 39, 41
BeiDou 173
bidirectional traceability 255, 256
binary scanning 127
black-box testing 114, 129, 130
BlazeMeter 232
Bluetooth 173
Bluetooth Low Energy (BLE) 19, 20, 173, 174
Board Support Package (BSP) 280
build scripts 228
build stage, CI 228, 229
build-versus-buy decision 187, 188

C

California Consumer Privacy Act (CCPA) 186
CANdela Diagnostic Data (CDD) 42
CAN Flexible Data-Rate (CAN FD) 176
capital expenditure (CapEx) 192
cellular modem module 213
Cellular V2X (C-V2X) 25
central controller 306
CERT C/C++ 125
certification testing 309
change management 94
CISQ CWE 125
Classic AUTOSAR 39
  Basic Software (BSW) 39
  components 40, 41

316 Index

development tools 42
diagnostic communication workflow 62-64
Microcontroller Abstraction
    Layer (MCAL) 39
Real-Time Engine (RTE) 39
reference link 39
versus Adaptive AUTOSAR 47
classic vehicle ECU diagnostics 183, 184
cloud 8
development 8
operations 8
cloud applications, and embedded software
development process 218
cloud backend 191
cloud deployment models 192
comparing 192
hybrid cloud 192, 193
private cloud 192
public cloud 192
cloud design considerations 179
connectivity management 180, 181
device management 180
cloud device gateway 199, 200
Cloud IoT Edge 200
Cloud-Native Application
    Security Top 10 104
cloud-native technologies 104
cloud service models 193
comparing 194
Infrastructure as a Service (IaaS) 193
Platform as a Service (PaaS) 193, 194
Software as a Service (SaaS) 193, 194
cloud service provider (CSP) 191
cockpit domain controller (CDC) 32
code review 113, 143
code stage, CI 226-228
commercially available
    off-the-shelf (COTS) 188

Common Vulnerabilities and
    Exposures (CVE) 126, 154
Common Weakness Enumerations
    (CWEs) 113, 154
Communication Management 45
Communication Manager (ComM) 41
communication/protocols, in DoIP
DoIP layer 55
IP layer 55
physical layer (Ethernet/WLAN) 55
TCP/UDP layer 55
UDS 55
Compact Disc (CD) 4
compatibility testing 232
competency management 94
compliance testing 232
configuration files 228
configuration management 94
Connected, Autonomous, Shared,
    Electric (CASE) 4
Autonomous 5
Connected 5
Electric 5
Shared 5
connected car services 16
connected mobility revolution 24
connected supply chain and
    manufacturing 27
smart parking solutions 24, 25
Vehicle-to-Everything (V2X)
    communication 25-27
connectivity management 209
Connectivity Management
    Platform (CMP) 209
connectivity management solution 179, 180
functions 180, 181
consequences 87, 90

Constrained Application Protocol (CoAP) 202
containerization 197
containers 196
continuous deployment/delivery (CD) 233
  deploy stage 233, 234
  monitor stage 243-247
  operate stage 239
  release stage 233
continuous integration (CI) 226
  build stage 228, 229
  code stage 226-228
  test stage 230
continuous integration/continuous delivery (CI/CD) 104
continuous integration/continuous deployment (CI/CD) 218
Controller Area Network (CAN) 4, 61, 78, 172, 176, 177, 210
copyleft licenses 149
cost-benefit analysis (CBA) 187
cross-site scripting (XSS) 128
Cybersecurity Assurance Level (CAL) 10, 146
  activities 116
  risk level, mapping 117
Cybersecurity Implementation 97
cybersecurity interface agreement 98
cybersecurity interface agreements for development (CIADs) 141, 145, 146
Cybersecurity Management System (CSMS) 9, 94
Cybersecurity Requirements Elicitation 97
Cybersecurity Risk Management 97
cybersecurity threats 84
cyclic redundancy code (CRC) 56

CycloneDX 153
  versus Software Package Data Exchange (SPDX) 154

D

damage potential 90
data identifier (DID) 52
data link layer (DLL) 40
data management
  for automotive IoT use cases 13
data persistence 212
DaVinci Adaptive AUTOSAR 47
  reference link 47
debugger 293
Dedicated Short-Range Communication (DSRC) 25
defense-in-depth approach 92
delta image 229
deploy stage, CD 233, 234
  deployment strategies 235, 236
  regions and availability zones 237, 238
Design Failure Mode and Effects Analysis (DFMEA) 267
  example 268
Desirability, Feasibility, and Viability (DFV) model 164, 165
development (Dev) teams 107
development interface agreement (DIA) 141
Development, Security, and Operations (DevSecOps) 8
development team 152
development tools 289
  debugger 293
  life cycle management tools 290
device management 202, 203
device management solution 179, 180
  functions 180

device manager 214
DevOps 107
DevSecOps 84, 107, 108, 121
DevSecOps life cycle 222, 223
  build stage 223
  code stage 222
  deploy stage 223
  monitor stage 223
  operate stage 223
  plan stage 222, 224, 225
  release stage 223
  test stage 223
Diagnostic Communication
    Manager (DCM) 41
diagnostic communication workflow
  in Classic AUTOSAR 62-64
Diagnostic Extract Template (DTEXT) 46
Diagnostic Manager 45, 64, 72
Diagnostic over Internet Protocol
    (DoIP) 53-55
  challenges 62
  example message flow 60, 61
  message format 56-59
Diagnostic Protocol Data Unit
    (D-PDU) 184, 214
diagnostic service management
  in Adaptive AUTOSAR 64, 65
Diagnostics over Internet
    Protocol (DoIP) 40, 41
diagnostic trouble codes (DTCs) 50
disaster recovery testing 232
Disciplined Agile 303
distributed denial-of-service (DDoS) 92, 207
distributed development approach 139
documentation management 94
domain controller 32, 36, 37
DOORS Next Generation (DNG) 292

dynamic application security testing
    (DAST) 114, 128, 143
  performing 129
  tools 129
dynamic test approach 130

E

edge computing 200
Elastic Block Store (EBS) 195
Elastic Container Service (ECS) 194
Elastic File System (EFS) 195
Elastic Stack (ELK Stack) 246, 247
electrical/electronic (E/E) 92
Electric Vehicles (EVs) 5
Electronic Control Units
    (ECUs) 4, 31, 49, 85, 130, 139, 170, 200
Elektrobit (EB) 42
embedded software development 273
  additional aspects 280, 281
  application notes 279
  automotive-focused aspects 281-283
  datasheets 278
  device drivers 279
  electrical engineering 274-276
  errata 278
  Hardware Abstraction Layer (HAL) 280
  operating systems 285-287
  power state management 283, 284
  schematics/block diagrams 277
embedded Subscriber Identity
    Module (eSIM) 172, 177
end of line (EOL) 233
end-user mobile device 171
  functions 171
enhanced driver experience and safety 16
Enterprise Kanban 303
epics 106

Ethernet Driver (Eth) 40
Ethernet Interface (EthIf) 41
Ethernet State Manager (EthSM) 41
Ethernet Switch (EthSwt) 40
Ethernet Transceiver (EthTrcv) 40
European Union (EU) 186
event-driven architecture (EDA) 196, 197
Execution Management 46
Executive Order 14028 (EO14028) 154
exploitation 132
eXtensible Markup Language (XML) 67, 124

F

Federal Communications
    Commission (FCC) 186
fishbone 270, 271
four golden signals 244
functional blocks, to build secure
    edge application on TCU
  cellular modem module 213
  data persistence 212
  global navigation satellite system
    (GNSS) module 212
  MQTT agent 213
  OTA updater 213
  power manager 212
  state sync 213
  system logger 212
  system manager 212
  TCU diagnostics 213
  Transport Layer Security (TLS) 213
functional safety 260, 261
functional safety development process 262
  software architectural design 264
  software integration and testing 266
  software project initiation 262

software safety requirements, verifying 266
software unit design and
    implementation 264, 265
software unit testing 265
specification of software safety
    requirements 263, 264
functional testing 230
fuzzed data 130
fuzz testing 114, 129, 130, 143
  performing 130, 131

G

Galileo 173
gateway design considerations 172
  CAN 176, 177
  gateway hardware 178
  GNSS receivers 172, 173
  sensors 177
  SIM/eSIM 177
  wired communication 175, 176
  wireless communication 173
gateway hardware 178
Gatling 232
gdb 293
General Data Protection
    Regulation (GDPR) 186
Generally Accepted Privacy
    Principles (GAPP) 103
generic practices (GPs) 254
generic Precision Time Protocol (gPTP) 283
gigabit per second (Gbps) 54
gigabytes (GB) 33
gigahertz (GHz) 33
Global Navigation Satellite System
    (GNSS) 172, 212
Globalnaya Navigatsionnaya Sputnikovaya
    Sistema (GLONASS) 172

Global Positioning System (GPS) 172
Google App Engine (GAE) 194
Google Cloud Artifact Repository 229
Google Cloud Platform (GCP) 192
Google Compute Engine (GCE) 194
Google Kubernetes Engine (GKE) 194
Grafana dashboard 246
gray-box approach 130

H

hardware-in-the-loop (HIL)
 testing 130, 218, 231
Hazard Analysis and Risk
 Assessment (HARA) 261
healthy software engineering culture
 tricks, for enabling 252
heating, ventilation, and air
 conditioning (HVAC) 54
high availability (HA) 205
high-performance computing
 (HPC) 184, 214
High-Speed CAN (HS-CAN) 177
hybrid cloud 192, 193
Hypertext Transfer Protocol (HTTP) 67, 202
hypervisors 287-289

I

IaaS offerings
 comparison, from various cloud
 providers 194, 195
identifying dependencies approach 127
identity and access management
 (IAM) 193, 209, 210
identity management (IdM) 210
incident management 242
 workflow 243

incremental builds 228
industrial IoT (IIoT) devices 175
InfluxDB 205
information gathering 132
information security management 100
Information Security Management
 System (ISMS) 10, 100
Infrastructure as a Service (IaaS) 193
input/output (I/O) extender 33
Integrated Development
 Environment (IDE) 293
interactive application security
 testing (IAST) 114
International Assessor Certification
 Scheme (intacs) 305
International Organization for
 Standardization (ISO) 186
Internet of Things (IoT) 3
Internet Protocol (IP) 54
intrusion detection systems (IDSs) 92
in-vehicle infotainment (IVI) 85
in-vehicle use case 182
IoT application architecture 198, 199
 API gateway 206-208
 cloud device gateway 199, 200
 connectivity management 209
 device management 202, 203
 edge computing 200
 IAM 209, 210
 over-the-air (OTA) solutions 203, 204
 rule engine 206
 stream processing 200, 202
 telemetry datastore 205
ISO 5230 10
ISO 8475 10
ISO 8477 10
ISO 21434 204
ISO 21448 10

ISO 24089 10, 204
ISO 26262 10
ISO 27001 10, 83, 100, 101
ISO/IEC 5230:2020 155
ISO/IEC 18974:2023 155
ISO/PAS 5112 10
ISO/SAE 8475 117
ISO/SAE 21434 10, 92, 93
ISO/SAE 21434 Cybersecurity
    Engineering 83
ISO/SAE 21434 organizational-level
    requirements 93
  management systems 94
  policies and processes 94
  roles and responsibilities 94
ISO/SAE 21434 project-level
    requirements 95
  continual monitoring and updates 96
  cybersecurity goals and concept 95
  cybersecurity plan 95
  cybersecurity specifications
    and requirements 95
  secure development 95
  TARA 95
  verification and validation 95

J

JavaScript Object Notation
    (JSON) 67, 124, 153
JSON Web Tokens (JWT) 207

K

key performance indicators
    (KPIs) 245, 282, 309
known attack patterns 128

known vulnerabilities 114, 126, 128
KPI testing 309

L

Lambda data processing architecture
  batch layer 201
  serving layer 201
  speed layer 201
Lane Departure Warning (LDW) systems 18
Large-Scrum Scale (LeSS) 303
legal and compliance team 152
license compliance 127
license information 126
license types, for OSS components 150
Linux Journey
  URL 287
LoadRunner 232
Local Interconnect Network (LIN) 61
Log and Trace Management 46
Long Term Evolution for Machines
    (LTE-M) 174, 175
Long Term Evolution (LTE) 174
low-power, wide-area (LPWA) 175
Low-Speed CAN (LS-CAN) 177
LwM2M 202
LwM2M Objects 202

M

machine learning (ML) 194
machine-to-machine (M2M)
    communications 175, 202
malformed input 130
MAN.7 97
management systems
  change management 94
  competency management 94

configuration management 94
documentation management 94
requirements management 94
tool management 95
Master In Slave Out (MISO) 37
Master Out Slave In (MOSI) 37
Media Access Control Security (MACsec) 92
medium Earth orbit (MEO) 172
megabytes (MB) 39
megahertz (MHz) 39
Message Queuing Telemetry Transport (MQTT) 130, 199
Microcontroller Abstraction Layer (MCAL) 39, 42
microelectromechanical systems (MEMS) sensors 177
microservices 196
MISRA C/C++ 125
mitigations 87, 91
product security solutions 92
secure development processes 92
Mobile Application Security Testing Guide 105
mobile apps 7
mobile IoT app 86
Mobile Network Operator (MNO) 181, 209
Mobile Top 10 104
Mobile Virtual Network Operator (MVNO) 181, 209
modern software development 8, 9
MongoDB 205
monitor stage, CD 243-247
MQTT agent 213
MQTT broker
working 199
multi-tenant architecture 234

N

Narrowband IoT (NB-IoT) 174, 175
National Telecommunications and Information Administration (NTIA) 155
focusing, areas 155
Navigation with Indian Constellation (NavIC) 173
Near Field Communication (NFC) 11, 19, 20, 173, 174
negative acknowledgment (NACK) 58
Network Management 41, 46
network operations center (NOC) 243
Nexus 303
NIST Cybersecurity Framework 83, 98
detect function 99
identify function 99
protect function 99
recover function 99
respond function 99
NIST Special Publication (SP) 86
non-DoIP nodes 61
non-functional testing 232
non-vehicle systems 98
Non-Volatile Memory Manager (NVMM) 41
non-volatile memory (NVM) 41
NoSQL databases 205
offerings, from cloud providers 205

O

OMA Lightweight M2M (OMA-LwM2M) 200
on-board diagnostics (OBD) 58
Open Authorization (OAuth) 207
OpenChain 155

Open Diagnostic data eXchange (ODX) 42, 68
Open Diagnostic eXchange (ODX) 214
Open Dialogue data eXchange (ODX) 183
Open Mobile Alliance (OMA) 202
open source software (OSS) 84, 137
  components 126
  used, for managing risks 146, 147
open source software (OSS), usage considerations
  license compliance 149-151
  operational risk 151-153
  security vulnerabilities 147-149
Open Systems Interconnection (OSI) 54
Open Test Sequence eXchange (OTX) 183, 214
Open Web Application Security Project (OWASP) 83, 103
  cheat sheets 103
  testing guides 104, 105
  top ten risks 104
operate stage, CD 239-243
operational expenditure (OpEx) 193
operations (Ops) teams 107
optimized fleet management 22
  driver performance monitoring 22
  predictive maintenance 23, 24
  real-time vehicle tracking and telematics 22
Organization for the Advancement of Structured Information Standards (OASIS) 199
Original Equipment Manufacturers (OEMs) 7, 19, 85, 139, 143, 203
OSS vulnerability management workflow 149
OTA updater 213
Over-The-Air (OTA) software updates 6, 62, 179, 229
Over-The-Air (OTA) solutions 203, 204
OWASP Defectdojo 124
OWASP Mobile Top 10 125
OWASP Top 10 125

P

pattern matching 127
penetration testing 114, 131, 143
  open source tools 133
  performing 132, 133
performance testing 232
permissive licenses 149
Persistent Storage Management 45
personalized in-car experience 18
personally identifiable information (PII) 103
phone as a key application 12, 19, 20
  local system 20
  remote system 21
plan stage, DevSecOps life cycle 224
  architecture and design 224
  collaboration and communication 225
  objectives and requirements, gathering 224
  resource planning 224
  risk management 225
  roadmap and prioritization 225
  security and compliance 224
Platform as a Service (PaaS) 193, 194
positioning, navigation, and timing (PNT) 172
power manager 212
pre-certification testing 309
predictive diagnostics 12
predictive maintenance 216, 217

Primary Virtual Machine (PVM) 288
private cloud 192
processes 251
product security solutions 92
project inventory 114
   cybersecurity assurance levels (CALs) 116, 117
   example 118-120
   project information and risk level 115, 116
Project Team 143
proximity use case 182
public cloud 192
Public Key Infrastructure (PKI) 214
publish/subscribe (pub/sub) messaging protocol 199

Q

Qualcomm
   URL 33
quality assurance (QA) teams 229
Quasi-Zenith Satellite System (QZSS) 173

R

Radio Equipment Directive (RED) 186
Rational Quality Manager (RQM) 292
Rational Teams Concert (RTC) 292
Real-Time Engine (RTE) 39
real-time operating system (RTOS) 38, 218
receiving data (RX) 37
Redis 205
Redis Source Available License (RSALv2) 151
region 237
regulatory compliance 185
   cybersecurity regulations 186
   data localization and storage regulations 186
   data privacy and protection regulations 186
   liability and responsibility regulations 186
   OTA updates and software management regulations 186
   standardization and interoperability regulations 186
   telematics and remote access 186
   vehicle safety standards 185
release stage, CD 233
remote diagnostics application 12, 66, 167, 168, 182, 213-216
   classic vehicle ECU diagnostics 183, 184
   service-oriented vehicle diagnostics 184, 185
remote updating 203
remote use case 182
Representational State Transfer (REST) 67, 72, 73
   principles 73
   reference link 73
Requests for Quotes (RFQs) 297
requirements management 94
responsible, accountable, support, inform, and consult (RASIC) chart 140-143
risk classification system 261
risk matrix 91, 116
   example 116
risks 87, 90
   managing, with open source software (OSS) 146, 147
risk treatment options 91
Risk Treatment Validation 97
Risk Treatment Verification 97
RS232 175
RS485 175
rule engine 206

S

safety mechanism 261
Safety Of The Intended Function (SOTIF) 10
SANS Top 25 125
Scalable service-Oriented MiddlewarE
  over IP (Some/IP) 45
Scale Agile Framework (SAFe) 303
Scaled Agile 301-303
  frameworks 303
scrum 107
Scrum@Scale 303
SD 45
SEC.1 97
SEC.2 97
SEC.3 97
SEC.4 97
secure development processes 92
Secure Onboard Communication
  (SecOC) 92
secure software development life
  cycle (SSDLC) 111
  activities 112
secure software development platform
  AppSec tooling 122, 124
  code repositories 121
  compliance record 122, 123
  continuous integration (CI) pipeline 122
  establishing 120
  necessity 120, 121
  overview 121, 122
  policies 122, 123
  project development teams 121
  purpose 120, 121
  remediation workflow 122
  requirements 122, 123
  software package 122
  vulnerability management 122-124

secure software supply chain risk
  management (SSCRM) 155
  risks, assessing 156
  risks, identifying 156
  risks, mitigating 156
  risks, mitigating activities and practices 157
security 310
  bottom-up 310
  top-down 310
security activities, ASPICE for Cybersecurity
  ACQ.2 98
  MAN.7 97
  SEC.1 97
  SEC.2 97
  SEC.3 97
  SEC.4 97
security information and event
  management (SIEM) 124
Security Module 45
security requirements 113
security (Sec) teams 107, 143, 152
security testing 232
security threats
  cybersecurity threats 84
  overview 84
  recent attacks, examples 85
Serial Clock (SCLK) 37
Serial Peripheral Interface (SPI) 36, 37, 211
server-based applications 195, 196
server-based computing 195
serverless applications 195, 196
serverless architecture 196
serverless computing 195
Server Side Public License (SSPLv1) 151
Service Discovery (Sd) 41
Service Identifier (SID) 50
service-level agreement (SLA) 102

service mesh 241
　control plane 241, 242
　data plane 241, 242
Service Oriented Architecture (SOA) 6, 7, 43
Service-Oriented Vehicle Diagnostics (SOVD) 67, 69-72, 78, 184, 185, 214
　diagnostic message example 74, 75
　documentation and demo 77
　interface example, as part of applications on server side 75, 76
　in-vehicle access 69
　proximity access 69
　remote access 69
　REST 73
　versus Unified Diagnostic Services (UDS) 78
shared responsibility model 193
shift-left 108
SIM 177
Simple Object Access Protocol (SOAP) 208
Simple Storage Service (S3) 195
simplified threat model, automotive IoT ecosystem
　assets 88
　consequences 90
　mitigations 91
　risks 90, 91
　threats 89
single-tenant architecture 233
Slave Select/Chip Select (SS/CS) 37
smart parking solutions 24, 25
snippet analysis 127
SoC hypervisor 37
Socket Adapter (SoAd) 41
Software as a Service (SaaS) 101, 193, 194
software bill of materials (SBOM) 126, 153
　Executive Order 14028 (EO14028) 154
　fields 153
　formats 153, 154

NTIA 155
OpenChain 155
software-centric architecture 6
software components (SWCs) 41
software composition 8
software composition analysis (SCA) 126
　performing 127
　tools 127, 128
software-defined vehicles (SDVs) 6, 53, 138, 301
software development ecosystem 294
　co-suppliers 296
　customers 295
　suppliers 296-299
Software Development Life Cycle (SDLC) 108, 155
　tools 290-292
software-first approach 6
Software Identification (SWID) 154
software-in-the-loop (SIL) testing 131, 231
Software Package Data Exchange (SPDX) 153
　versus CycloneDX 154
Software Risk Manager 124
software supply chain 137-140
software transparency 155
Software Updates Management System (SUMS) 10
SOVD2UDS adapter 72
SOVD library 72
SOVD server/gateway 72
sprints 106
sprint tasks 106
stability testing 309
standards and regulations 9
　ASPICE 10
　ASPICE for Cybersecurity 10
　ISO 5230 10

ISO 8475 10
ISO 8477 10
ISO 21448 10
ISO 24089 10
ISO 26262 10
ISO 27001 10
ISO/PAS 5112 10
ISO/SAE 21434 10
UN R155 9
UN R156 10
Statement of Work (SOW) 298
state sync 213
Static Analysis Results Interchange Format (SARIF) 124
static application security testing (SAST) 113, 124, 143
  performing 125
  tools 126
stories 106
stream processing 200-202
STRIDE 86
Structured Query Language (SQL) injection 128
subscription management 179
supplier 143
Supplier Request and Selection 98
System and Organization Controls 2 (SOC 2) 83, 101
  principles 102, 103
system components, automotive IoT system 169
  end-user mobile device 171
  vehicle cloud platform 170, 171
  vehicle telematics gateway 169, 170
system design 163
  process overview 164, 165

system engineering
  interfaces 165
system logger 212
system manager 212
System-on-Chip (SoC) 33, 212

T

Targeted Attack Feasibility (TAF) 10
TCP/IP Stack (TcpIp) 41
TCU diagnostics 213
technical safety concept 261
technical safety requirement 261
Telematics Control Unit (TCU) 85, 210
telemetry datastore 205
test stage, CI 230
  functional testing 230, 231
  non-functional testing 232
  performance testing 232
The Update Framework (TUF) 204
Threat Analysis and Risk Assessment (TARA) 86, 113, 143
Threat Model 113, 143
threats 87, 89
tiers 139
time series databases 205
time to market (TTM) 188
Tire Pressure Monitoring Systems (TPMSs) 23
tool management 95
Top 10 CI/CD Security Risks 104
Top 10 Web Application Security Risks 104
transmitting data (TX) 37
Transport Layer Security (TLS) 92, 213
Trust Services Criteria (TSC) 101

U

ultra-low-power, wide-area (ULPWA) 175
Ultra-Wide Band (UWB) 5, 19, 20, 173, 174
Unified Diagnostic Services
(UDS) 49, 50, 55, 67, 68, 182, 214
   message structure 50-53
   versus Service-Oriented Vehicle
      Diagnostics (SOVD) 78
Uniform Resource Identifiers (URIs) 73
United Nations Economic Commission
    for Europe (UNECE) R155 and
    R156 UN regulations 204
Universal Asynchronous Receiver-
    Transmitter (UART) 36, 37, 211
unknown or new attack patterns 130
unknown vulnerabilities 128, 130
UN R155 9
UN R156 10
Update and Configuration Management 46
Uptane 204
user experience (UX) 163
user management 179
UX-driven design (UXDD) 163, 165
   benefits 166, 167
   characteristics 166

V

Vector
   URL 42
Vector SOVD console
   reference link 77
vehicle architecture 31
   central computer, with multiple
      domain-specific SoCs 34
   central computer with single SoC 35
   centralized zonal domain architecture 32, 33
   distributed architecture 32
vehicle cloud platform 170
   functions 170, 171
vehicle domain controller (VDC) 32
vehicle identification number (VIN) 50, 85
Vehicle Interface Processor (VIP) 210
vehicle-in-the-loop (VIL) testing 231
vehicle IoT app 86
vehicle management module 179
vehicle management use case 12
vehicle manager 213
vehicle OTA updates 179
vehicle telematics gateway 169, 210-213
   functions 170
vehicle telemetry 179
Vehicle-to-Everything (V2X)
      communication 5, 25-27, 176
vendor security assessments 143-145
virtual machines (VMs) 38, 193, 288
Virtual Network (VNet) 195
Virtual Private Cloud (VPC) 195
virtual reality (VR) 174
V-model 105, 106
vulnerability analysis 132
Vulnerability Exploitability
    eXchange (VEX) 154
vulnerability scanning 114, 143

W

watchlist 148
web application firewall (WAF) 208, 240
Web Security Testing Guide 104
white-box approach 114, 130
Wi-Fi 174
wired communication 175, 176

wireless communication 173
    long range 175
    short range 173, 174
wireless local area network (WLAN) 54

Z

zone controller 33

www.packtpub.com

Subscribe to our online digital library for full access to over 7,000 books and videos, as well as industry leading tools to help you plan your personal development and advance your career. For more information, please visit our website.

Why subscribe?

- Spend less time learning and more time coding with practical eBooks and Videos from over 4,000 industry professionals
- Improve your learning with Skill Plans built especially for you
- Get a free eBook or video every month
- Fully searchable for easy access to vital information
- Copy and paste, print, and bookmark content

Did you know that Packt offers eBook versions of every book published, with PDF and ePub files available? You can upgrade to the eBook version at packtpub.com and as a print book customer, you are entitled to a discount on the eBook copy. Get in touch with us at customercare@packtpub.com for more details.

At www.packtpub.com, you can also read a collection of free technical articles, sign up for a range of free newsletters, and receive exclusive discounts and offers on Packt books and eBooks.

Other Books You May Enjoy

If you enjoyed this book, you may be interested in these other books by Packt:

The Azure IoT Handbook

Dan Clark

ISBN: 978-1-83763-361-6

- Get to grips with setting up and deploying IoT devices at scale
- Use Azure IoT Hub for device management and message routing
- Explore Azure services for analyzing streaming data
- Uncover effective techniques for visualizing real-time streaming data
- Delve into the essentials of monitoring and logging to secure your IoT system
- Gain insights into real-time analytics with Power BI
- Create workflows and alerts triggered by streaming data

Hands-on ESP32 with Arduino IDE

Asim Zulfiqar

ISBN: 978-1-83763-803-1

- Understand the architecture of ESP32 including all its ins and outs
- Get to grips with writing code for ESP32 using Arduino IDE 2.0
- Interface sensors with ESP32, focusing on the science behind it
- Familiarize yourself with the architecture of various IoT network protocols in-depth
- Gain an understanding of the network protocols involved in IoT device communication
- Evaluate and select the ideal data-based IoT protocol for your project or application
- Apply IoT principles to real-world projects using Arduino IDE 2.0

Packt is searching for authors like you

If you're interested in becoming an author for Packt, please visit `authors.packtpub.com` and apply today. We have worked with thousands of developers and tech professionals, just like you, to help them share their insight with the global tech community. You can make a general application, apply for a specific hot topic that we are recruiting an author for, or submit your own idea.

Share Your Thoughts

Now you've finished *Building Secure Automotive IoT Applications*, we'd love to hear your thoughts! Scan the QR code below to go straight to the Amazon review page for this book and share your feedback or leave a review on the site that you purchased it from.

https://packt.link/r/1-835-46550-1

Your review is important to us and the tech community and will help us make sure we're delivering excellent quality content.

www.ingramcontent.com/pod-product-compliance
Lightning Source LLC
Chambersburg PA
CBHW080756300426
44114CB00020B/2743